Design Solutions for Adaptive Hypermedia Listening Software

Vehbi Turel
Bingol University, Turkey

IGI Global
PUBLISHER of TIMELY KNOWLEDGE

A volume in the Advances in Educational Technologies and Instructional Design (AETID) Book Series

Published in the United States of America by
IGI Global
Engineering Science Reference (an imprint of IGI Global)
701 E. Chocolate Avenue
Hershey PA, USA 17033
Tel: 717-533-8845
Fax: 717-533-8661
E-mail: cust@igi-global.com
Web site: http://www.igi-global.com

Library of Congress Cataloging-in-Publication Data

Names: Turel, Vehbi, 1965- editor.
Title: Design solutions for adaptive hypermedia listening software / Vehbi Turel, editor.
Description: Hershey, PA : Engineering Science Reference, [2021] | Includes bibliographical references and index. | Summary: "This book of contributed chapters aims to enable companies, institutions, teachers and researchers to design and create technically and pedagogically sound and efficient interactive adaptive hypermedia listening environments for language learners in any language in general and in English in specific"-- Provided by publisher.
Identifiers: LCCN 2021009398 (print) | LCCN 2021009399 (ebook) | ISBN 9781799878766 (h/c) | ISBN 9781799878773 (s/c) | ISBN 9781799878780 (eISBN)
Subjects: LCSH: Voice computing. | Interactive multimedia. | Software architecture.
Classification: LCC QA76.9.V65 D47 2021 (print) | LCC QA76.9.V65 (ebook) | DDC 006.2/48392--dc23
LC record available at https://lccn.loc.gov/2021009398
LC ebook record available at https://lccn.loc.gov/2021009399

This book is published in the IGI Global book series Advances in Educational Technologies and Instructional Design (AETID) (ISSN: 2326-8905; eISSN: 2326-8913)

British Cataloguing in Publication Data
A Cataloguing in Publication record for this book is available from the British Library.

For electronic access to this publication, please contact: eresources@igi-global.com.

Advances in Educational Technologies and Instructional Design (AETID) Book Series

Lawrence A. Tomei
Robert Morris University, USA

ISSN:2326-8905
EISSN:2326-8913

Mission

Education has undergone, and continues to undergo, immense changes in the way it is enacted and distributed to both child and adult learners. In modern education, the traditional classroom learning experience has evolved to include technological resources and to provide online classroom opportunities to students of all ages regardless of their geographical locations. From distance education, Massive-Open-Online-Courses (MOOCs), and electronic tablets in the classroom, technology is now an integral part of learning and is also affecting the way educators communicate information to students.

The **Advances in Educational Technologies & Instructional Design (AETID) Book Series** explores new research and theories for facilitating learning and improving educational performance utilizing technological processes and resources. The series examines technologies that can be integrated into K-12 classrooms to improve skills and learning abilities in all subjects including STEM education and language learning. Additionally, it studies the emergence of fully online classrooms for young and adult learners alike, and the communication and accountability challenges that can arise. Trending topics that are covered include adaptive learning, game-based learning, virtual school environments, and social media effects. School administrators, educators, academicians, researchers, and students will find this series to be an excellent resource for the effective design and implementation of learning technologies in their classes.

Coverage

- Instructional Design
- Digital Divide in Education
- Hybrid Learning
- Educational Telecommunications
- Web 2.0 and Education
- Bring-Your-Own-Device
- Social Media Effects on Education
- Higher Education Technologies
- Curriculum Development
- Adaptive Learning

The Advances in Educational Technologies and Instructional Design (AETID) Book Series (ISSN 2326-8905) is published by IGI Global, 701 E. Chocolate Avenue, Hershey, PA 17033-1240, USA, www.igi-global.com. This series is composed of titles available for purchase individually; each title is edited to be contextually exclusive from any other title within the series. For pricing and ordering information please visit http://www.igi-global.com/book-series/advances-educational-technologies-instructional-design/73678. Postmaster: Send all address changes to above address.

Titles in this Series

For a list of additional titles in this series, please visit: http://www.igi-global.com/book-series/advances-educational-technologies-instructional-design/73678

Handbook of Research on Empowering Early Childhood Educators With Technology
Jade Burris (West Chester University of Pennsylvania, USA) Dina Rosen (Kean University, USA) and Donna Karno (University of Maine at Farmington, USA)
Information Science Reference • © 2021 • 400pp • H/C (ISBN: 9781799868880) • US $245.00

Shifting to Online Learning Through Faculty Collaborative Support
Caroline M. Crawford (University of Houston-Clear Lake, USA)
Information Science Reference • © 2021 • 375pp • H/C (ISBN: 9781799869443) • US $195.00

Handbook of Research on Analyzing IT Opportunities for Inclusive Digital Learning
Patricia Ordóñez de Pablos (The University of Oviedo, Spain) Mohammad Nabil Almunawar (Universiti Brunei Darussalam, Brunei) Kwok Tai Chui (The Open University of Hong Kong, Hong Kong) and Maniam Kaliannan (Business School, Nottingham University, Malaysia)
Information Science Reference • © 2021 • 403pp • H/C (ISBN: 9781799871842) • US $245.00

Evolving Multicultural Education for Global Classrooms
Richard Keith Gordon (Seisa University, Japan & California State University, USA) Kawser Ahmed (University of Winnipeg, Canada & Conflict and Resilience Research Institute, Canada) and Miwako Hosoda (Seisa University, Japan)
Information Science Reference • © 2021 • 287pp • H/C (ISBN: 9781799876496) • US $195.00

eLearning Engagement in a Transformative Social Learning Environment
Caroline M. Crawford (University of Houston-Clear Lake, USA) and Melissa A. Simons (Independent Researcher, USA)
Information Science Reference • © 2021 • 340pp • H/C (ISBN: 9781799869566) • US $195.00

Assistive Technologies for Differently Abled Students
Sangeeta Dhamdhere-Rao (Modern College of Arts, Science and Commerce, India) and Frederic Andres (National Insitute of Informatics, Japan)
Information Science Reference • © 2021 • 305pp • H/C (ISBN: 9781799847366) • US $195.00

Developments in Virtual Learning Environments and the Global Workplace
Stephanie Swartz (Mainz University of Applied Sciences, Germany) Belem Barbosa (University of Porto, Portugal) Izzy Crawford (Robert Gordon University, UK) and Susan Luck (Pfeiffer University, USA)

IGI Global
PUBLISHER of TIMELY KNOWLEDGE

701 East Chocolate Avenue, Hershey, PA 17033, USA
Tel: 717-533-8845 x100 • Fax: 717-533-8661
E-Mail: cust@igi-global.com • www.igi-global.com

Table of Contents

Detailed Table of Contents

Due to the nature of listening, multiple media tools have been utilized in second language (L2) listening instruction for a long time. This chapter recruited the most recently published empirical studies in addition to the meta-analysis conducted in 2015 (Kang) and analyzed available evidence on the effects of different multiple media tools on L2 listening comprehension. The results revealed a medium-to-large effect of multiple media tools on listening comprehension in between-group designs (Cohen's d = .70), which was almost identical to the result observed five years ago. The effects of individual multiple media tools have also been statistically synthesized. Moderator analysis could help L2 educators and test developers make decisions on applying different multiple media tools in the fields of L2 instruction and assessment. Specifically, subtitles, captions, as well as self-regulated listening and slow speed were recommended to teachers and test developers as a means to improve learners' listening comprehension.

The design and development of adaptive hypermedia (AH) listening environments consist of different stages. Each stage in design and development of AH listening applications is vital in the design and development process so that cost-effective AH listening software can be designed and developed in a true sense. Pedagogically, epistemologically, and technically, a wide range of design principles and guidelines have to be heeded at each design and development stage so that effective and efficient AH listening environments can be designed and developed.

Use of movies and TV shows for foreign language learning purposes pervades informal settings where viewers enjoy the multiplicity of soundtracks and of subtitles. However, this is not an integral part of formal language education policies. Hoping to inform policy making processes, this study aimed to investigate the most efficient use of audio/subtitle combinations for specific purposes. As a result, reversed subtitling was found to be effective for vocabulary learning whereas, generally, bimodal subtitling was considered to benefit accent-related problems. The participants self-reported preferrence for non-subtitled version in case of listening comprehension in general, which was due to the subtitle effect in the case of students at lower levels. Considering that mostly interlingual subtitled content is the only version in most platforms, these findings make it necessary to prefer a platform presenting a variety of alternative combinations in terms of soundtrack and subtitles.

Podcasts, one type of mobile-assisted language learning, have been promoted as promising tools in second language (L2) learning. Yet, quality indicators of podcast programs have not been generalized. Applying previous empirical results on predictors of listening comprehension, the purpose of this chapter was to identify English language learning podcast programs in iTunes that were preferred by language learners and teachers and build up a list of the indicators that can be considered for selecting and designing high-quality English language learning podcasts. Participants were 26 intermediate English as second language (ESL) learners at a southwestern American university and two of their listening and speaking teachers. The results revealed that the ESL students and their teachers both preferred podcasts that had interesting topics, natural speed, and clear pronunciation.

Reduced forms (RFs) spoken by native English speakers have been challenging on the part of the second language (L2) learners. This chapter aims to address suprasegmental features to Turkish preparatory language school students in relation to L2 listening comprehension. Considering the limited research on RFs in learning English as a L2 context, this pre-test post-test control group design study aimed to explore whether the instruction of five RFs in sentential level results in any difference in listening comprehension test performance. The five forms entail contraction, assimilation, flap, elision, and linking. A total of 343 were recruited, and RFs instruction was delivered through the web page designated for the study for five weeks, and the performance of the eight groups was measured twice throughout the study. The findings indicated that sentence level of RFs instruction through web-based learning facilitates the listening comprehension of RFs.

In the field of second language (L2) perception, there is a common adherence to quantitative methods to examine reduced forms (RFs). This chapter extends the field by reporting on an investigation that analyzed L2 listeners' perceptions of RFs in English from a qualitative perspective. RFs instruction through web-based activities was delivered to a total of 80 learners of English of varying proficiency for five weeks. Twenty participants reflected on their performance on RFs listening tasks and provided justifications for their perceptions of the target RFs. Qualitative analysis revealed that the RFs that influenced L2 learners' perceptions of RFs were linking, pause phenomena, and assimilation. The results of using such qualitative methodology highlights the important role that RFs plays in perception judgements in syllable-timed languages such as Turkish, a factor which has not always been given much prominence in previous L2 fluency quantitative research.

This mixed-method 14-week case study investigated the impact of peer feedback on Turkish English as a foreign language (EFL) prospective teachers' attitudes to and performance in a blended academic writing course in the tertiary setting. The peer feedback provision process involved online and face-to-face written peer feedback on weblogs, enhanced with a six-hour training phase for peer feedback. Weekly student journals, group feedback conferences, the end-of-the-semester survey, and the pre- and post-writing tests were used in the data collection. The qualitative data were analyzed through content analysis and the quantitative data via descriptive statistics and an independent paired sample t-test. The study indicated that the use of face-to-face and online peer feedback together was perceived as beneficial and effective by EFL pre-service teachers and that it led to a relatively successful performance in academic writing. It is suggested that participants unfamiliar with peer revision be provided with training and continuous scaffolding/guidance to ensure the effectiveness of the process.

With its engaging spectacle and discussion of key issues, the US presidential election debates draw high viewership of not only Americans but also people of other nationalities through not only US-based television channels but also other mainstream channels around the world. Since many of the viewers are likely to be the learners of English as the most widely used Lingua Franca, who listen for as well as to the content, they should be helped with their processing of what they listen to, given the listening comprehension challenges. The demanding factors of speech rate and vocabulary necessitate the use of transcripts as an aid for listening comprehension. Transcripts facilitate the required vocabulary buildup and allow for semantic mapping through frequency counts of recurrent key words or word strings uncovering linguistic patterns and laying foundation for the enriching listening practices. Therefore, this chapter focuses on using transcripts accordingly.

Preface

USING DIGITAL TECHNOLOGY TO DESIGN AND DEVELOP ADAPTIVE HYPERMEDIA LISTENING SOFTWARE

As we, educators and materials developers, live in a digital age, it is obligatory for us to meet the needs of today's digital learners who are in general fluent and competitive in digital technologies. It is due to these facts that today's learners are known as 'digital natives', whose learning styles and preferences are different from the learners of just a few decades ago. Therefore, today's learners expect adaptive hypermedia (AH) learning materials to be used more often and widely both in teaching and learning. In order to be able to meet such demands, we, as educators and materials developers, need to and have to respond to such learning demands and differences. This can only be achieved through accommodating today's learners with correct and efficient adaptive hypermedia learning environments at all stages of teaching and learning.

From the point of view of second / foreign language learning (SLL /FLL) and teaching, fortunately, tremendous amount of research has been conducted as well as published in the field of (a) using educational technology for SLL/FLL and (b) the design and development of AH language learning environments in a very wide range of different peer-reviewed internationals journals. However, no compiled book has ever been written or edited on the 'design and development of AH listening environments' to date. Moreover, no 'instructional design model for design and development of AH listening environments' has ever been suggested although many instructional design models for design and development of hypermedia are suggested.

This book, *Design Solutions for Adaptive Hypermedia Listening Software,* offers a unique way for designing and developing efficient and cost effective AH listening applications for the learners of any language (e.g. English, Kurdish, Arabic, Hebrew, Chinese, Spanish, Japanese, Russian, French, German, Persian, Italian, Portuguese, Hindi, Urdu, Greek, Turkish etc.) that is taught/learnt as a second/foreign language. This book is an updated valuable resource in addition to *Intelligent Design of Interactive Multimedia Listening Software.*

The eleven chapters of this book, - which are authored by specialists and experts in the field of listening and the design and development of AH listening applications as a part of SLL/FLL from a wide range of different universities -, cover (a) SLL/FLL and AH listening-applications overview theories, hypotheses, approaches and trends, (b) efficient design principles and guidelines for design of different learning elements of AH listening applications such as instructional design models, optimum combinations, tasks, activities, instructions, captions, tests, unfamiliar items, aural texts, speech rate, and other relevant design issues, and (c) pertinent resources. These chapters of the book present the results of original and empirical research covering these areas. In the same way, the chapters of the book report results of both quantitative and qualitative empirical data on the actual design and development of a wide range of digital learning elements (or their optimum combinations) of AH listening environments.

Design Solutions for Adaptive Hypermedia Listening Software provides undergraduate and post-graduate language and computer students; language and digital education teachers, tutors, lecturers and researchers; educational institutions; commercial software companies and software developers, and individual software developers with the most up to date compiled research, findings and practical design principles and guidelines in the field of AH listening applications. This unique compiled book enables all potential linguistic and digital education scholars, students, institutions, software companies and software developers to better understand efficient and effective hypermedia design solutions and development principles and guidelines. Furthermore, this book empowers them to design and create efficient cost-effective AH listening applications for any language as a part of SLL/FLL.

HOW THIS BOOK ORGANISED

The eleven chapters in this book cover three key areas that are vital in the design and development process of AH listening environments for any language as a part of SLL/FLL in a true sense. These key areas are: (1) meta-analyses, optimum combinations, and design models, (2) instructions, feedback, and design issues, and (3) film adaptations, listening, and testing.

Meta-Analyses, Optimum Combinations, and Design Models

In Chapter 1, "The Effect of Using Technology in L2 Listening: A Meta-Analysis of Studies from 1980 to 2020", Kang recruited the most recently published empirical studies in addition to the meta-analysis conducted in 2015 (Kang, 2015) and analyzed available evidence on the effects of different multiple media tools on L2 listening comprehension. The results revealed a medium-to-large effect of multiple media tools on listening comprehension in between-group designs (Cohen's d = .70), which was almost identical to the result observed five years ago. The effects of individual multiple media tools have also been statistically synthesized. Moderator analysis could help L2 educators and test developers make decisions on applying different multiple media tools in the fields of L2 instruction and assessment. Specifically, subtitles captions, as well as self-regulated listening and slow speed, were recommended to teachers and test developers as a means to improve learners' listening comprehension.

In Chapter 2, "An Instructional Design Model for Design and Development of Adaptive Hypermedia Listening Environments", *Turel* suggested an instructional design model for design and development of AH listening applications (i.e. AHLE Model – Adaptive Hypermedia Listening Environments Model). Each stage of the suggested instructional design model is explained thoroughly and the principles and guidelines that need to be taken into heed at each stage are explained clearly. From the point of view of designing and developing AH listening environments, an instructional design model is a framework (i.e. set of principles and guidelines) that enable designers and developers of AH listening environments to design and develop pedagogically cost effective listening applications so that target instructional goals can be achieved.

In Chapter 3, "How Best to Use Audio/Subtitle Combinations in the Use of Films for Non-Native Learners of English", *Zengin,* investigated the most efficient use of audio/subtitle combinations for specific purposes. As a result, reversed subtitling was found out to be effective for vocabulary learning whereas generally bimodal subtitling was considered to benefit accent-related problems. The participants self-reported preference for non-subtitled version in case of listening comprehension in general, which was due to the 'subtitle effect in the case of students at lower levels. Considering that mostly inter-lingual subtitled content is the only version in most platforms, these findings make it necessary to prefer a platform presenting a variety of alternative combinations in terms of soundtrack and subtitles.

Instructions, Feedback, and Design Issues

In Chapter 4, "Quality Indicators for Designing English Learning Podcasts", *Kang* focuses on Podcasts, which are one type of mobile assisted language learning tools, and have been promoted as promising tools in second language (L2) learning. Yet, quality indicators of podcast programs have not been generalized. Applying previous empirical results on predictors of listening comprehension, the purpose of this chapter was to identify English language learning podcast programs in iTunes that were preferred by language learners and teachers and build up a list of the indicators that can be considered for selecting and designing high-quality English language learning podcasts. Participants were 26 intermediate English as second language (ESL) learners at a southwestern American university and two of their listening and speaking teachers. The results revealed that the ESL students and their teachers both preferred podcasts that had interesting topics, natural speed, and clear pronunciation.

In Chapter 5, "The Effect of Audio and Video Modality on Perception of Reduced Forms: The Role of Web-Based Instruction", *Gulozer,* and *Kocoglu* address supra-segmental features to Turkish preparatory language school students in relation to L2 listening comprehension. Reduced forms (RFs) spoken by native English speakers have been challenging on the part of the second language (L2) learners. Considering the limited research on RFs in learning English as a L2 context, this pre-test post-test control group design study aimed to explore whether the instruction of five RFs in sentential level results in any difference in listening comprehension test performance. The five forms entail contraction, assimilation, flap, elision and linking. A total of 343 were recruited and RFs instruction was delivered through the web page designated for the study for five weeks and the performance of the eight groups was measured twice throughout the study. The findings indicated that sentence level of RFs instruction through web-based learning facilitates the listening comprehension of RFs.

In Chapter 6, "Boosting L2 Listening Comprehension Through Web-Based Listening Activities on Reduced Forms", *Gulozer* extends the research on 'reduced forms' (RFs) by reporting on an investigation that analyzed L2 listeners' perceptions of RFs in English from a qualitative perspective. RFs instruction through web-based activities was delivered to a total of 80 learners of English of varying proficiency for five weeks. Twenty participants reflected on their performance on RFs listening tasks and provided justifications for their perceptions of the target RFs. Qualitative analysis revealed that the RFs that influenced L2 learners' perceptions of RFs were linking, pause phenomena and assimilation. The results of using such qualitative methodology highlights the important role that RFs plays in perception judgements in syllable-timed languages such as Turkish, a factor which has not always been given much prominence in previous L2 fluency quantitative research.

In Chapter 7, "The Impact of Online and Face-to-Face Peer Feedback on Pre-Service Teachers in EFL Academic Writing", *Kaçar* investigated the impact of peer feedback on Turkish English as a Foreign Language (EFL) prospective teachers' attitudes to and performance in a blended academic writing course in the tertiary setting in a mixed-method 14-week case study. The peer feedback provision process involved online and face-to-face written peer feedback on weblogs, enhanced with a six-hour training phase. Weekly student journals, group feedback conferences (intra-feedback), the end-of-the-semester survey, and the pre- and post-writing test were used to collect data. The qualitative data were analyzed through content analysis and the quantitative data via descriptive statistics and an independent paired sample t-test. The study indicated that EFL pre-service teachers perceived the use of face-to-face and online peer feedback together as beneficial. The use of face-to-face and online peer feedback together led to relatively improved performance in academic writing. It is suggested that participants unfamiliar with peer revision be provided with training and continuous guidance to ensure effective peer feedback provision.

Film Adaptations, Listening, and Testing

In Chapter 8, "Significance of Transcript Availability for Viewing TV Coverage of US Elections", *Zengin* focuses using transcripts for enriching listening practices. With its engaging spectacle and discussion of key issues, the US presidential election debates draw high viewership of not only Americans but also people of other nationalities through not only US-based television channels but also other mainstream channels around the world. Since many of the viewers are likely to be the learners of English as the most widely used Lingua Franca, who listen for as well as to the content, they should be helped with their processing of what they listen to, given the listening comprehension challenges. The demanding factors of speech rate and vocabulary necessitate the use of transcripts as an aid for listening comprehension. Transcripts facilitate the required vocabulary build-up and allow for semantic mapping through frequency counts of recurrent key words or word strings uncovering linguistic patterns and laying foundation for the enriching listening practices. Therefore, this chapter focuses on using transcripts accordingly.

In Chapter 9, "Benefits of Film Adaptations of Literary Works for English Language Learning", *Öztürk,* and *Zengin* presents the benefits of film adaptations for language learners. Despite their potential in language learning, films have not been covered adequately in academic research except some attempts to develop strategies. One of these strategies has been to compare film scripts and their translations, to identify chunks, technical terms and creative language examples and to list them with their translations, and to do creative dialogue writing activities especially by modeling these scripts where chunks are used extensively. Believing in the importance of evaluating the film scripts with their translations, this chapter investigated the processes the researcher/teacher's high school students went through in learning and using the target language starting with the use of the film adaptations of the literary works, writing their scripts modelling the film scripts, acting their own scripts and shooting their performances. The mixed method research design was used with the triangulation of the qualitative and quantitative data. 68 high school students' views were investigated with Likert scale (1-5) questionnaire and their comments about the items.

In Chapter 10, "Exploring Teachers' Beliefs about Listening in a Foreign Language", *Irgin* investigated the beliefs and practices of 96 language teachers in Turkey. A mixed-method research design was employed to answer the questions to what extent teachers' beliefs and practices reflect the issues stated in the literature on second language listening, what their beliefs and listening instructional practices are, whether their beliefs and practices converge, and if so, what factors underpin them. The data was collected through a questionnaire. The results of the descriptive statistics and coding approach indicate that the teachers' beliefs about instructional listening in a foreign language are inconsistent with their practices of teaching listening to their pupils. The teachers' beliefs reflect the effective completion of listening tasks referring to the pre-, during and post-listening procedures; however, their practices do not emphasize on the use of metacognitive listening strategies and metacognitive awareness to make learners autonomous.

In Chapter 11, "Examination of Speaking Test Performance from an Interactional Perspective', *Tokdemir Demirel,* and *Başer* explore the use of structured group tasks in the testing of speaking skills in the context of a tertiary level first year speaking course. Participants included undergraduate university students from the department of English translation and interpreting at a state university located in Central Turkey. The structured task used in the study guided students into a group discussion during the speaking exam. The discussions in the exam sessions were video-recorded and transcribed. The students were also given an open-ended questionnaire to reveal their opinions about the ins and outs of the speaking test task tried out in the study. The questionnaire consisted of two parts. The first part required students to fill out their demographic information and the second part inquired students' opinions about the group-speaking task. The results of the analysis and questionnaire responses revealed additional opportunities provided by the use of structured tasks in speaking exams.

In conclusion, not only does this unique compiled book impact the field of AH listening software very positively in general, but it also contributes significantly to the design and development of AH listening software in specific cases. Compiled with experimental and practical design principals and guidelines, as well as an instructional design model for design and development of adaptive hypermedia listening environments (i.e. AHLE Model – Adaptive Hypermedia Listening Environments Model), this is a book that no undergraduate and post-graduate language and computer students; language and digital education teachers, tutors, lecturers and researchers; language and digital educational institutions; commercial software companies and software developers, and individual software developers can afford to be without.

Vehbi Turel
Bingol University, Turkey

Introduction

LISTENING, ADAPTIVE HYPERMEDIA AND OPTIMUM DESIGN

Listening

Listening refers to "a set of cognitive interactions involved in oral language processing" (Rost, 1994, p. 3778). It is also "the process of deriving meaning from sound" (ibid). The Chambers Encyclopaedic English Dictionary (CEED) defines listening as "to give attention so as to hear something to follow advice". Listening can also be defined as, "… following and understanding…. Following … is closely connected with the intelligibility of a message. That is, can a listener repeat the message aloud, sub-vocally, or in the mind's eye?" (Vanderplank, 1988, p. 32). Moreover, Bacon (1992, p. 388) states that "the act of listening involves an interaction of input, task, and individual variables".

The above definitions of listening provide us with some key features: 'a set of cognitive interactions in processing'; 'the process of deriving meaning …', 'giving attention'; 'following, understanding and repeating/producing'; 'the interaction of input, task and individual variables'. These terms need to be discussed in detail. Then, what does listening mean? Is it a term related to only language studies? Is it always a process of deriving meaning from sound? Is it always a process of following only advice? To answer these questions correctly, firstly a comparison should be made, to some extent, between first language (L1) and second language (L2) listening. Inevitably, there are many -similarities in L1 and L2 listening such as following speech, trying to understand it, and, in some cases, answering. Apart from these similarities, there are some more aspects of foreign/second language (FL/SL) listening, not all of which are indicated clearly in the above definitions. Such features can vary depending on (a) the characteristics of language learners, (b) the characteristics of listening input, (c) listening purposes and (d) learning objectives. Language learners, for instance, sometimes:

- Only listen to acquire the correct pronunciation, stress and intonation, as these features are important elements of listening texts and possess a role in understanding (Goh, 2000, p. 59; Anderson-Hsieh, 1992, p. 52; Ur, 1992, p. 13; Richards, 1983, p. 226)
- Only focus on grammatical structures (i.e. syntax) to find out how such structures are used in the target language, as syntax plays an important role in comprehension and foreign/second language learning (FLL/SLL; Conrad, 1989, p. 14). Let us, for instance, say that language learners hear such a sentence in a listening text: 'Minority people *could have*, and *should have*, *been* given equal rights by the modern states'. If language learners do not know what "*could have*", "*should have*" (i.e. unrealised/undone modal auxiliaries in past) and also "*been*" (i.e. passive voice), mean, then

naturally and logically they cannot be expected to comprehend what they hear. It is because of this Conrad (1989, p. 14) says: "Knowledge of TL [target language] syntax seems to be an important factor in increasing the amount of linguistic material that can be retained by short memory".

- Pay attention to unfamiliar items (i.e. lexis), as they might think that it is important at that particular time to understand that particular unfamiliar lexis or learn new language items. Moreover, an unfamiliar lexis is another element that might cause difficulty for language learners in understanding listening texts (Turel, 2014b; Chung & Huang, 1998, pp. 559-60; Underwood, 1993, p. 17).
- Follow each sentence and try to translate it into L1 (i.e. providing equivalents of L2 in L1). This is mostly done in the early stages of language learning process (Eastman, 1987, p. 198; O'Malley, Chamot, Stewner-Manzanares, Kupper & Russo, 1985a, p. 40), which does not help comprehension and acquisition a lot (Goh, 2002, p. 192; Goh, 2000, pp. 62, 67, 69). Not only do native speakers ordinarily not use translation strategy, but also this strategy is not accepted as a highly efficient strategy (O'Malley et al., 1985a, p. 39), as translation slows down processing and takes language learners' attention away from clues that might assist comprehension (Goh, 2002, p. 192).
- Listen to understand the surface meaning (main idea) of the listening input (i.e. involves actual understanding of listening input, initially main ideas). This is, maybe, the most global type of listening function that language learners carry out. This is common in L1 listening, as well. In fact, in real life when we listen to listening texts in L1, we sometimes only focus on the main ideas owing to the circumstances because either we do not want to comprehend in detail or due to other reasons such as not being interested in the topic. It might be because of such similarities that the proponents of the procedural skills hypothesis (i.e. the nature of individual skill in native speakers - NSs - and non-native speakers – NNSs - is the same although the ability is acquired gradually) claim "there is no fundamental difference in language processing between NNS and NS…" (Pienemann, 2001, p. 335).
- Listen to understand the deeper meaning of the listening input (i.e. syntax, social- and culture-based meanings, the illocutionary forces/a speaker's intention), which is also done in L1 listening
- Listen/view to understand not only what they hear but also the non-verbal aspects of the segments, as the meaning of listening input does not always consist of only words, which is also the case in L1 listening. Non-verbal aspects of listening texts and situations might sometimes play a vital role in comprehension. To this end, Oxford (1993, p. 207) even claims that in some cases "as much as 93% of the total meaning of any interaction comes from non-verbal, often visual, clues", an idea which might not be agreed on by everyone.

In other words, for language learners, listening*:*

- Is not a process of deriving meaning *every time*
- Is not a process of *hearing* something because *hearing*, as defined in Collins Cobuild Dictionary (CCD), is "a sense that makes it possible for us to be aware of sound." As understood here, hearing is a sense which enables us to be aware of sound regardless of whether we want to hear or not such as *noise*, *chatting*, *shouting*, *crying*, *swearing*, *music* and *so on*. In listening, language learners consciously follow speech, someone talking, news and so on to, in general, grasp meaning and sometimes for other indicated purposes.
- Is not *only* a process of following. It is more a process of understanding which is carried out by using world - (social and cultural) knowledge, comprehension clues, listening strategies, and con-

textual references (Weissenrieder 1987, p. 23; Long, 1989, p. 32; O'Malley, Chamot & Kupper, 1989, pp. 431-32).

- Is not a process of tuning in to a listening text or someone unconsciously and unintentionally. Conversely, listening requires careful attention and it is mostly on purpose, otherwise it will be *hearing*, not *listening*
- All of these, of course, are not exactly what language learners always do when they listen to/view listening texts. These are only some of the approaches that language learners follow when they listen to listening input. It is because of these different listening purposes that the types of listening are categorised into different groups such as (1) empathic listening (requires focusing on facts and emotional content), (2) appreciative listening (includes listening to enjoy), (3) critical listening (focuses on analysing and indicating the importance of arguments), and (4) relational listening (requires a holistic approach by paying attention to the whole material; (Oxford, 1993, p. 208-89;Vo, 2015; Coakley & Wolvin, 1995).

Comprehension

Comprehension is defined, in the CCD, as "full knowledge and understanding of the meaning of something", while Lund (1991, p. 196) says "... the construction of meaning using both the decoded language and the comprehender's prior knowledge". According to Faerch and Kasper (1986, p. 264-65) when input and knowledge match each other, then comprehension occurs, but this matching is not perfect as a rule because of the gaps which regularly occur either in the recipient's or in the interlocutor's sources. They say that the comprehension process relies on three types of information:

"... linguistic and other communicative input, the recipient's linguistic and other (socio-cultural, world) knowledge, and contextual information deriving from the situational context and the linguistic co-text" (ibid: 264).

The CEED defines comprehension as "the process of power of understanding; the scope or range of one's knowledge or understanding". Chapelle (1998, p. 28) goes further and groups comprehension into (1) 'semantic comprehension' and (2) 'syntactic comprehension'. 'Semantic comprehension' is considered the more difficult process (Carroll, 1977, p. 505). Comprehension can also be defined as our general capacity to think, and solve problems by using all available (verbal, visual, situational, contextual etc.) input depending on our transactional (information seeking) objectives (Figure 1).

According to Anderson's (1995, p. 379) comprehension framework, comprehension consists of perception (i.e. the encoding of the input), parsing (i.e. the transformation of words into mental representation) and utilisation (i.e. relating mental representations to existing knowledge/elaboration, which enables to store in long-term memory).

Figure 1. Listening comprehension (Turel, 2015b)

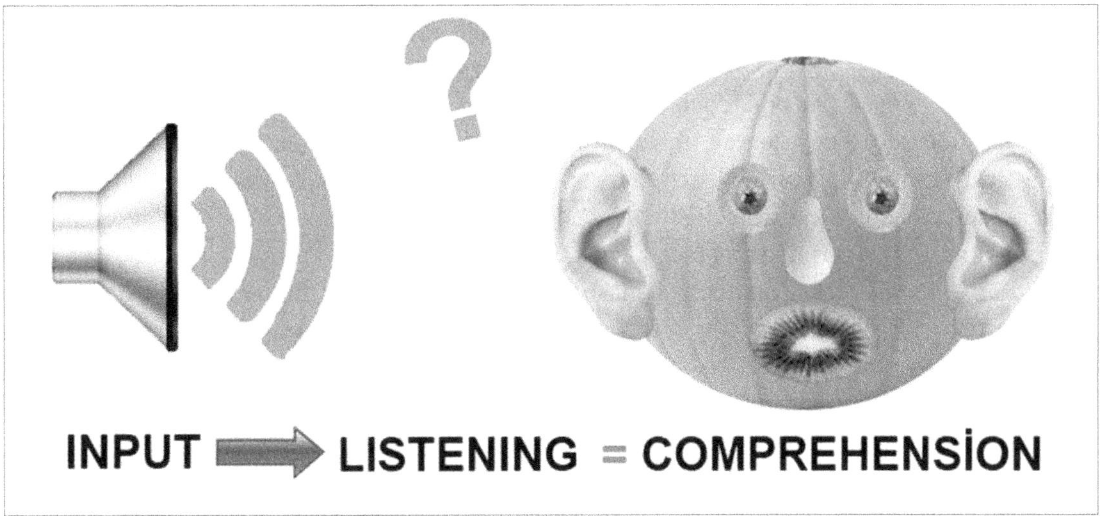

UNDERSTANDING

According to Vanderplank, understanding "… involves much wider notions such as prior or pragmatic knowledge, ability to follow argument, contextual relevance, insider information, etc." (1988, p. 32). Rost states that understanding is "...an optimal interaction between comprehension of the oral input and interpretation of that input in the situational context" (1994, p. 3799). These significant definitions shed invaluable light on 'comprehension' term. The definitions of 'comprehension' also reveal different variables such as input (Faerch & Kasper, 1986, p. 264-5), language learners' linguistic and world knowledge (ibid), contextual information, meaning of 'input', construction of meaning, target culture (Turel, 2014c; Turel & Kilic, 2014; Platt & Brooks, 2002, p. 369; Kohn, 2001, p. 254; McLoughlin & Oliver, 2000, p. 59; Vygotsky, p. 1978), syntactic and lexical comprehension (Hegelheimer & Chapelle, 2000, p. 42; Smith, 2003, p. 52), encoding, transformation of words into representations, elaboration, long-term memory, ability to follow, optimal interaction and interpretations (Brown, 1978, p. 59; Eastman, 1991, p. 185). All these variables play an important role in understanding. For us, in terms of listening skills, comprehension and understanding, as can also be derived from the definitions, mean more or less the same thing, but in order to make them clearer we need to discuss them a bit further.

It is a de facto that it is not always possible for language learners to fully comprehend or understand listening input. This might also often be the case in L1 listening. Brown and Yule, to this end, emphasise: "…a 100 percent notion. What native-listeners operate with are partial, reasonable interpretations of what they are listening to…" (1992, p. 57).

Language learners do not always want to fully comprehend, as a result of their transactional goals. The degree of the process of power of comprehension is largely dependent on the learners' transactional (information seeking) objectives; interpersonal aims; abilities; linguistic (language), social cultural and world knowledge. Partly to this end, Porter and Roberts (1981, p. 42) state: "… we understand a TL [target language] at different levels according to our ability to use linguistic clues and situational and paralinguistic information". Let us say that we listen to a news programme particularly to find out what has happened as a result of an incident, not where and when it happened because of our transactional objectives. Then, inevitably, we will not follow and focus on when and where that particular incident has occurred, as we are not interested in full understanding, and only interested in something specific. To this end, Oxford (1993, p. 206) says: "Not everything to which a student is exposed becomes 'intake', only the part that is significant and to which the student is paying attention".

Comprehension is, in the same way, not always a process of using information and solving problems. If language learners are supposed to listen to listening texts and answer questions about them, find out something specific or solve the problem, then it can be said that it is a process of using information and solving problems. Let us imagine that language learners listen to a radio programme about which they do not have any pre-knowledge and only want to know what it is about. In the same way, the learners are not also requested to complete any task. In such a case, firstly, the learners cannot use any pre-information and solve the problems, as they lack it. Secondly, the learners' transactional objectives do not also necessitate them to understand in detail and to use this information to solve problems, answer questions or analyse it. Is this a process of using information and solving problems?

In terms of language learners, therefore, it can be said that comprehension is, in general, a process of deriving meaning from the listening texts they tune in to. This process is also conscious and on purpose, which not only targets deriving meaning, but also targets acquiring the other aspects of the target language in any possible way. The degree of this process is largely dependent on language learners' level; their age, level of intelligence (i.e. fast/slow language learners; Gardner, 1985, p. 23), gender (ibid: 43); their learning style preferences (Carson & Longhini, 2002, p. 408; McLoughlin, 1999, pp. 222-23; Brickell, 1993, p. 103) and the strategies they use (Hurd, Beaven & Ortega, 2001, p. 342; Cohen, 1998, p. 5; Chamot & O'Malley, 1994, p. 371); their transactional objectives (Oxford, 1993, p. 206); world knowledge and linguistic knowledge (Porter & Roberts, 1981, p. 42); the difficulty/ease of the listening texts (Anderson & Lynch, 1989, pp. 6, 56; Rixon, 1992, p. 65; Underwood, 1993, p. 17); and contextual information (i.e. visuals, knowledge about the structure of the listening text/formal schemata -e.g. news programme or event: pre-thematic - the beginning -, thematic -main -, and posthematic context - the end of the story; Faerch & Kasper, 1986, p. 264). Regarding individual differences, Bacon (1992, pp. 399-400) points out: "Individual learner variables such as attitude and motivation, background knowledge, perceptual style, previous language-learning experience, and learning strategies all contribute to how a listener will interact with input". Due to these factors, language learners approach listening input in different ways, which results in comprehension/understanding or interpretation in diverse ways. In terms of many aspects, this process is also valid for L1 comprehension.

LISTENING COMPREHENSION

Listening comprehension is defined as "... being able to understand native speech at normal speed in unstructured situations" (Geddes & White, 1978, p. 38). Regarding language learners' listening comprehension, Brown (1978, p. 59) and Eastman (1991, p. 185) state that language learners have to be prepared and encouraged to target a reasonable, logical and acceptable interpretation rather than a 'correct' interpretation. However, when someone looks at the definition of listening comprehension, it might be significantly important in terms of which approach listening comprehension is defined, as each approach in listening gives priority to different aspects of listening comprehension. Byrnes (1984, pp. 317-78), for example, groups listening comprehension into three approaches, and in each approach the focus is different: (a) A linguistic approach: this approach indicates how listeners arrive at a structural description of a listening text (sentence) based on the phonological, syntactic and semantic aspects. (2) A conceptual approach: it focuses on how a listener puts a conceptual structure into a linguistic input (a non-linguistic onto a linguistic-structure). (3) A communicative approach: this approach sees listening comprehension as a result of a transactional interaction between a speaker and a listener. It is achieved successfully particularly when the listeners identify the illocutionary forces (i.e. the force of a sentence/a speaker's intention). It focuses on meaning, not on form, and functions as a part of an integrated-skill approach.

Moreover, there are many factors that affect listening comprehension. These are: (a) learners' characteristics (i.e. their age, L1 listening ability, L2 proficiency/level, vocabulary development in L2, - Vandergrift, 2006 -; level of intelligence, learning-style-preferences, learning effectiveness, transactional objectives, psychological readiness and willingness, - which is also considered one of the main components of autonomy, e.g. Littlewood, 1996, p. 428, 1997, p. 82 -; their experience in the target language, familiarity with the target culture and topic, world and social-cultural knowledge and so on, e.g. Turel, 2014c; Turel &Kilic, 2014), (b) input related factors (i.e. length and amount of input, cultural differences, subject matter, coherence, types of language input, unfamiliar items, discourse marks, the difficulty or ease of the listening input/level of input,, the presence or absence of comprehension clues, tasks, information organisation, lexis, syntax) and (c) sound related factors (i.e. speech rate; voice and background noise; pause and hesitation phenomena; stress, intonation, and rhythm patterns; the number of speakers; expertness and the gender of speakers; accent; Turel, 2014b; Leveridge & Yang, 2014; Yang & Chang, 2014; Chen, 2011). In other words, when language learners listen to/view or are exposed to listening input, their personal characteristics, and input and sound related factors play a vital role in full-/half-/mis-comprehension of listening input. In short, listening comprehension is the process of many interweaved multifaceted cognitive processes (Diao, Chandler, & Sweller, 2007; Chang, Tseng, & Tseng, 2011; Meinardi, 2009; Goh & Taib, 2006; Wipf, 1984).

In terms of learners (and teachers) of a language as an FL/SL, it can be said that listening comprehension is a way of partial or total matching. This matching is between what language learners tune in to (i.e. linguistic and other communicative input/what the interlocutor says) and the resources the learners have. The resources can be: world, social-cultural and linguistic knowledge; abilities; strategies –cognitive and meta-cognitive -; contextual (processing) references in the form of textual references (i.e. Mary and John - they), referential connections (i.e. the verbal system and the visual representation system) or situational references (i.e. subsequent - the current similar event - and prior references - the previous similar event); skills and the linguistic co-text (i.e. the linguistic environment/context). These resources help learners grasp and arrive at an acceptable, logical and reasonable interpretation of the listening input. The degree of the power of this matching is largely dependent on the degree of using the above-mentioned sources effectively. The better language learners can use these sources and apply effective strategies, the better they can fulfil listening comprehension, which might mean that the listening comprehension process is complete and (close to) perfect. In short, listening comprehension in foreign/second language acquisition (FLA/SLA) occurs through the phases of language learning process shown in Figure 2 below.

Figure 2. The framework of the phases of listening comprehension in FLA/SLA process (Turel, 2015b)

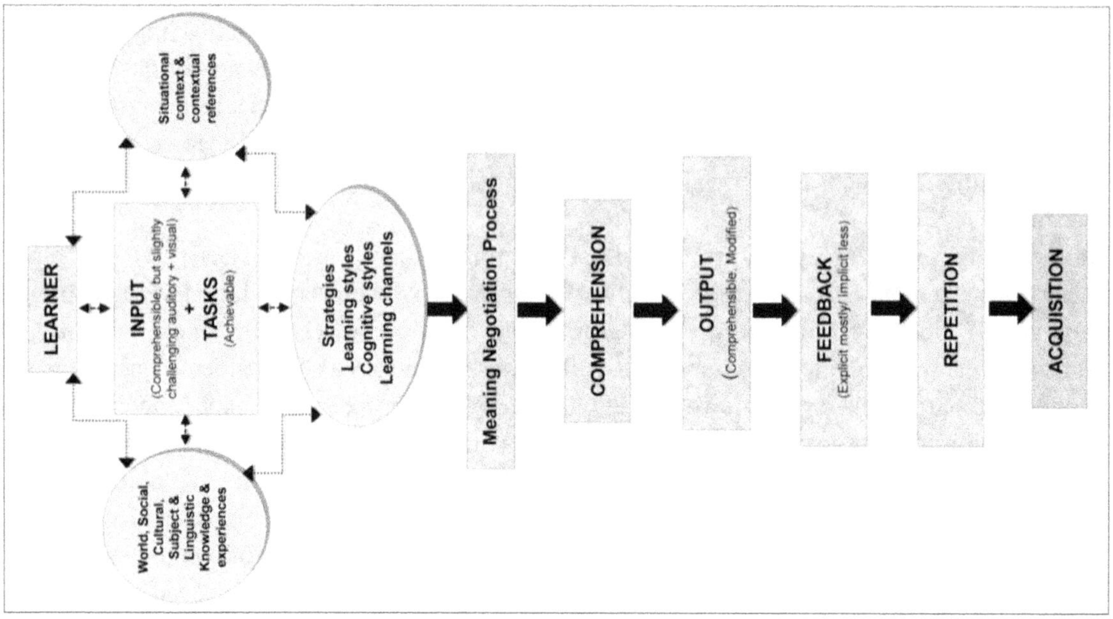

What does Figure 2 mean? For language learners to be able to comprehend the provided listening input, they:

- Need to be provided with *comprehensible input*. The assumption underlying this is that if language learners comprehend, then they can acquire FL/SL (see Vidal, 2003, p. 79 about the role of comprehension in vocabulary gain). Carroll (1977, p. 500) also says: "… when comprehension is successful, the person presumably has learned something he has not known before and has thus achieved a new kind of competence,' Additionally, Long (1983, p. 138) points out: "It is widely assumed… that samples… heard but not understood by a would-be acquirer… serve no useful purpose in the SLA process. Only comprehensible input will do". There is, however, a point about 'comprehension' that needs to be emphasised here. Comprehension that takes place through a combination of semantic and syntactic processing can make the linguistic characteristic of the input 'intake' and this holds the potential for developing the language learners' linguistic system (Chapelle, 1998, p. 23). In addition to *comprehensible input*, Krashen (1982, p. 21) manages to draw our attention to another vital point: "We acquire … only when we understand language that contains structure that is 'a little beyond' where we are now". In fact, Krashen is not the one who mentions this first. Casambre (1962, p. 166), for example, says: "'… a certain degree of difficulty is necessary to make subjects learn and retain the material with greater permanence. Pedagogically, it means that 'spoon-feeding' should be avoided".
- Use linguistic input (i.e. titles, grammatical structures, words, phrases, clauses, prefixes and suffixes, stress and intonation and international cognates). Here the difficulty or easiness of the listening text plays an important role. If they, for example, feature many unfamiliar items such as nouns, verbs, phrasal verbs, idioms, grammatical structures (Underwood, 1993, p. 17; O'Malley et al., 1989, p. 428; Markham & Latham, 1987, 168) and strong-regional dialects (Turel, 2004, p. 87),

then inevitably these will hinder listening comprehension (Rivers, 1981, p. 164; Boyle, 1984, p. 35; Anderson & Lynch, 1989, pp. 6, 56).

- Utilise available comprehension aids (i.e. supplementary aids) such as facial gestures, diagrams, imaginary drawings, cartoons, charts, still or motion pictures, maps, key words, captions and the like. While their presence can in general increase and improve listening comprehension, similarly their absence can also, to a great extent, handicap (Turel, 2014d; Jones, 2003, 2004, 2006, 2009; Jones & Plass, 2002; Arnold & Brooks, 1976, pp. 711-16; Omaggio, 1979, pp. 112-15; Mueller, 1980, pp. 335-40).
- Use their world and social-cultural knowledge of subject-matter, and their target language information, and then match it with what they hear. This can be a great aid in terms of enhancing listening comprehension (Long, 1990, pp. 65-80; Weissenrieder, 1987, p. 23; Markham & Latham, 1987, p. 168). The degree of using such informational sources to their advantage is dependent on the amount of such knowledge language learners have, and the capacity and capability of using it (i.e. the use of affective strategies, Turel, 2004, pp. 107-08).
- Use situational contexts and contextual references (i.e. subsequent and prior reference). On encountering listening, for instance, a listening text related to 'seeing a doctor', language learners can roughly guess what a doctor-and-a-patient conversation might consist of. Such a situational context can help language learners to narrow their scope and focus, which can result in better grasping and understanding. Language learners also use contextual references providing that they have. In the above example, language learners can make use of their previous experiences related to doctor-patient context (prior reference) to understand the 'seeing a doctor' listening text (the subsequent reference).
- Apply listening strategies such as (a) meta-cognitive strategies – advance organisers, directed attention, selective attention, self-management, advance preparation, self-monitoring, delayed production, self-evaluation, self-reinforcement-. (b) cognitive strategies – repetition, resourcing, directed physical response, translation, grouping, note taking, deduction, recombination, imagery, auditory representation, key-word, contextualization, elaboration, transfer, inferencing, question for clarification-. (c) social strategies, (d) affective strategies (O'Malley et al., 1985a; O'Malley, Chamot, Stewner-Manzanares, Russo & Kupper, 1985b; O'Malley & Chamot, 1990; Bacon, 1992).
- These processing stages, which are known as 'meaning-negotiation process' lead to *comprehension*, which can result in productive practice (i.e. *output*) in the target language (Ortega, 1997, p. 87). Produced comprehensible (modified) output can aid language learners to become aware of linguistic problems that may further help them to modify/develop their output. As a result, language learners might be forced into a more syntactic-processing mode (Chapelle, 1997, pp. 25-6), which is called *ideal comprehensible output*. This is considered an important contributor to language development both at linguistic and semantic level (Chapelle, 1998, p. 28). For instance, let us assume that a language learner listens to a native speaker (NS) who says what they bought the day before. After listening, the language learner says: 'I see, you buy**ed** a book yesterday' and the NS says: 'Yes, I **bought** a book'. In this example, after a successful comprehension process, the language learner produces output, which is comprehensible at semantic level, but problematic only at linguistic (i.e. grammar/syntax) level, and therefore receives (implicit) feedback. This can force the language learner into a more syntactic process (especially when output is corrected explicitly or more correctly when feedback as well as the produced problematic output manages to

attract the learners' attention to syntactic, lexical and/or semantical features that are problematic). Not only can this ultimately help language learners realise their mistakes, but it can also provide them with the opportunity to modify/develop the produced problematic output as a result of productive practice and received (implicit/explicit) feedback. It is because of this that Chapelle (1998, p. 24) says error correction affords the opportunity to "focus on form" (Long, 1988)' and "focus on form is expected to be beneficial when it occurs during the process of attempting to construct meanings". This process, which is referred to as 'negotiation of meaning' (Harrington & Levy, 2001, p. 15; Hegelheimer & Chapelle, 2000, p. 42; Long, 1996), however, mostly occurs during comprehension/communication breakdown when language learners are confused about meaning or syntax (Hegelheimer & Chapelle, 2000, p. 42; Smith, 2003, p. 52). Once, for example, the author of this study heard a child at the age of Key stage 2 (KS2) calling another child at the age of KS1 'gay'. The author got confused (the meaning-negotiation process began), as a child could not be a gay. After negotiating with an NS, he found out the word 'gay' was used as a kind of 'swear word' (the meaning-negotiation process finished). In another case, the author of this study heard phrase 'Uncle Hitler' from a TV programme. When he used that term for practising purposes while speaking to an NS, that person looked at him strangely, an attitude (related to lexis) the assumption of which the author of this study has not figured out yet (i.e. unfinished meaning-negotiation process). Having comprehension problems can draw language learners' attention to 'certain' aspects of the target language. This can lead them to request help in the form of accessing feedback, asking someone (Williams, 2001, p. 337; Smith, 2003, pp. 39-40) or looking at a resource book. This 'meaning negotiation process' can result in comprehension, which is very likely to lead to (correctly modified) output that is more likely to be learnt (i.e. *acquisition*), as language learners have to make *more effort* to process, comprehend and produce. Thus, the learners are unlikely to forget according to the depth processing theory, which suggests that without enough effort (deep processing) information will be forgotten (Craik & Lockhart, 1972).

- In sum, there is an active interplay between all phases/elements illustrated in Figure 1, where each phase/knowledge source has a two-way access to every other source in the task of analysing and interpreting the sensory *input* (i.e. both linguistic and non-linguistic) while using available sources and applying effective strategies to match with the input. This results in *comprehension* after the *meaning-negotiation process*, which leads to productive *output* that results in foreign language *acquisition* (in a non-target language community, SLA in a target language community). Some of these elements - words, phrases, grammatical structures and the like - are kept temporarily in short-term memory, which is limited in storage capacity and therefore can only hold a limited amount of information for a limited amount of time; while semantic cues are acquired and sustained in long-term memory. It is even like this in L1 comprehension. We, for example, do not remember words, phrasal-structures, and sentences, wholly, even partially, after a period of time, but we do mostly remember semantic representations. This is limited and subconscious, and also depends on some factors such as reviewing (i.e. effort), which is considered an effective factor in FLA/SLA according to some theories (i.e. the social learning theory, the conditioning theory; Robinson, 1989, pp. 119-130, 1991, pp. 158 - 59; Bandura, 1977; Skinner, 1953) and findings (Schmidt, 1990, p. 147).
- All of these require adaptive hypermedia (AH) designers and developers to be more careful, as a wide range of digital elements can be combined and delivered on the same digital platform instaneously, and thus they are more likely to make the provided input unchallenging or cause cogni-

tive overload. As a result, there are some implications of these for the design and development of AH listening environments, which are discussed below in details.

- The first implication is that listening input to be used in AH listening environments need to be *comprehensible*. The provision of comprehensible input in AH listening environments is also a requirement of the comprehension input hypothesis and theory, which state that language learners acquire only when they understand, and therefore "consider intake synonymous with comprehensible input"(Krashen, 1984, p. 21; Ellis, 1997, p. 47; Tschirner, 2001, p. 311). Likewise, it is also a requirement of the social learning theory, which "posits that people's judgements about their potential ability to perform well or to cope in a situation actually affect their efforts…" (Robinson, 1991, p. 157). The person perception theory, which requires us to avoid focussing language learners' attention on their weaknesses so that they do not develop negative judgements about their ability to perform (ibid: 157), and the social-psychological theory and the socio-educational model, which focus on the role of attitudes and motivation in FLA/SLA (Gardner, 1985, p. 158; Masgoret & Gardner, 2003, pp. 158-59, 124, 127) also require us to provide comprehensible listening texts in AH listening environments as a part of SLL/FLL. Comprehensible input can also be considered as a requirement of the cognitive load theory, which "places a primary emphasis on working memory capacity limitations as a factor in instructional design" and suggests that "too many elements of information may overwhelm working memory, decreasing the effectiveness of instruction" (Kalyuga, 2000, p. 161; Sweller, 1999). Language learners have limited processing capacity and thus proper allocation of cognitive information is critical to learning (Kalyuga, 2000, p. 161; Hedberg, Harper & Brown, 1993, p. 6; Sweller, 1999). Failure to take into account working memory considerations can cause cognitive overload particularly in AH listening environments where many elements are combined and presented on the same digital platform, and are optimum and personalised. In short, input that features many unfamiliar lexis (i.e. words), syntax (i.e. grammar), accent and so on, and thus is beyond language learners' current level can cause risk of overload on language learners' limited working memory.
- The second implication is that the *comprehensible linguistic input* used in AH listening environments should be *slightly challenging*, which is -as mentioned above - also a requirement of the comprehension input hypothesis. Unchallenging input, like very difficult one, can demotivate language learners, especially autonomous learners, as they are by themselves and this might cause them to get bored and quit very easily. Moreover, *slightly challenging* input require language learners to put more *effort* and revise, which is necessary for FLA/SLA according to the depth-processing theory, which suggests that without enough effort (deep processing) input will be forgotten (Craik & Lockhart, 1972). In short, as a Kurdish Proverb says, we need to: "Plough neatly, deeply but do not hurt the oxen".

This (i.e. providing *comprehensible*, but *slightly challenging* input that requires *enough effort,* but does *not develop negative judgements…*) is what need to be achieved in AH listening environments, where much more can be achieved better in comparison to conventional materials. In sum, designers and developers of AH listening environments need to make sure that the provided comprehensible input is *slightly* beyond the current level of the language learners.

- The third implication is that the designers and developers of AH applications need to draw the language learners' *attention* to both syntactical (i.e. recognition) and semantic (i.e. selection/meaning) characteristics of the comprehensible input in AH listening environments. Although comprehensible input is necessary, this alone is not sufficient for FLA/SLA. Comprehensible input, for the author of this study, does not mean the input the surface meaning (i.e. syntax) of which is understood, but also input the deeper meaning (i.e. grammar structures, culture-based meanings, the illocutionary forces/a speaker's intention) of which is understood. Drawing learners' attention to both syntactical and semantic characteristics of the comprehensible input is a requirement of the noticing hypothesis, which suggests that paying/drawing attention to (specific) forms in the input is necessary for learning new linguistic features of a language (Schmidt, 1993, 1995; Nicholas, Lightbown & Spada, 2001, p. 721; Leow, 2001, pp. 114, 144). This can be implemented effectively in particular in an AH environment, where a wide range of digital elements can be combined and presented much more efficiently (Turel, 2015a; Turel, 2014a; Turel, 2014b; Turel, 2014c; Turel, 2012). In terms of noticing, while some claim that forms are perceived first (Tschirner, 2001, p. 312), others claim that we go 'for meaning first' (Krashen, 1984, p. 21). The author of this study personally thinks that the priority is changeable depending on language learners' level and transactional objectives. Low-level language learners might go mostly for forms first, while upper-level language learners might go mostly for meaning. In the same way, when language learners have heard a form or structure (i.e. unplanned form; Williams, 2001, p. 327) which they have been looking forward to for ages (ibid: 326), then forms are likely to be perceived first. In like manner, when language learners are interested in 'what's going on', in this case semantic is perceived first (although the priority might even change a few times at one single exposure). Some might call it 'a personal anectode', but, this is, at least, what has happened to the author of this study, who *has* learnt both a second language and a foreign language for ages.

- Finally, in addition to providing *slightly challenging comprehensible input* in an AH environment where language learners' *attention* is drawn to *syntactical and semantic characteristics* of the input, the designers and developers of AH listening environments need also to provide language learners with *the opportunities to produce comprehensible (modified) output* (Ortega, 1997, p. 87). According to the comprehensible output hypothesis, in order for input to lead to full productive competence, language learners need to be given the opportunity to produce (comprehensible) output. The opportunity to produce comprehensible output further promotes noticing, contributes to language learning through 'hypothesis testing' and serves as a metalinguistic function - i.e. the opportunity to think about and analyse the -produced- forms and structures - (Shehadeh, 2002, p. 608; Tschirner, 2001, p. 311; Swain, 1985, p. 249, 1995)

Listening Skills

Listening skills are natural and global forever learning skills. Language learning is normally carried out by using senses of vision, hearing/listening, speaking and the others (i.e. tasting, touching and smelling), which are called 'senses of human being' as well as reading and writing. We can, for example, require language learners to touch or hold cotton when teaching the words 'soft' or 'light'; a stone when teaching 'hard' and 'heavy' or require them to taste 'honey' when teaching 'sweet' or smell perfume/a flower while teaching 'smell', 'nice smell' or 'bad smell'. These, doubtless, have their role in the language learning process. In terms of listening, the first two senses (i.e. vision and hearing) might play more important roles in comparison to 'speaking', as listening relies on both auditory and visual input and requires the use of sight and hearing when listening to someone talking or listening texts. Of course, when the listening text is audio-lingual, the sense of sight cannot be used. Instead, imagination can be used. Likewise, the other communicative performance skills (i.e. speaking and writing) enable 'productive output' which is not only necessary for the listening process to be complete, but is also vital in the language learning process. The author of this study, for instance, once watched a TV programme in which one NS said: "Finally, I've figured out what kind of jerk I am", while he was an inpatient in the hospital during his PhD. study. He (partly) comprehended the sentence and wanted to practice the 'new' word 'jerk' as a conscious language learner, and thus he said to one of the NS inpatients: "You know, finally I've figured out what kind of jerk you are", the objective of which was just to practise. Although the patient got really very angry, this unfortunate productive output in the form of speaking skill helped the author of the study to learn the new word truly (although it has other meanings, as well), as it became clear that you could use that word for yourself, but not always for others whenever you want (negotiation of meaning). This simple concrete example shows the role of careful listening (i.e. more accurately the role of comprehension and paying attention to syntactical and semantic features), as well as speaking in the form of reproducing the comprehended input in the form of comprehensible (but problematic) productive output in FLA/SLA.

When someone looks at the percentage of language learners who are auditory and visual (Reid, 1987, pp. 96-97; Dunn & Dunn, 1979, pp. 238-44), the importance of listening skills in FLA/SLA becomes clearer. This means that most language learners use their senses of sight and hearing, which are most required for listening. The implication is that more emphasis on teaching/learning of listening skills in AH listening environments might better contribute to FLA/SLA. In other words, more and better AH listening environments as a part of FLL/SLL need to be designed and developed both for classroom and autonomous study. In such applications, in particular in those created for autonomous study, the teaching of listening can be delivered in the form of instructions, global and local help, feedback and tasks (Turel, 2012, pp. 35-49). These are some of the possibilities that are considered for strategy training in classroom and autonomy (Hurd et al., 2001, pp. 346, 350; Kohonen, 1991, p.109; Mangiafico, 1996, p.112) and facilitators of learning (Sturtridge, 1987, p. 9).

Maybe it is because of the value and importance of listening in language learning process that before commencing to say basic meaningful words, babies listen for roughly one to one-and-half years and then begin to utter meaningful chunks. Though this is not an FLL process, there still remains invaluable facts in terms of the role listening can play in the FLL process. The question, on the other hand, is that although listening is so valuable in terms of (a) the senses devoted, (b) its frequent use in daily life, (c) its role in learning L1 and (e) FLL/SLL, it is hard to say that it is exploited appropriately. As a result, not only do language learners lack effective listening skills, as "most-learners do not come into the learning situation with the knowledge and skills to determine content and learning processes which will enable them to reach their objectives in learning another language" (Nunan, 1997, p. 201), but they do not also utilise such listening skills to learn effectively. Fortunately, the listening skill has been considered as more central in the FLL/SLL and teaching/learning process in the last two decades. It has been obvious that this skill is vitally necessary in this process and needs to be taught more overtly (Rost, 1990; Brown & Yule, 1992; Ur, 1992).

It has also been clear language learners of all abilities and levels can develop and extend their listening skills, as such skills are teachable/learnable. To this end, some researchers state language learners can improve their language performance by being trained to use beneficial strategies (Rost & Ross, 1991; O'Malley et al., 1985a; Tezza, 1962). In an experiment, the learning strategies of treatment groups, who were trained for 50 minutes daily for 8 days, improved, and they outperformed the control group

(O'Malley et al., 1985b, p. 574). Rubin and Thomson (1992, p. 10) also state: "...students can be taught to control their use of listening strategies". In the same way, Cohen (1984, pp. 101-12) states that foreign language learners can use strategies that have been shown to be successful to quicken learning. Additionally, Ehrmann (1963, p. 18-20) reported that listening training improved the aural comprehension of Hebrew although some other studies partly differ from those obtained by the above researchers (Rubin & Thomson, 1992; O'Malley et al., 1985a; O'Malley et al., 1985b; Tezza, 1962, pp. 39, 46).

It appears that listening skills:

- are a fundamental skill in the process of FLL/SLL, and relies on the senses of 'sight' and 'vision' depending on listening mode. So, when creating AH listening software, the focus should be on both 'visuals' and 'auditory' input. In the same way, all types of listening materials (i.e. audio-only, audio + supplementary contextual visuals, audio + animation, talking-heads video + supplementary contextual visuals, video-only) need to, and should be provided in AH listening environments (Turel, 2011).
- are teachable and learnable. Teaching/learning listening skills does not end in the classroom. Thus, (autonomous) AH listening environments should "function as a surrogate teacher" (Frankel, 1987, p. 53) as the whole point of skills development is to develop autonomy, which is the main way of life-long learning. This is the requirement of the autonomous learning theory, which "demands that learners take control of their learning" (Voller, 1997, p. 106). Improving language learners' skills and strategies, which considered "the main issue" in autonomy (Benson, 1997, p. 19) should be the objective of the AH listening applications. AH listening applications should aim to help language learners to develop their ability to learn independently (i.e. to equip them with the listening skills and techniques they need) through guiding, task involvement and encouraging them to take responsibility for their own learning and to apply affective strategies, while improving their listening.
- Listening training can help language learners to acquire strategies and use them (Hurd et al. 2001: 346; Goh, 2000, p. 73; Field, 1998, p. 117) although all levels of language learners might not benefit to the same degree. Low-level language learners can, for instance, benefit more, which means that AH listening applications can be more effective in their training (Turel, 2012). Through tasks, language learners can, for instance, be required to use certain cognitive strategies such as 'infer missing or unfamiliar words using contexts, co-text and prior knowledge'; 'predict contents before listening using context and prior knowledge'; 'use prior knowledge to elaborate and complete interpretation'; 'take notes of content words', relate one part of the listening text to another', or 'visualise scenes' (Goh, 2000, p. 72) at different stages (Turel, 2012). Likewise, language learners can be requested to employ certain metacognitive strategies such as to 'pay attention to visuals'; 'establish purpose for listening'; 'listen selectively according to purpose' or pay attention to certain input (Goh, 2000, p. 72) or evaluate their progress (Turel, 2012). Such requirements are in line with constructivist approaches to FLL/SLL, as according to the constructivist learning theory FLL/SLL is a dynamic process where language learners are the architects, rather than the passive recipients of knowledge.
- In short, listening skills are necessary and teachable, and play a role in FLL/SLL process (Rost, 1990, p. 154). These skills are not always, however, taught and acquired properly. This means that the skills are not used to the advantage of language learners fully. This might be due to many reasons such as the lack of (1) listening courses in the curriculum, (2) properly trained (listening)

teachers, (3) appropriate teaching, (4) appropriate materials, or (5) due to the overcrowded classes and so on. For example, although the importance of listening in FLL/SLL was mostly first emphasised a few decades ago, in many countries listening is not practically included in the curriculum adequately yet. Likewise, many teachers are either not trained properly or do not teach as they are supposed to (TES May 21, 2004, p. 3). The teachers are not always aware of the language learners' needs or do not always care satisfactorily. Likewise, for instance, although the role of visuals in FLL/SLL is well known (Furnham et al., 2002, pp. 191-210; Ginther, 2002, pp. 133 - 67; Dodson & Schacter, 2002, pp. 782: 802; Gyselinck et al., 2002, pp. 665-85; Amrhein et al., 2002, 843-57; Al-Seghayer, 2001, p. 203; Secules et al., 1992, pp. 480 - 90), they are not used enough by teachers as well as in AH listening applications. The author of this study was shocked when he realised the big gap between the implications of findings in the field of FLL/SLL and CALL, and what really often goes on in classrooms. Learning materials, whether they are conventional or AH environments, are also not always convenient (Turel, 2014). In one case, the language learning materials were mostly about robbery, burglary, and stealing. The author of this study felt as if he were attending a 'robbery course'. When he had private conversations with some other language learners who were attending the same pre-sessional course, he realised that they felt the same, but nobody felt safe enough to tell their opinion. This is often the case in our civilised (!) world particularly in a century when flattering is favoured most and people dare not often to tell their honest opinion (TES April 30, 2004, p. 1) although the benefits of truth extends far beyond individuals and the century we live in, and its contribution to humanity, as a whole, is tremendous and invaluable.

Adaptive Hypermedia

When one nowadays speaks of 'AH' mostly a few things comes to mind: (a) it is the combination and instantaneous delivery of different digital elements such as sound, video, text, pictures, graphics and so forth on the same digital platform. (b) It is navigational and interactive. (c) It provides user control and ease of use, and (d) it offers 'personalized learning'. The ultimate objective of educational AH is to design and develop pedagogically sound and epistemologically flexible personalized digital learning environments which both supports a wide range of learners who are diverse in terms of abilities, disabilities, levels, interests, backgrounds, and other characteristics, and is also to enable learners to make maximum use of the available personalized interactive, navigational, and simultaneous learning environments (Turel, 2018, pp. 2357-68; 2015a, pp. 2495-96; Shute & Zapata- Rivera, 2012; Brusilovsky, 1996:),

Turel (2018, 2015a, p. 167) defines AH as a design, combination and delivery of a wide range of optimum digital elements on the same digital platform which as a whole provides a multidimensional, multi-sensory environment in which multiple rich, efficient, instant, comprehensible, optimum and meaningful input and feedback can be presented to the learners of all subjects at all stages of learning (Figure 3; see Brett, 1998, p. 81; the Dictionary of Multimedia, 1997, p. 151; Ashworth, 1996, p. 81; Thierry, 1996, p. 7 for further definitions of pertinent terms such as multimedia, interactive multimedia, and multiple media). It appears that AH is a powerful digital tool, as it enables materials developers to combine and deliver a wide range of different digital elements on the same digital platform in very beneficial and meaningful ways, which cannot be provided by any other conventional tools. The links in AH can be in the form of hyperlinks, hypertext, buttons, hotspots or hotwords so that the ultimate users can retrieve information and/or navigate. In other words, certain words or units provide links to other documents or move ultimate users to another position in the environment or display information pertinent to the particular links on the current positions. This provided information can be in the form of sound, text, picture, graphics, video and so on, or optimum combinations. AH enables target learners to make personal preferences, and it records these personal preferences, their individual needs and learning goals, and then uses them throughout interaction with the target learners in order to meet their personal needs so that they can learn better (Turel, 2018, pp. 2357-68; 2015a, p. 2497). This is relatively a new direction in the field of digital learning environments (Brusilovsky, 2012, p. 46) and can consists of different models (Kahraman, Sagiroglu & Colak, 2013, p. 60).

Figure 3. Combination and delivery of a wide range of digital elements on the same digital platform (Turel, 2015b)

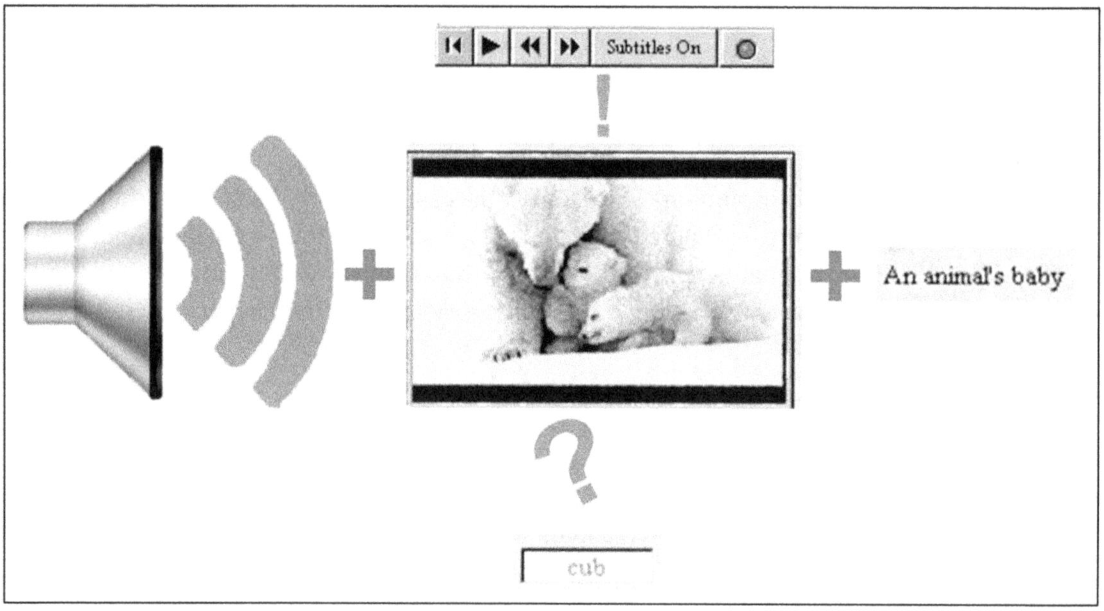

Adaptive Hypermedia in FLL/SLL

In terms of FLL/SLL and teaching, there are some outstanding differences between AH applications and conventional materials. These outstanding differences are: (a) the adaptive optimum combinations and delivery of a wide range of different digital elements on the same digital platform. This cannot be provided with any other media type. (b) AH applications are navigational and interactive. (c) Language learners have better control over AH applications. (d) It is easier for language learners to use AH applications, as ‘everything’ is on the same digital platform as well as instantaneous due to the presence of hyperlinks in the form of hotspots, hypertext and buttons. In the same way, AH language applications have many different combined elements, which can make (a) input more easily comprehensible and (b) learning enjoyable, all of which can result in FLL/SLL better (Turel, 2018; 2015a; Turel, 2014e, pp. 292-309; Stevens, 1995, pp. 289-99; Stepp-Greany, 2002, p. 172). For instance, in Soboleva and Tronenko’s (2002, p. 495) study, language learners appreciated different aspects of CALL, whether it is delivered on the Web or on CD-ROM format, and they found it interesting, rich, simulative, helpful, interactive and so on. These results are also supported by other studies (e.g. Turel, 2014a). In Borras and Lafayette's (1994, p. 67) study, the majority preferred the computer laboratory to the language laboratory. The positive attitudes towards AH are also agreed on by language teachers in that they think, "the use of computers would motivate the student to study (81.47 - 0%) …" (Carbola-Calero, 2001, p. 11). The results of other conducted investigations revealed that language learners remembered better from the digital platform treatment then the traditional treatment (Crosby, Stelovsky & Ashworth, 1994, pp. 3-13). A study by Brett (1997, pp. 39-53) also reports that language learners' success rate while using digital platforms is greater than traditional ones (ibid: 45-6). A research study by Deville, Kelly, Paulussen, Vandecasteele & Zimmer (1996, p. 81), which was conducted on a group of beginner language learners who accessed interactive and navigational digital listening environments, revealed that language learners generally performed better in listening comprehension. Likewise, interactive and navigational digital platforms are also effective for other skills (i.e. writing, reading, teaching of grammar and culture) (Soboleva & Tronenko, 2002, p. 494; Adair-Hauck, Willingham-McLain & Youngs, 1999, p. 294) although speaking was "… clearly beyond computers' working possibilities" a decade ago (Fleta et al., 1999, p. 55). It is now feasible even to integrate oral responding tasks in listening tasks with the maturity of natural language processing technology. The TOEFL integrated listening and speaking tasks are, for example, good examples although they were once only used for assessment purposes (Xi, 2010, pp. 291-300).

Since FLL/SLL is a multi-channel phenomenon, the combination of adaptive optimum different elements provides a multidimensional, multi-sensory, interactive, simultaneous, navigational, and adaptive (personalised) environments in which language can be presented in different ways (Turel, 2011). This can meet the needs of different individual learners who have different learning style preferences (Carson & Longhini, 2002, p. 408; McLoughlin, 1999, pp. 222-23; Brickell, 1993, p. 103), although in terms of listening skills only two of the sensory channels- visual and acoustic - are especially privileged. For instance, while the provision of audio-only meets the needs of auditory learners (Ridgway, 2000, p.182), video can meet the needs of visual learners (Adair-Hauck et al. 1999, p. 289; Peter, 1994, p. 202; Hart, 1992, p. 5; Tschirner, 2001, p. 310). Likewise, functional interactivity facilitates the negotiation of meaning, which is necessary in FLL/SLL, as learners can access personalised hyperlinks, glossaries, feedback, captions, and so on immediately, and find out what and why they have not understood, and the underlying assumptions (Turel, 2015a). All these pedagogically and epistemologically positive features of AH facilitate listening comprehension, draw learners' attention to 'their weaknesses' and certain aspects (i.e. syntactical and sematic features) of input (Nicholas et al., 2001, p. 721; Leow, 2001, pp. 114, 144; Schmidt, 1993, 1995) and result in depth processing (Craik & Lockhart, 1972), all of which are necessary conditions for FLL/SLL.

The combinations of adaptive optimum different digital elements through links on the same digital platforms provide navigational and functional interactivity and multi-dimensional and sensory support (Herron, Dubreil, Corrie & Cole, 2002, p. 37; Leffa, 1992, p. 66). For instance, when learners need to, they can access personalised instant dictionaries, grammar structures, captions, feedback and the like. All of them are combined and are on the same digital platform. Learners can re-listen to / re-view the same listening texts without losing time. All of these are good opportunities to exercise control over FLL information (Tschimer, 2001, pp. 312-13; Brickell, 1993, p. 103). These pedagogically effective and efficient aspects of AH can enable learners to find out the difficulties, the right solutions, what the rules are; to analyse the mistakes that have been made; and to find out why they have made such mistakes by assessing their answers, recording and scoring them, pointing out and explaining mistakes.

This outstanding capability of AH provides more real-world learning contexts and more authentic and personalised interactive tasks (Turel, 2018; 2015a).

AH environments provide learners with the opportunity to learn FL/SL at their own pace, without fear of making mistakes in the presence of a teacher or other learners. AH environments are tension free environments in which learners can use computers individually, as learners have no teachers and friends present. In particular, during self-study, AH environments are private and flexible workplaces where learners can take risks; work in their own place (Turel, 2018; 2015a; Tschirner, 2001, p. 307), in their own time, at the pace they need, and in the way they enjoy. The assumptions underlying these are that (a) AH environments give learners the control (Soboleva & Tronenko, 2002, p. 493; Trinder, 2002, p. 75), and (b) provide them with different optimum personalised choices, tasks and feedback. In other words, AH environments enable learners to decide what to study, when to study, how to study, how long to study, how often to study, and even where to study in a considerably easier environment, as wide range of adaptive digital optimum elements are on the same platform and everything is simultaneous. These can make learners feel more comfortable and might result in promoting development of self-confidence and provoking working hard. It is due to these reasons that learners do not complain about the fear of making mistakes while using adaptive digital learning environments (Deville et al., 1996, p. 83). Conversely, learners express their comfort of working with such environments (Trinder, 2002, p. 75; Fleta et al., 1999, p. 55; Brett, 1997, pp. 46-7; Peter, 1994, pp. 157-58; Hoven, 1999, p. 91).

Since AH offers learners many choices (i.e. sound sequences, video clips, video/sound clips with personalised optional captions, vocabulary and grammar links, pertinent comprehension tests with immediate, remedial and innovative feedback, instructions, visuals and the like), it is highly motivating. When learners make mistakes, this does not even de-motivate them because they have the opportunity to receive instant and meaningful feedback, and practise as many times as they want to, and need to while working with digital environments (Soboleva & Tronenko, 2002, p. 483; Ayres, 2002, pp. 247-48; Lyall & NcNamara, 2000, p. 9). This is an important factor in FLL/SLL especially during autonomous study because "how to engage the interest of the learner and so sustain his motivation to learn" is considered "a related problem for the writer of self-study materials" (Frankel, 1987, p. 53). Most importantly, not only is motivation "directly linked to achievement" according to the socio-educational model (Masgoret & Gardner, 2003, p. 129), which suggests that integrativeness and attitudes towards the learning situation cause learners' motivation to learn an FL/SL and motivation is responsible for achievement (Masgoret & Gardner, 2003, p. 124; Gardner, 1985, p. 158), but motivation is also common to all models of FLL/SLL (i.e. the acculturation model, the conscious reinforcement model, the intergroup model, the interactionist SLA model, the LMR-plus Model, the monitor model, the social context model, the social psychological model, the elaboration Theory model – Gardner, 1985, pp. 142, 125-66).

Technically, pedagogically and psychologically, it is easier for the designers and developers of AH to combine and deliver different digital optimum elements on the same digital platform more effectively in hypermedia environments. For instance, hypermedia (i.e. digital learning environments) provides a non-linear editing facility. Thus, materials writers can cut and move digitised audio/video in any order (Tschirner, 2001, p. 307). Moreover, digital sound has more advantages in comparison to tape-cassettes for learners. These are: "(1) The option of instantly accessing sound in non-linear form, (2) The facilities of stepping and isolating, (3) Instant record, replay and comparison, (4) The ability to synchronise text and graphics" (Willets, 1988, p. 4).

Being able to feature in, combine and deliver different digital optimum elements on the same adaptive digital platform, AH can guide learners more effectively. For example, when learners make a mistake, they receive personalised instant and meaningful feedback (Turel, 2018; 2015a; 2012, pp. 35-49). This can enable them to find out why they have made the particular mistakes, how they can overcome such difficulties in future occasions, and improve new strategies. Due to these reasons, digital environments are considered efficient for self-study use (Soboleva & Tronenko, 2002, p. 483; Brett, 1996, p. 206). Additionally, if learners are instructed about which strategies they need to follow in which situations, then they can be directed and guided (Turel, 2018; 2015a; 2012; Debski & Gruba, 1999, pp. 219-20; Barnett, 1993, p. 303). This is important because learners can be taught to "use appropriate comprehension strategies" (Goh, 2000, p. 71). This is mostly needed during self-study, as learners are by themselves (Debski & Gruba, 1999, p. 219-20).

AH can prepare learners more effectively and adequately for listening input. The assumptions are that, for instance, AH empowers the designers and developers of listening materials to (1) provide learners with different personalised optimum digital elements such as unknown items, grammatical rules, special features of text type, short audio messages, short video clips, sample sentences, graphics, animations, visuals, simplified written versions of the text, information about the speakers, their roles, how they interact, the content and so on at the pre-listening stages, which can not only be implemented in different and efficient ways, but also prepares learners very well to the while-listening stages (Turel, 2015c).

AH can qualify learners to overcome potential sources of difficulties of listening input such as unfamiliar items, proper names, cultural difficulties, grammar structures, fast speech, and unfamiliar accents (Turel & McKenna, 2015; Turel, 2018; 2015a; 2014c; Turel & Kilic, 2014). The meanings of unfamiliar items can be provided through personalised hypertext, hotspot or other links such as dictionaries. These can be explained in target language through synonyms or antonyms (Turel & McKenna, 2015). Likewise, learners' attention can be drawn to cognates, false cognates and polysemous words, which are useful for vocabulary acquisition and helpful for understanding, and improving listening as well as the other skills (Vidal, 2003, p. 80, Hammer & Monod, 1978, p. 32; Nakic, 1981, pp. 11-12), at the pre-listening stages. Additionally, simple sentences and short paragraphs featuring unknown items can be provided. In some cases, the unfamiliar items can be explained throughout visuals, audio or video clips. Their equivalents in L1 can be given if LLs are monolingual or if we know the L1 of the target customers. Unfamiliar proper names can also be given in advance so that learners will not have difficulty. Such names can be given through personalised optimum pictures and cartoons that have labels and instructions. Cultural differences and unknown grammar structures can be explained and illustrated through simple personalised samples, pictures, audio or video clips. The assumptions behind cultural differences, for instance, can be elucidated and presented in different ways (Turel & McKenna, 2015; Turel, 2014d, pp. 268-293; Turel & Kilic, 2014, pp. 245-267).

Fast-speech and unfamiliar accents can be overcome in AH in different ways: (1) They can be overcome by providing captions, as they make fast speech comprehensible (Turel, 2015d). (2) Fast speech can be overcome by giving learners the control of speech-rate or alternatively by providing personalised slow versions in addition to the authentic versions (Turel & McKenna, 2015). Captions can also help learners to come to grips with proper names (Turel, 2015d). Not only do these help learners to overcome the difficulties, but they also help learners comprehend, gain confidence, and prepare for future occasions, which are very beneficial (Fox, Romano-Hvid & Sheffield, 1992, p. 48; Deville et al., 1996, p. 82).

In conventional materials, feedback is normally given in learner's books or answer key books. When learners have difficulty, they can access them. Although this is useful, feedback in conventional materials is very limited, as (a) it is not instant and (b) it consists of restricted elements such as text, pictures and graphics. In AH, feedback is (a) immediate and personalised, (b) can consist of different optimum digital elements (i.e. audio, video, visuals, text, animation or optimum combinations) which meet both learners' visual and acoustic needs resulting in FLA/SLA and (c) can be conditional (Turel, 2012, pp. 35-49). Such simultaneous and personalised feedback can help learners to (a) find out what and why they could not understand and (b) overcome the difficulties causing them not to comprehend, which facilitates meaning negotiation, draws attention and raises conscious/metacognitive awareness. This can guide and lead learners to develop new and effective strategies, which is one of the targets that material writers want to, and need to, fulfil especially in autonomous materials (ibid: 35-49). To this end, a famous scholar, Kahlil Gibran, in his book entitled 'The Prophet'; has a poem, the part of which runs thus: "Then said the teacher: 'Speak to us of teaching.' ... 'If he is indeed wise, he does not bid you enter the house of his wisdom, but rather leads you to the threshold of your own mind".

All learners of all subjects do not have the same background and abilities. While some have high abilities and know quite a lot about target language and target culture, others may not. As AH enables materials writers to make use of different digital optimum elements at personalised level, which can make input comprehensible at different levels and create personalised gradual and different tasks, each individual can find what is most appropriate for them or a way of working which is most convenient for them. For instance, learners with high proficiency can be provided with personalised listening input (i.e. listening texts without captions), while those with low proficiency can be provided with captions. In the same way, if the speech rate is too fast for certain learners, then they can be provided with slowed down or the slow versions in AH. In short, AH provides learners with personalised digital learning environments. That is why; it is called 'adaptive' hypermedia.

Hypermedia can be used for classroom use and individual use (Turel, 2018, 2015a) like all types of digital learning environments (Soboleva & Tronenko, 2002; Gillespie & McKee, 1999), as it has much to offer. The results of research by Mangiafico (1996, p. 106) suggest that the use of the same digital learning program for both classroom use and individual use can be equally beneficial in enhancing students' FLL/SLL listening comprehension. Although the author of this study thinks and believes that AH is beneficial for both classroom and individual use, he believes that it can be more helpful for individual-use, as it is 'adaptive' (depending on the nature of AH applications).

The effective and efficient design of AH environments (i.e. personalised optimum combination of a wide range of digital learning features on the same digital platform), on the other hand, require the designers and developers of AH environments to consider the requirements of a wide range of theories such as the dual coding theory and the generative theory of multimedia, which suggest that presenting two concurrent elements to teach one element is effective, as it helps guide learners' cognitive processes (Turel, 2018, 2015a; Moreno & Mayer, 2002; Al-Seghayer, 2001, p. 226; Mayer, 1997; Chun & Plass, 1996, p. 515; Mayer & Sims, 1994; Mayer & Anderson, 1992; Wittrock, 1990; Paivio, 1986) although this is not always the case (Amrhein et al., 2002, pp. 843-57; Kalyuga, 2000, pp. 2-3). In the same way, this aspect of hypermedia also requires the designers and developers of AH to consider the requirements of the cognitive load theory. Therefor, it is essential to design every (optimum) digital elements of AH listening environments efficiently (i.e. personalised optimum combinations or 'right balance' for 'the right learner') (see Figure 4 for optimum combinations).

Figure 4. Optimum design of different digital elements on the same digital platform

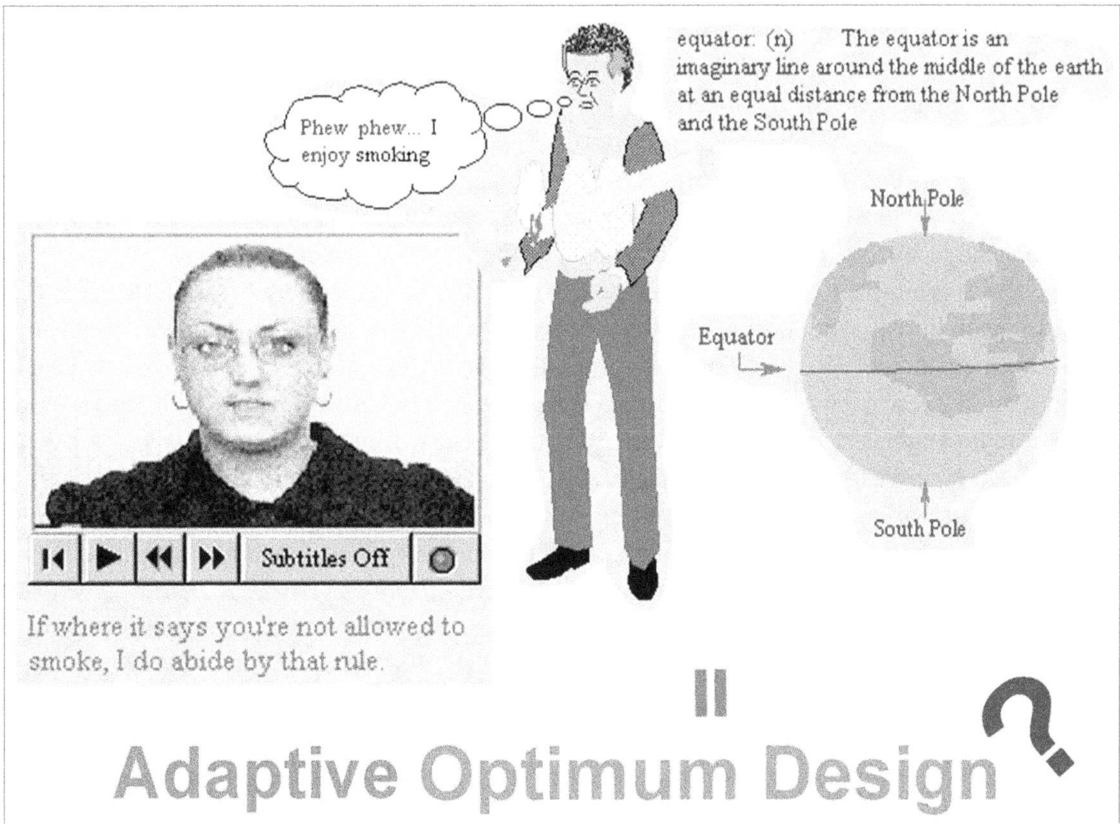

To sum up, educational digital technology provides an invaluable digital environment that enables the designers and developers of AH listening applications to overcome listening difficulties and provide language learners with conditions that facilitate listening comprehension as a part of FLL/SLL. These can make an undeniable contribution to listening skill development as a part of the FLL/SLL process.

THE LIMITATIONS OF ADAPTIVE HYPERMEDIA IN TEACHING LISTENING

Though computer based educational technology was not so not perfect two decades ago (i.e. the quality of video clips, for instance, would diminish when they were compressed; Soboleva & Tronenko 2002, pp. 488, 496), since then technological developments in the field of educational technology have occurred very rapidly. As a result, very high quality and efficient digital computer technology is nowadays available. Therefore, AH designers and developers can design and develop technologically as well as pedagogically very highly sophisticated qualities.

The main problem in designing and developing technologically and pedagogically sophisticated AH is not the technological dimension any more. This is what, at least, the author of this study thinks so. Currently, there are many digital learning applications on the market that are not entirely sophisticated technically, pedagogically and psychologically (Turel, 2014a, pp. 167-183; Ferney & Waller, 2001, p.

156; Trinder, 2002, pp. 69-84). Such applications were even called 'shovelware' in the past (Clifford, 1998, pp. 2-8), as was the case more than three decades ago (Le Mon, 1988, p. 39). Some even featured spelling errors although they were written by NSs (TES Teacher 2004, p. 18). The problems fundamentally stem from digital material designers and developers (i.e. materials writers), not the technology itself, as in many cases many materials writers are not expert either technically or educationally. Only a team that consists of both pertinent educators and technologists can create both technically and pedagogically very sophisticated and ideal AH listening applications. If technically and pedagogically sophisticated AH applications on the market are surveyed carefully, it will be seen that teams made up of educators and technologists, not either, have created them. This is the case for many digital learning projects, as well (Phillips, Pospisil & Richardson, 2001, pp. 96-114; Grob & Wolf, 2001, p. 234). To be able to create in real sense cost-effective/professionally sophisticated AH (listening) applications, the active participation of all experts needed (depending on the type of the AH application we want to create) is a must. These experts are: instructional designers, (specialist) teachers, programmers, graphic designers, audio engineers, photographers, artists, voice actors, film directors/specialists, musicians, animators, (target) learners and so on (Turel, 2018, 2015a; Turel & McKenna, 2013, pp. 188–209). All truly professional works are always achieved by a team of pertinent experts, not by single individual, "where one person acts as a curriculum expert, content expert, programmer and instructional designer" (Hedberg et al., 1993, p. 4), The involvement of such experts is vital and has to be in true sense (Turel, 2018, p. 2364; Nicholson & Ngai, 1996, p. 3). Target learners' involvement, for instance, "produces more useable and effective" digital learning applications (Nikolova, 2002, p.112; Kennedy & McNaught, 1997, p. 6; MacGregor, 1993, p. 3; Eraut, 1988), although the findings in the field of digital learning environments and CALL are very often learners' preferences, views, ideas, progress etc. Equally, the lack of a specialist programmer hinders not only the use of the maximum potential of the computer tools or authoring programmes, but also results in lack of the minimum requirements (Grob & Wolff, 2001, p. 249; Lyall & McNamara, 2000, p. 8). The involvement of these experts needs to be in true sense (Turel, 2018, 2015a). In the past, there were even some digital learning programmes that created by experts, but were developed "with minimal reference to the educational research available, both within a particular discipline and about student learning in general" (Kennedy & McNaught, 1997, p. 1). AH applications also need to be evaluated, which happens while creating (i.e. formative evaluation) and after (i.e. summative evaluation), and revised accordingly (Colpaert, 2002, p. 439; Phillips et al., 2001, p. 5; Sims, 2000, p. 4; Adair-Hauck et al., 1999, p. 274). In other words, the stages of AH design and development are categorized into six separate stages (i.e. feasibility, setting up a team of experts, designing, programming, testing, and evaluating) and at each stage a wide range of principles and guidelines need to be borne in mind so that cost effective AH language listening environments can be designed and developed (Turel & McKenna, 2013, pp. 188–209).

In conclusion, it is the limitations of AH designers and developers that are the problem in the development of AH applications rather than the limitations of the computer based educational technology. This unique compiled book, which is a revised and further developed version of *Intelligent Design of Interactive Multimedia Listening Software* (Turel, 2015b), will further enable all linguistic and computer based education scholars, students, institutions; software companies and teams of digital learning material developers not only to better understand efficient and effective AH design and development principles and guidelines, but also will empower them to design and create efficient and cost effective AH listening applications for any language as a part of FLL/SLL.

Vehbi Turel
Bingol University, Turkey

REFERENCES

Adair-Hauck, B., Willingham-McLain, L., & Youngs, B. E. (1999). Evaluating the integration of technology and second language learning. *CALLICO Journal, 17*(2), 269–306. doi:10.1558/cj.v17i2.269-306

Al-Seghayer, K. (2001). The effect of multimedia annotation modes on L2 vocabulary acquisition: A comparative study. *Language Learning & Technology, 5*(4), 202–232.

Amrhein, P. C., McDaniel, M. A., & Waddill, P. (2002). Revisiting the picture-superiority effect in symbolic comparisons: Do pictures provide privileged access? *Journal of Experimental Psychology. Learning, Memory, and Cognition, 28*(5), 843–857. doi:10.1037/0278-7393.28.5.843 PMID:12219794

Anderson, A., & Lynch, T. (1989). *Listening*. OUP.

Anderson, J. R. (1995). *Cognitive psychology and its implications* (4th ed.). Freeman.

Anderson-Hsieh, J. (1992). Using electronic visual feedback to teach suprasegmentals. *System, 20*(1), 51–62. doi:10.1016/0346-251X(92)90007-P

Arnold, D. J., & Brooks, P. H. (1976). Influence of contextual organising material on children's listening comprehension. *Journal of Educational Psychology, 68*(6), 711–716. doi:10.1037/0022-0663.68.6.711

Ashworth, D. (1996). Hypermedia and CALL'. In M. C. Pennington (Ed.), *The power of CALL* (pp. 79–95). Athelstan.

Ayres, R. (2002). Learner attitudes towards the use of CALL. *Computer Assisted Language Learning, 15*(3), 241–249. doi:10.1076/call.15.3.241.8189

Bacon, S. M. (1992). Authentic listening in Spanish: How learners adjust their strategies to the difficulty of the input. *Hispania, 75*(2), 398–411. doi:10.2307/344077

Bandura, A. (1977). *Social learning theory*. Prentice-Hall.

Barnett, L. (1993). Teacher off: Computer technology, guidance and self-access. *System, 21*(23), 295–304. doi:10.1016/0346-251X(93)90021-8

Benson, P. (1997). The philosophy and politics of learner autonomy. In P. Benson & P. Voller (Eds.), *Autonomy & independence in language learning* (pp. 18–34). Addison Wesley Longman Limited.

Borras, I., & Lafayette, R. C. (1994). Effects of Multimedia courseware sub-titling on the speaking performance of college students of French. *Modern Language Journal, 78*(1), 61–75. doi:10.1111/j.1540-4781.1994.tb02015.x

Boyle, J. P. (1984). Factors affecting listening comprehension. *ELT Journal, 38*(1), 34–38. doi:10.1093/elt/38.1.34

Brett, P. (1996). Using multimedia: An investigation of learners' attitudes. *Computer Assisted Language Learning*, *9*(2), 191–212. doi:10.1080/0958822960090209

Brett, P. (1997). A comparative study of the effects of the use of multimedia on listening comprehension. *System*, *25*(1), 39–53. doi:10.1016/S0346-251X(96)00059-0

Brett, P. (1998). An intuitive, theoretical and empirical perspective on the effectiveness question for multimedia. In K. Cameron (Ed.), *Multimedia CALL: Theory and Practice* (pp. 81–92). Elm Bank Publication.

Brickell, G. (1993). Navigation and learning style. *Australian Journal of Educational Technology*, *9*(2), 103–114. Retrieved January 1, 2015, from http://www.ascilite.org.au/ajet/ajet9/brickell.html

Brown, G. (1978). Listening comprehension. In *ETIC Occasional Paper: The Teaching of Comprehension* (pp. 48-64). The British Council.

Brown, G., & Yule, G. (1992). *Teaching the spoken language: An approach based on the analysis of conversational English*. CUP.

Brusilovsky, P. (1996). Methods and techniques of adaptive hypermedia. *User Modeling and User-Adapted Interaction*, *6*(2-3), 87–129. doi:10.1007/BF00143964

Brusilovsky, P. (2007). Adaptive Navigation Support. In P. Brusilovski, A. Kobsa, & W. Nejdl (Eds.), *The Adaptive Web: Methods and Strategies of Web Personalization* (Vol. 4321, pp. 263–290). Springer-Verlag. doi:10.1007/978-3-540-72079-9_8

Brusilovsky, P. (2012). Adaptive Hypermedia for Education and Training. In P. J. Durlach & A. M. Lesgold (Eds.), *Adaptive Technologies for Training and Education* (pp. 46–68). CUP., doi:10.1017/CBO9781139049580.006

Brusilovsky, P., Eklund, J., & Schwarz, E. (1998). Web-based education for all: A tool for development adaptive courseware. *Computer Networks and ISDN Systems*, *30*(1), 291–300. doi:10.1016/S0169-7552(98)00082-8

Brusilovsky, P., & Millán, E. (2007). User models for adaptive hypermedia and adaptive educational systems. In P. Brusilovski, A. Kobsa, & W. Nejdl (Eds.), *The Adaptive Web: Methods and Strategies of Web Personalization* (Vol. 4321, pp. 3–53). Springer-Verlag. doi:10.1007/978-3-540-72079-9_1

Byrnes, H. (1984). The role of listening comprehension: A theoretical base. *Foreign Language Annals*, *17*(4), 317–329. doi:10.1111/j.1944-9720.1984.tb03235.x

Carbola-Calero, M. V. F. (2001). The EFL teacher and the introduction of multimedia in the classroom. *Computer Assisted Language Learning*, *14*(1), 3–14. doi:10.1076/call.14.1.3.5785

Carroll, J. B. (1977). On learning from being told. In M. C. Wittrock (Ed.), *Learning and instruction* (pp. 496–512). McCutchan.

Carson, J. G., & Longhini, A. (2002). Focussing on learning styles and strategies: A diary study in an immersion setting. *Language Learning*, *52*(2), 401–438. doi:10.1111/0023-8333.00188

Casambre, A. J. (1962). *The effect of certain variables in informative speaking on listener comprehension* (Unpublished Ph.D. thesis). Ohio State University.

Cauldwell, R. T. (1996). Direct encounters with fast speech on CD-AUDIO to teach listening. *System*, *24*(4), 521–528. doi:10.1016/S0346-251X(96)00046-2

CCD (Collins Cobuild Dictionary). (1987). HarperCollins Publishers.

CEED (The Chambers Encyclopaedic English Dictionary). (1994). Larousse Plc.

Chamot, A. U., & O'Malley, J. M. (1994). Language learner and learning strategies. In N. C. Ellis (Ed.), *Implicit and explicit learning of languages* (pp. 371–392). Academic Press.

Chang, C. C., Tseng, K. H., & Tseng, J. S. (2011). Is single or dual channel with different English proficiencies better for English listening comprehension, cognitive load and attitude in ubiquitous learning environment? *Computers & Education*, *57*(4), 2313–2321. doi:10.1016/j.compedu.2011.06.006

Chapelle, C. (1997). CALL in the year 2000: Still in search of research paradigms? *Language Learning & Technology*, *3*(1), 19–43.

Chapelle, C. A. (1998). Multimedia CALL: Lessons to be learned from research on instructed SLA. *Language Learning & Technology*, *2*(1), 22–34.

Chen, H. J. (2011). Developing and evaluating SynctoLearn, a fully automatic video and transcript synchronization tool for EFL learners. *Computer Assisted Language Learning*, *24*(2), 117–130. doi:10.1080/09588221.2010.526947

Chun, M. D., & Plass, J. L. (1996). Facilitating reading comprehension with multimedia. *System*, *24*(4), 503–519. doi:10.1016/S0346-251X(96)00038-3

Chung, J. M., & Huang, S. C. (1998). The effects of three aural advance organisers for video viewing in a foreign language classroom. *System*, *26*(4), 553–565. doi:10.1016/S0346-251X(98)00037-2

Clifford, R. (1998). Mirror, mirror, on the wall: Reflections on computer assisted language learning. *CALICO Journal*, *16*(1), 1–11. doi:10.1558/cj.v16i1.1-9

Coakley, C. G., & Wolvin, A. D. (1986). Listening in the native language. In B. H. Wing (Ed.), *Listening, reading, and writing: Analysis and application. Northeast Conference Reports: Northeast Conference on the Teaching of Foreign Languages* (pp. 11-42). Middlebury, VT: The Conference.

Cohen, A. (1984). Studying second-language learning strategies: How do we get information? *Applied Linguistics*, *5*(2), 101–112. doi:10.1093/applin/5.2.101

Cohen, A. D. (1998). *Strategies in learning and using a second language*. Addison Wesley Longman Limited.

Colpaert, J. (2002). The world of CALL. *Computer Assisted Language Learning*, *15*(5), 437–439. doi:10.1076/call.15.5.437.13469

Conrad, L. (1989). The effects of time-compressed speech on native and EFL listening comprehension. *Studies in Second Language Acquisition*, *11*(1), 1–16. doi:10.1017/S0272263100007804

Craik, F. I. M., & Lockart, R. S. (1972). Levels of processing: A framework for memory research. *Journal of Verbal Learning and Verbal Behavior*, *11*(6), 671–684. doi:10.1016/S0022-5371(72)80001-X

Crosby, M. C., Stelovsky, J., & Ashworth, D. (1994). Hypermedia as a facilitator for retention: A case study using Kanji City Issues. *Computer Assisted Language Learning*, *7*(1), 3–13. doi:10.1080/0958822940070102

Debski, R., & Gruba, P. (1999). A qualitative survey of tertiary instructor attitudes towards project-based CALL. *Computer Assisted Language Learning*, *12*(3), 219–239. doi:10.1076/call.12.3.219.5715

Deville, G., Kelly, P., Paulussen, H., Vandecasteele, M., & Zimmer, C. (1996). The use of a multi-media support for remedial learning of English with heterogeneous groups of 'false beginners'. *Computer Assisted Language Learning*, *9*(1), 75–84. doi:10.1080/0958822960090105

Diao, Y., Chandler, P., & Sweller, J. (2007). The effect of written text on comprehension of spoken English as a foreign language. *The American Journal of Psychology*, *120*(2), 237–261. PMID:17650920

Dodson, C. S., & Schacter, D. L. (2002). When false recognition meets metacognition: The distinctiveness heuristic. *Journal of Memory and Language*, *46*(4), 782–802. doi:10.1006/jmla.2001.2822

Dunn, R., & Dunn, K. J. (1979). Learning style/teaching styles: Should they … can they … be matched? *Educational Leadership*, *36*, 238–244.

Eastman, J. K. (1987). Remedial training in listening comprehension. *System*, *15*(2), 197–201. doi:10.1016/0346-251X(87)90067-4

Eastman, J. K. (1991). Learning to listen and comprehend: The beginning stages. *System*, *19*(3), 179–187. doi:10.1016/0346-251X(91)90042-N

Ehrmann, E. L. (1963). Listening comprehension in teaching of a foreign language. *Modern Language Journal*, *47*(1), 18–20. doi:10.2307/319940

Ellis, R. (1997). *Second Language Acquisition*. OUP.

Eraut, M. (1988). What has happened to learning design? In H. Mathias, N. Rushby, & R. Budgett (Eds.), *Aspects of Educational Technology XXI: Designing New Systems and Technologies for Learning*. Kogan Page.

Faerch, C., & Kasper, G. (1986). The role of comprehension in second language learning. *Applied Linguistics*, *7*(3), 257–274. doi:10.1093/applin/7.3.257

Ferney, D., & Waller, S. (2001). Reflections on multimedia design criteria for the international language learning community. *Computer Assisted Language Learning*, *14*(2), 145–168. doi:10.1076/call.14.2.145.5781

Field, J. (1998). Skills and strategies: Towards a new methodology for listening. *ELT Journal*, *52*(2), 110–118. doi:10.1093/elt/52.2.110

Fleta, B. M., Sabater, C. P., Salom, L. G., Guillot, C. P., Monreal, C. S., & Turney, E. (1999). Evaluating multimedia programs for language learning: A case study. *ReCALL*, *11*(3), 50–57.

Fox, J., Romano-Hvid, B. C., & Sheffield, J. S. (1992). *New perspectives in Modern Language Learning*. Employment Department-Group UK.

Frankel, F. (1987). Self-study materials: Involving the learner. In M. Geddes & G. Sturtridge (Eds.), *Individualisation* (pp. 52–60). Modern English Publications ltd.

Furnham, A., De Siena, S., & Gunter, B. (2002). Children's and adults' recall of children's news stories in both print and audio-visual presentation modalities. *Applied Cognitive Psychology*, *16*(2), 191–210. doi:10.1002/acp.777

Gardner, R. C. (1985). *Social psychology and second language learning: The role of attitudes and motivation*. Edward Arnold.

Geddes, M., & White, R. (1978). The use of semi-scripted simulated authentic speech and listening comprehension. *Audio-Visual Language Journal*, *16*(3), 137–145.

Gillespie, J., & McKee, J. (1999). Resistance to CALL: Degrees of student reluctance to use CALL and ICT. *ReCALL*, *11*(1), 38–46. doi:10.1017/S095834400000207X

Ginther, A. (2002). Context and content visuals and performance on listening comprehension stimuli. *Language Testing*, *19*(2), 133–167. doi:10.1191/0265532202lt225oa

Goh, C., & Taib, Y. (2006). Metacognitive instruction in listening for young learners. *ELT Journal*, *60*(3), 222–232. doi:10.1093/elt/ccl002

Goh, C. C. M. (2000). A cognitive perspective on language learners' listening comprehension problems. *System*, *28*(1), 55–75. doi:10.1016/S0346-251X(99)00060-3

Goh, C. C. M. (2002). Exploring listening comprehension tactics and their interaction patterns. *System*, *30*(2), 185–206. doi:10.1016/S0346-251X(02)00004-0

Grob, A., & Wolff, D. (2001). A multimedia tool to develop learner autonomy. *Computer Assisted Language Learning*, *14*(3-4), 233–249. doi:10.1076/call.14.3.233.5794

Gyselinck, V., Cornoldi, C., Dubois, V., De Beni, R., & Ehrlich, M. (2002). Visuospatial memory and phonological loop in learning from multimedia. *Applied Cognitive Psychology*, *16*(6), 665–685. doi:10.1002/acp.823

Hammer, P., & Momod, M. (1978). The role of English/French cognates in listening comprehension. *Audio-visual Language Journal.*, *16*(13), 29–32.

Harrington, M., & Levy, M. (2001). CALL begins with a 'C': Interaction in computer-mediated language learning. *System*, *29*(1), 15–26. doi:10.1016/S0346-251X(00)00043-9

Hedberg, J. G., Harper, B., & Brown, C. (1993). Reducing cognitive load in multimedia navigation. *Australian Journal of Educational Technology*, *9*(2), 157–181. Retrieved January 3, 2015, from http://www.ascilite.org.au/ajet/ajet9/hedberg.html

Hegelheimer, V., & Chapelle, C. A. (2000). Methodological issues in research on learner-computer interactions in CALL. *Language Learning & Technology*, *4*(1), 41–59.

Herron, C., Dubreil, S., Corrie, C., & Cole, S. P. (2002). A classroom Investigation: Can video improve intermediate-level French language students' ability to learn about a foreign culture? *Modern Language Journal*, *86*(1), 36–53. doi:10.1111/1540-4781.00135

Hodges, H. (1982). Madison prep-alternatives through learning styles. In J. W. Keefe (Ed.), *Student learning styles and brain behaviour: Programs, instrumentation, research* (pp. 28–31). National Association of Secondary School Principals.

Hoven, D. (1999). A model for listening and viewing comprehension in multimedia environments. *Language Learning & Technology*, *3*(1), 88–103.

Hurd, S., Beaven, T., & Ortega, A. (2001). Developing autonomy in a distance language learning context: Issues and dilemmas for course writers. *System*, *29*(3), 341–355. doi:10.1016/S0346-251X(01)00024-0

Jones, L. (2003). Supporting listening comprehension and vocabulary acquisition with multimedia annotations: The students' voice. *CALICO Journal*, *21*(1), 41–65. doi:10.1558/cj.v21i1.41-65

Jones, L. (2004). Testing L2 vocabulary recognition and recall using pictorial and written test items. *Language Learning & Technology*, *8*(3), 122–143. Retrieved January 9, 2015, from http://llt.msu.edu/vol8num3/jones

Jones, L. (2006). Effects of collaboration and multimedia annotations on vocabulary learning and listening comprehension. *CALICO Journal*, *24*(1), 33–58. doi:10.1558/cj.v24i1.33-58

Jones, L. (2009). Supporting student differences in listening comprehension and vocabulary learning with multimedia annotations. *CALICO Journal*, *26*(2), 267–289. doi:10.1558/cj.v26i2.267-289

Jones, L., & Plass, J. (2002). Supporting listening comprehension and vocabulary acquisition in French with multimedia annotations. *Modern Language Journal*, *86*(4), 546–561. doi:10.1111/1540-4781.00160

Kahraman, H. T., Sagiroglu, S., & Colak, I. (2013). A novel model for web-based adaptive educational hypermedia systems: SAHM (supervised adaptive hy- permedia model. *Computer Applications in Engineering Education*, *21*(1), 60–74. doi:10.1002/cae.20451

Kalyuga, S. (2000). When using sound with a text or picture is not beneficial for learning. *Australian Journal of Educational Technology*, *16*(2), 161–172. doi:10.14742/ajet.1829

Kennedy, D. M., & McNaught, C. (1997). Design elements for interactive multimedia. *Australian Journal of Educational Technology*, *13*(1), 1–22. Retrieved January 7, 2015, from http://www.ascilite.org.au/ajet/ajet13/kennedy.html

Kern, R. C. (1995). Restructuring classroom interaction with networked computers: Effects on quantity and characteristics of language production. *Modern Language Journal*, *79*(4), 457–476. doi:10.1111/j.1540-4781.1995.tb05445.x

Kohn, K. (2001). Developing multimedia CALL: The *telos language partner* approach. *Computer Assisted Language Learning*, *14*(3-4), 251–267. doi:10.1076/call.14.3.251.5788

Kohonen, V. (1991). Reviews: Leslie Dickenson: Self-Instruction in language learning, CUP 1987. *Applied Linguistics*, *12*(1), 107–112. doi:10.1093/applin/12.1.107

Krashen, S. D. (1982). *Principles and practice in second language acquisition*. Pergamon.

Krashen, S. D. (1984). *Principles and practice in second language acquisition*. Pergamon.

Le Mon, R. (1988). Cultural Differences affecting the international transfer of computerised language learning. *System, 16*(1), 37–40. doi:10.1016/0346-251X(88)90008-5

Leffa, V. J. (1992). Making foreign language texts comprehensible for beginners: An experiment with an electronic glossary. *System*, *20*(1), 63–73. doi:10.1016/0346-251X(92)90008-Q

Leow, R. P. (2001). Attention, awareness, and foreign language behaviour. *Language Learning*, *51*(Supplement 1), 113–155. doi:10.1111/j.1467-1770.2001.tb00016.x

Leveridge, A. N., & Yang, J. C. (2014). Learner perceptions of reliance on captions in EFL multimedia listening comprehension. *Computer Assisted Language Learning*, *27*(6), 545–559. doi:10.1080/09588221.2013.776968

Littlewood, W. (1996). Autonomy: An autonomy and a framework. *System*, *24*(4), 427–435. doi:10.1016/S0346-251X(96)00039-5

Littlewood, W. (1997). Self-access: Why do we want it and what can it do? In P. Benson & P. Voller (Eds.), *Autonomy & Independence in language learning* (pp. 79–91). Addison Wesley Longman Limited.

Long, D. R. (1989). Second language listening comprehension: A schema-theoretic perspective. *Modern Language Journal*, *73*(1), 32–40. doi:10.1111/j.1540-4781.1989.tb05307.x

Long, D. R. (1990). What you don't know can't help you: An exploratory study of background knowledge and second language listening comprehension. *Studies in Second Language Acquisition*, *12*(1), 65–80. doi:10.1017/S0272263100008743

Long, M. H. (1983). Native speaker/non-native speaker conversation and the negotiation of comprehensible input. *Applied Linguistics*, *4*(2), 127–141. doi:10.1093/applin/4.2.126

Long, M. H. (1988). Instructed interlanguage development. In L. Beebe (Ed.), *Issues in second language acquisition: Multiple perspectives* (pp. 115–141). Newbury House.

Long, M. H. (1996). The role of the linguistic environment in second language acquisition. In W. C. Ritchie & T. K. Bhaatoa (Eds.), *Handbook of second language acquisition* (pp. 413–468). Academic Press.

Lund, R. (1991). A comparison of second language listening and reading comprehension. *Modern Language Journal*, *75*(2), 196–204. doi:10.1111/j.1540-4781.1991.tb05350.x

Lyall, R., & McNamara, S. (2000). Learning tool or potplant stand? Students' opinions of learning from a CAL program in a distance education context. *Australian Journal of Educational Technology, 16*(2), 126-46. Retrieved January, 4, 2015, from http://www.ascillite.org.au/ajet/ajet16/lyall.html

MacGregor, R. C. (1993). Are we recognising the organisational impact on educational software design? *Australian Journal of Educational Technology*, *9*(1), 59–68. doi:10.14742/ajet.2097

Mangiafico, L. F. (1996). *The relative effects of classroom demonstration and individual use of interactive multimedia on second listening comprehension* (Unpublished Ph.D. Thesis). Vanderbilt University, Faculty of Graduate School.

Markham, P. L., & Latham, M. (1987). The influence of religion-specific background knowledge on the listening comprehension of adult second language students. *Language Learning*, *37*(2), 157–170. doi:10.1111/j.1467-1770.1987.tb00563.x

Masgoret, A. M., & Gardner, R. C. (2003). Attitudes, motivation, and second language learning: A meta-analysis of studies conducted by Gardner and associates. *Language Learning*, *53*(1), 123–163. doi:10.1111/1467-9922.00212

Mayer, E. (1997). Multimedia learning: Are we asking the right questions? *Educational Psychologist*, *32*(1), 1–19. doi:10.120715326985ep3201_1

Mayer, R. E., & Anderson, R. B. (1992). The instructive animation: Helping students build connections between words and pictures in multimedia learning. *Journal of Educational Psychology*, *84*(4), 444–452. doi:10.1037/0022-0663.84.4.444

Mayer, R. E., & Sims, V. K. (1994). For whom is a picture worth a thousand words? Extension of a dual coding theory of multimedia learning. *Journal of Educational Psychology*, *86*(3), 389–401. doi:10.1037/0022-0663.86.3.389

McLoughlin, C. (1999). The implications of the research literature on learning styles for the design of instructional material. *Australian Journal of Educational Technology*, *15*(3), 222–241. doi:10.14742/ajet.1859

McLoughlin, C., & Oliver, R. (2000). Designing environments for cultural inclusivity: A case study of indigenous online learning at tertiary level. *Australian Journal of Educational Technology*, *16*(1), 58–72. Retrieved January 13, 2015, from http://www.ascilite.org.au/ajet/ajet16/mcloughlin.html. doi:10.14742/ajet.1822

Meinardi, M. (2009). Speed bumps for authentic listening material. *ReCALL*, *21*(03), 302–318. doi:10.1017/S0958344009990048

Moreno, R., & Mayer, R. E. (2002). Verbal redundancy in multimedia learning: When reading helps listening. *Journal of Educational Psychology*, *94*(1), 156–163. doi:10.1037/0022-0663.94.1.156

Mueller, G. (1980). Visual contextual cues and listening comprehension: An experiment. *Modern Language Journal*, *64*(3), 335–340. doi:10.1111/j.1540-4781.1980.tb05202.x

Nakic, A. (1981). The use of authentic sound materials for beginners. In The teaching of listening comprehension (pp. 7-14). Tonbridge Printers Ltd.

Nicholas, H., Lightbown, P. M., & Spada, N. (2001). Recasts as feedback to language learners. *Language Learning*, *51*(4), 719–758. doi:10.1111/0023-8333.00172

Nicholson, A. Y. W., & Ngai, J. Y. K. (1996). Managing the development and production of interactive multimedia courseware in education. *Australian Journal of Educational Technology*, *12*(1), 35–45. doi:10.14742/ajet.2034

Nikolova, O. R. (2002). Effects of students' participation in authoring of multimedia materials on student acquisition of vocabulary. *Language Learning & Technology*, *6*(1), 100–122.

O'Malley, J. M., & Chamot, A. U. (1990). *Learning strategies in second language acquisition*. CUP. doi:10.1017/CBO9781139524490

O'Malley, J. M., Chamot, A. U., & Kupper, L. (1989). Listening comprehension strategies in second language acquisition. *Applied Linguistics*, *10*(4), 418–437. doi:10.1093/applin/10.4.418

O'Malley, J. M., Chamot, A. U., Stewner-Manzanares, G., Kupper, L., & Russo, R. (1985a). Learning strategies used by beginning and intermediate ESL students. *Language Learning*, *35*(1), 21–46. doi:10.1111/j.1467-1770.1985.tb01013.x

O'Malley, J. M., Chamot, A. U., Stewner-Manzanares, G., Russo, R. P., & Kupper, L. (1985b). Learning strategy applications with students of English as a Second language. *TESOL Quarterly*, *19*(3), 557–584. doi:10.2307/3586278

Omaggio, A. C. (1979). Pictures and second language comprehension: Do they help? *Foreign Language Annals*, *12*(2), 107–116. doi:10.1111/j.1944-9720.1979.tb00153.x

Ortega, L. (1997). Process and outco mes in networked classroom interaction: Defining the research agenda for L2 computer-assisted classroom discussion. *Language Learning & Technology*, *1*(1), 82–93.

Oxford, R. L. (1993). Research update on teaching L2 listening. *System*, *21*(2), 205–211. doi:10.1016/0346-251X(93)90042-F

Paivio, A. (1986). *Mental representation: A dual coding approach*. OUP.

Peter, M. (1994). *Investigating into the design of educational multimedia: Video, interactivity and narrative* (Unpublished Ph.D. Thesis). Open University.

Phillips, R., Pospisil, R., & Richardson, J. L. (2001). The use of QTVR image database for teaching veterinary radiology and diagnostic ultrasound to distance education students. *Australian Journal of Educational Technology*, *17*(1), 96–114. doi:10.14742/ajet.1775

Pienemann, M. (2001). Testing the procedural skill hypothesis. *Computer Assisted Language Learning*, *14*(3-4), 321–337. doi:10.1076/call.14.3.321.5791

Platt, E., & Brooks, F. B. (2002). Task engagement: A turning point in foreign language development. *Language Learning*, *52*(2), 365–400. doi:10.1111/0023-8333.00187

Porter, D., & Roberts, J. (1981). Authentic listening activities. *ELT Journal*, *36*(1), 37–47. doi:10.1093/elt/36.1.37

Reid, J. M. (1987). The learning style preferences of ESL students. *TESOL Quarterly*, *21*(1), 87–111. doi:10.2307/3586356

Richards, J. C. (1983). Listening comprehension: Approach, design, and procedure. *TESOL Quarterly*, *17*(3), 219–240. doi:10.2307/3586651

Ridgway, T. (2000). Listening strategies - I beg your pardon? *ELT Journal*, *54*(2), 179–185. doi:10.1093/elt/54.2.179

Rivers, V. M. (1981). *Teaching Foreign Language Skills*. The University of Chicago Press. doi:10.7208/chicago/9780226518855.001.0001

Rixon, S. (1992). *Developing listening skill*. Macmillan Publishers Ltd.

Robinson, G. L. (1989). The CLCCS CALL study: Methods, error feedback, attitudes, and achievement. In W. M. Flint Smith (Ed.), *Modern technology in FLE: Applications and project* (pp. 119–133). National Textbook Co.

Robinson, G. L. (1991). Effective feedback strategies in CALL: Learning theory and empirical research. In P. Dunkel (Ed.), *Computer-assisted language learning and testing: Research issues and practice* (pp. 155–167). Newbury House.

Rost, M. (1990). *Listening in language learning*. Longman Group UK Ltd.

Rost, M. (1994). Second Language Processing: Listening. In R. E. Asher (Ed.), *The Encyclopaedia of Language and Linguistic*. Pergamon Press.

Rost, M., & Ross, S. (1991). Learner use of strategies in interaction: Typology and teachability. *Language Learning*, *41*(2), 235–273. doi:10.1111/j.1467-1770.1991.tb00685.x

Rubin, J., & Thomson, I. (1992). Materials selection in strategy instruction for Russian listening comprehension. *ERIC DOC*, (NO: ED349796), 2–12.

Schmidt, R. W. (1990). The role of consciousness in second language learning. *Applied Linguistics*, *11*(2), 129–158. doi:10.1093/applin/11.2.129

Schmidt, R. W. (1993). Awareness and second language acquisition. *Annual Review of Applied Linguistics*, *13*, 206–226. doi:10.1017/S0267190500002476

Schmidt, R. W. (1995). Consciousness and foreign language learning: A tutorial on the role of attention and awareness in learning. In R. W. Schmidt (Ed.), Attention and awareness in foreign language learning (Technical Report No. 9) (pp. 1-63). Honolulu, HI: University of Hawai'i, Second Language Teaching and Curriculum Center.

Secules, T., Herron, C., & Tomasello, M. (1992). The effects of video context on foreign language learning. *Modern Language Journal*, *76*(4), 480–490. doi:10.1111/j.1540-4781.1992.tb05396.x

Shehadeh, A. (2002). Comprehensible output, from occurrence to acquisition: An argument for acquisitional research. *Language Learning*, *52*(3), 597–647. doi:10.1111/1467-9922.00196

Shute, V. J., & Zapata-Rivera, D. (2012). Adap- tive educational systems. In J. D. Paula & M. L. Alan (Eds.), *Adaptive technologies for training and education* (pp. 7–27). Cambridge University Press., doi:10.1017/CBO9781139049580.004

Sims, R. (2000). An interactive conundrum: Constructs of interactivity and learning theory. *Australian Journal of Educational Technology*, *16*(1), 45–57. doi:10.14742/ajet.1821

Skinner, B. F. (1953). *Science and human behaviour*. Macmillan.

Smith, B. (2003). Computer-mediated negotiated interaction: An expanded model. *Modern Language Journal*, *87*(1), 38–57. doi:10.1111/1540-4781.00177

Soboleva, O., & Tronenko, N. (2002). A Russian multimedia learning package for classroom use and self – study. *Computer Assisted Language Learning*, *15*(5), 483–499. doi:10.1076/call.15.5.483.13470

Stepp-Greany, J. (2002). Student perceptions on language learning in a technological Environment: Implications for the new millennium. *Language Learning & Technology*, *6*(1), 165–180.

Stevens, V. (1995). A study of student attitudes toward CALL in a self-access student resource centre. *System*, *19*(3), 289–299. doi:10.1016/0346-251X(91)90053-R

Sturtridge, G. (1987). Individualised learning: What are the options for the classroom teacher? In M. Geddes & G. Sturtridge (Eds.), *Individualisation* (pp. 8–14). Modern English Publications ltd.

Swain, M. (1985). Communicative competence: Some rules of comprehensible input and comprehensible output in its development. In S. M. Gass, & C. G. Madden (Eds.), Input in second language acquisition (pp. 235-53). Rowley, MA: Newbury House Publishers.

Swain, M. (1995). Three functions of output in second language learning. In G. Cook & B. Seidlhofer (Eds.), *Principle and practice in applied linguistics: Studies in honor of H.G. Widdowson* (pp. 125–144). OUP.

Sweller, J. (1999). *Instructional design*. ACER.

Taylor-leech, K. (1994). Authoring in Guide. *California Law Review*, (November), 4–8.

Tezza, J. (1962). *The effects of listening training on audio-lingual learning* (Unpublished Ph.D. Thesis). University of Pittsburgh.

The Dictionary of Multimedia. (1997). In S. Collin, Ed.). Peter Collin Publishing.

Thierry, C. (1996). Learning a second language for specific purposes within a hypermedia framework. *Computer Assisted Language Learning*, *9*(1), 3–43. doi:10.1080/0958822960090102

Toyoda, E. (2001). Exercise of learner autonomy in project-oriented CALL. *CALL-EJ Online, 2*(2). Retrieved January 7, 2015, from http://www.lerc.ritsumei.ac.jp/callej/5-2/toyoda.html

Trinder, R. (2002). Forum: Multimedia in the Business English classroom: The learners' point of view. *Computer Assisted Language Learning*, *15*(1), 69–84. doi:10.1076/call.15.1.69.7291

Tschirner, E. (2001). Language acquisition in the classroom: The role of digital video. *Computer Assisted Language Learning*, *14*(3-4), 305–319. doi:10.1076/call.14.3.305.5796

Tumposky, N. (1987). The learner on his own. In M. Geddes & G. Sturtridge (Eds.), *Individualisation* (pp. 4–7). Modern English Publications ltd.

Türel, V. (2004). *Design of multimedia software: Investigating the design of some elements of interactive multimedia listening software for autonomous intermediate language learners* (Unpublished Ph.D. Thesis). The University of Manchester. Manchester, UK.

Turel, V. (2011). Learners' attitudes to repetitious exposure in multimedia listening software. *The EUROCALL Review, 19*, 57-83. Retrieved January 13, 2015, from http://www.eurocall-languages.org/wordpress/wp-content/uploads/2014/01/review19.pdf

Turel, V. (2012). Design of feedback in interactive multimedia language learning environments. *Linguistik Online, 54*(4), 35-49. Retrieved January 7, 2015, from http://www.linguistik-online.de/54_12/tuerel.pdf

Turel, V. (2014a). Learners' perceptions towards interactive multimedia environments. *Hacettepe Üniversitesi Eğitim Fakültesi Dergisi, 29*(3), 167-183. http://www.efdergi.hacettepe.edu.tr/201429-3VEHB%C4%B0%20T%C3%9CREL.pdf

Turel, V. (2014b). Factors affecting listening. In C. M. Akrivopoulou & N. Garipidis (Eds.), *Human Rights and the Impact of ICT in the Public Sphere: Participation, Democracy, and Political Autonomy* (pp. 310–331). IGI-Global. doi:10.4018/978-1-4666-6248-3.ch018

Turel, V. (2014c). Design of cultural differences in hypermedia environments. *International Journal of Human Rights and Constitutional Studies, 2*(2), 150–170. doi:10.1504/IJHRCS.2014.062766

Turel, V. (2014d). The use and design of supplementary visuals for the enhancement of listening skills in hypermedia. In C. M. Akrivopoulou & N. Garipidis (Eds.), *Human Rights and the Impact of ICT in the Public Sphere: Participation, Democracy, and Political Autonomy* (pp. 268–293). IGI-Global. doi:10.4018/978-1-4666-6248-3.ch016

Turel, V. (2014e). Priority of listening materials for autonomous intermediate language learners. In C. M. Akrivopoulou & N. Garipidis (Eds.), *Human Rights and the Impact of ICT in the Public Sphere: Participation, Democracy, and Political Autonomy* (pp. 292–309). IGI-Global. doi:10.4018/978-1-4666-6248-3.ch017

Turel, V. (2015a). Hypermedia and its' role in learning. In M. Khosrow-Pour (Ed.), Encyclopaedia of Information Science and Technology (3rd ed.). (Vols. 1–10; pp. 2495–2505). IGI-Global. doi:10.4018/978-1-4666-5888-2.ch243

Turel, V. (2015b). Intelligent design of interactive multimedia listening software. Hershey, PA: IGI Global. doi:10.4018/978-1-4666-8499-7

Turel, V. (2015c). Design of the pre-listening tasks in hypermedia environments. *International Journal of Human Rights and Constitutional Studies.*, *3*(2), 181–201. doi:10.1504/IJHRCS.2015.070601

Turel, V. (2015d in press). Intelligent Design of Captions in Interactive Multimedia Listening Environments. In V. Turel (Ed.), *Intelligent design of interactive multimedia listening software*. IGI-Global. doi:10.4018/978-1-4666-8499-7.ch003

Turel, V. (2018). Adaptive Hypermedia in Education. In M. Khosrow-Pour (Ed.), Encyclopaedia of information science and technology (4th ed.). (Vols. 1–10; pp. 2357–2368). IGI-Global., doi:10.4018/978-1-5225-2255-3

Turel, V., & Kilic, E. (2014). The inclusion and design of cultural differences in Interactive multimedia environments. In C. M. Akrivopoulou & N. Garipidis (Eds.), *Human Rights and the Impact of ICT in the Public Sphere: Participation, Democracy, and Political Autonomy* (pp. 245–267). IGI-Global., doi:10.4018/978-1-4666-6248-3.ch015

Turel, V., & McKenna, P. (2013). Design of Language Learning Software. In B. Zou & ... (Eds.), *Computer- assisted foreign language teaching and learning: Technological advances* (pp. 188–209). IGI Global. doi:10.4018/978-1-4666-2821-2.ch011

Turel, V., & McKenna, P. (2015 in press). Design of multimedia listening software: Unfamiliar iems, glossary, aural texts and speech rate. In V. Turel (Ed.), *Intelligent design of interactive multimedia listening software*. IGI-Global. doi:10.4018/978-1-4666-8499-7.ch007

Underwood, M. (1993). *Teaching listening*. Longman Group UK Ltd.

Ur, P. (1992). *Teaching listening comprehension*. CUP.

Vandergrift, L. (2006). Second language listening: Listening ability or language proficiency? *Modern Language Journal*, *90*(1), 1, 6–18. doi:10.1111/j.1540-4781.2006.00381.x

Vanderplank, R. (1988). Implications of Difference in Native and Non-Native Speaker Approaches to Listening. *British Journal of Language Teaching*, *26*(1), 32–41.

Vidal, K. (2003). Academic listening: A source of vocabulary acquisition? *Applied Linguistics*, *24*(1), 56–89. doi:10.1093/applin/24.1.56

Vo, Y. (in press). Interactive Multimedia and Listening. In V. Turel (Ed.), Intelligent design of interactive multimedia listening software. Hershey, PA, USA: IGI-Global.

Voller, P. (1997). Does the teacher have a role in autonomous language learning? In P. Benson & P. Voller (Eds.), *Autonomy & Independence in Language Learning* (pp. 98–113). Addison Wesley Longman Limited.

Vygotsky, L. (1978). *Mind in society: The development of higher psychological processes*. Harvard University Press.

Weissenrieder, M. (1987). Listening to the news in Spanish. *Modern Language Journal*, *71*(1), 18–27. doi:10.1111/j.1540-4781.1987.tb01049.x

Willets, J. (1988). Listening to your computer: Digital sound in CALL. *MUESLI News*, (January), 4–4.

Williams, J. (2001). The effectiveness of spontaneous attention to form. *System*, *29*(3), 325–340. doi:10.1016/S0346-251X(01)00022-7

Wipf, J. (1984). Strategies for teaching school language listening comprehension. *Foreign Language Annals*, *17*(4), 345–348. doi:10.1111/j.1944-9720.1984.tb03240.x

Wittrock, C. (1990). Generative process of comprehension. *Education Psychologist*, *24*(4), 345–376. doi:10.120715326985ep2404_2

Xi, X. (2010). Automated scoring and feedback systems: Where are we and where are we heading? *Language Testing*, *27*(3), 291–300. doi:10.1177/0265532210364643

Yang, J. C., & Chang, P. (2014). Captions and reduced forms instruction: The impact on EFL students' listening comprehension. *ReCALL*, *26*(1), 44–61. doi:10.1017/S0958344013000219

Chapter 1
The Effect of Using Technology in L2 Listening: A Meta-Analysis of Studies From 1980 to 2020

Tingting Kang
Lafayette College, USA

ABSTRACT

Due to the nature of listening, multiple media tools have been utilized in second language (L2) listening instruction for a long time. This chapter recruited the most recently published empirical studies in addition to the meta-analysis conducted in 2015 (Kang) and analyzed available evidence on the effects of different multiple media tools on L2 listening comprehension. The results revealed a medium-to-large effect of multiple media tools on listening comprehension in between-group designs (Cohen's d = .70), which was almost identical to the result observed five years ago. The effects of individual multiple media tools have also been statistically synthesized. Moderator analysis could help L2 educators and test developers make decisions on applying different multiple media tools in the fields of L2 instruction and assessment. Specifically, subtitles, captions, as well as self-regulated listening and slow speed were recommended to teachers and test developers as a means to improve learners' listening comprehension.

INTRODUCTION

Listening comprehension is considered to be a very crucial language skill because it appears at "the early stages of L2 learning" (Nation & Newton, 2009, p. 37), enables learners to receive language input, and facilitates the emergence of other language skills (Vandergrift & Goh, 2012). According to Burley-Allen (1995), the average time spent on basic skills during the daily communication process is 35% for speaking, 16% for reading, 9% for writing, and 40% for listening. However, listening has also been described as "the least understood and most overlooked of the four skills (listening, speaking, reading, and writing)" (Nation & Newton, 2009, p. 37). The continuous speech-flow makes the listening comprehension

DOI: 10.4018/978-1-7998-7876-6.ch001

process hard to capture and analyze. For instance, in the real spoken language, listeners cannot slow it down or break it down into manageable chunks. Therefore, effectively teaching and learning L2 listening is the greatest concern for most of the L2 educators.

Several models have been developed to explain how listening process functions, including the bottom-up model, the top-down model, and the interactive model (Flowerdew & Miller, 2005). In the bottom-up model, listeners start with the small units in the acoustic message, such as individual sounds, to achieve listening comprehension. The top-down model suggests that learners rely on prior contextual knowledge and utterance interpretations in processing a text. Allowing for the individual variations in listening processing, the interactive model synthesizes the bottom-up and top-down models. With regards to listening activities, Rost (2011) discussed six widely used instructional designs in L2 listening, namely, intensive listening, selective listening, interactive listening, extensive listening, responsive listening, and autonomous listening.

In light of the possible factors in listening passages that influence listening success, Buck (2001), and Vandergrift and Goh (2012) provided very comprehensive summaries. Specifically, the statistically significant predictors of listening comprehension in listening passages that have been empirically tested are vocabulary, syntax/grammar, topics, speech rate, and multiple media tools. Table 1 summarizes these predictors and the corresponding empirical studies. According to Stæhr (2009), advanced EFL learners' vocabulary knowledge was significantly correlated with their listening comprehension skills and predicted 51% of the variance in their listening scores. In terms of the syntactic or grammatical knowledge, listeners at lower proficiency levels tended to pay more attention to syntactic cues instead of global semantic cues, whereas advanced learners paid more attention to the semantic cues than the syntactic cues in listening passages (Conrad, 1985). Macaro, Vanderplank, and Graham (2005) conducted a meta- analysis on the effectiveness of prior knowledge in listening comprehension. The results showed that prior knowledge was positively related with listening comprehension; however, some studies also suggested that prior knowledge could lead to inaccurate comprehension. With regards to speech rate, as Griffiths (1990b) identified, speech rates faster than 200 words per minutes/3.8 syllables per second would weaken lower intermediate learners' listening comprehension. Specifically, the lower intermediate learners' listening comprehension scores were significantly higher when the passages delivered at slow speech rate (127 words per minutes/2.5 syllables per second) than the average speech rate (188 words per minutes/2.75 syllables per second) or the fast speech rate (250 words per minutes/5 syllables per second; Griffiths, 1992).

Lastly, due to the acoustic nature of listening, multiple media tools have been utilized in L2 listening ever since the sound could be captured through technology. Such multiple media tools include audiocassette players, radio, video, language laboratory, and the like. Multiple media tools have the advantages of facilitating L2 learners' extensive and intensive listening practice, promoting learning motivation, and getting access to authentic listening materials. Different researchers have empirically tested the effects of these multiple media tools in the literature. Specifically, the effects of visual in listening tests have been widely discussed; for example, according to Wagner (2010; 2013), audiovisual input could result in significantly higher listening comprehension scores than the audio-only input.

Table 1. Predictors of listening comprehension

Factors	Empirical Studies
vocabulary	Bonk, 2000; Mecartty, 2000; Stæhr, 2009; Trofimovich & Isaacs, 2012
syntax/grammar	Conrad, 1985; Mecartty, 2000
prior knowledge/topic	Buck & Tatsuoka, 1998; Freedle & Kostin, 1999; Long, 1990; Macaro, Vanderplank, & Graham, 2005
speech rate	Buck & Tatsuoka, 1998; Griffiths, 1990ab; Griffiths, 1992; Jensen & Vinther, 2003; Zhao, 1997
multimedia	Parry & Meredith, 1984; Shin, 1998; Wagner, 2010; Wagner 2013

Unlike traditional literature reviews, meta-analysis approach enables researchers to have a quantifiable sense of previous studies. Previous meta-analysis on computer assistant language learning (CALL) embraced topics such as the comprehensive term, CALL (e.g., Grgurović, Chapelle, & Shelley, 2013; Zhao, 2003), CALL-based glosses to enhance reading comprehension and vocabulary acquisition (e.g., Abraham, 2008; Taylor, 2009; Yun, 2011), technology-based writing instruction (e.g., Goldberg, Russell, & Cook, 2003; Little, Clark, Tani, & Connor, 2018), computer assisted vocabulary instruction (e.g., Chiu, 2013), distance learning (e.g., Allen, Mabry, Mattrey, Bourhis, Titsworth, & Burrell 2004; Bernard, 2014), MALL (mobile assisted language learning; e.g., Chen, Chen, Jia, & An, 2020; Sung, Chang, & Yang, 2015), web-based instruction (e.g., Chang & Lin, 2013), digital game (e.g., Chiu & Reynolds, 2012; Kao, 2014), data-driven approach in language learning (e.g., Boulton & Cobb, 2017; Mizumoto & Chujo, 2015), and small group and individual learning with technology (e.g., Lou, Abrami, & d'Apollonia, 2001). Figure 1 summarizes the meta-analytic effects of CALL from these studies that reported Cohen's d values (M = .67; SD = .39). Thus far, only one meta-analysis focused on the effectiveness of using multiple media tools on L2 listening, which was published 5 years ago (Kang, 2015; Cohen's d = .69).

This chapter aims to recruit the most recently published empirical studies on L2 multiple media listening tools, to discover the effectiveness of these multiple media tools on listening comprehension, and, further, use the moderator analysis to help L2 educators and test developers make decisions on applying different multiple media tools in the fields of L2 instruction and assessment and help L2 researchers explore the research gap for future research. In the results section, incorporating the results from a meta-analysis study, some of the most frequently researched multiple media tools were discussed by answering the following questions:

1. What is the overall effectiveness of multiple media tools in L2 listening instruction?
2. How is the effectiveness of multiple media tools in L2 listening instruction affected by moderators including contexts (e.g., second language vs. foreign language, country, and institution), treatments (e.g., MALL, delayed posttest, and instruction duration), and measures (e.g., closed-ended questions vs. open-ended questions and reliability reported)?

Figure 1. Overall meta-analytic effects of CALL (Cohen's d)

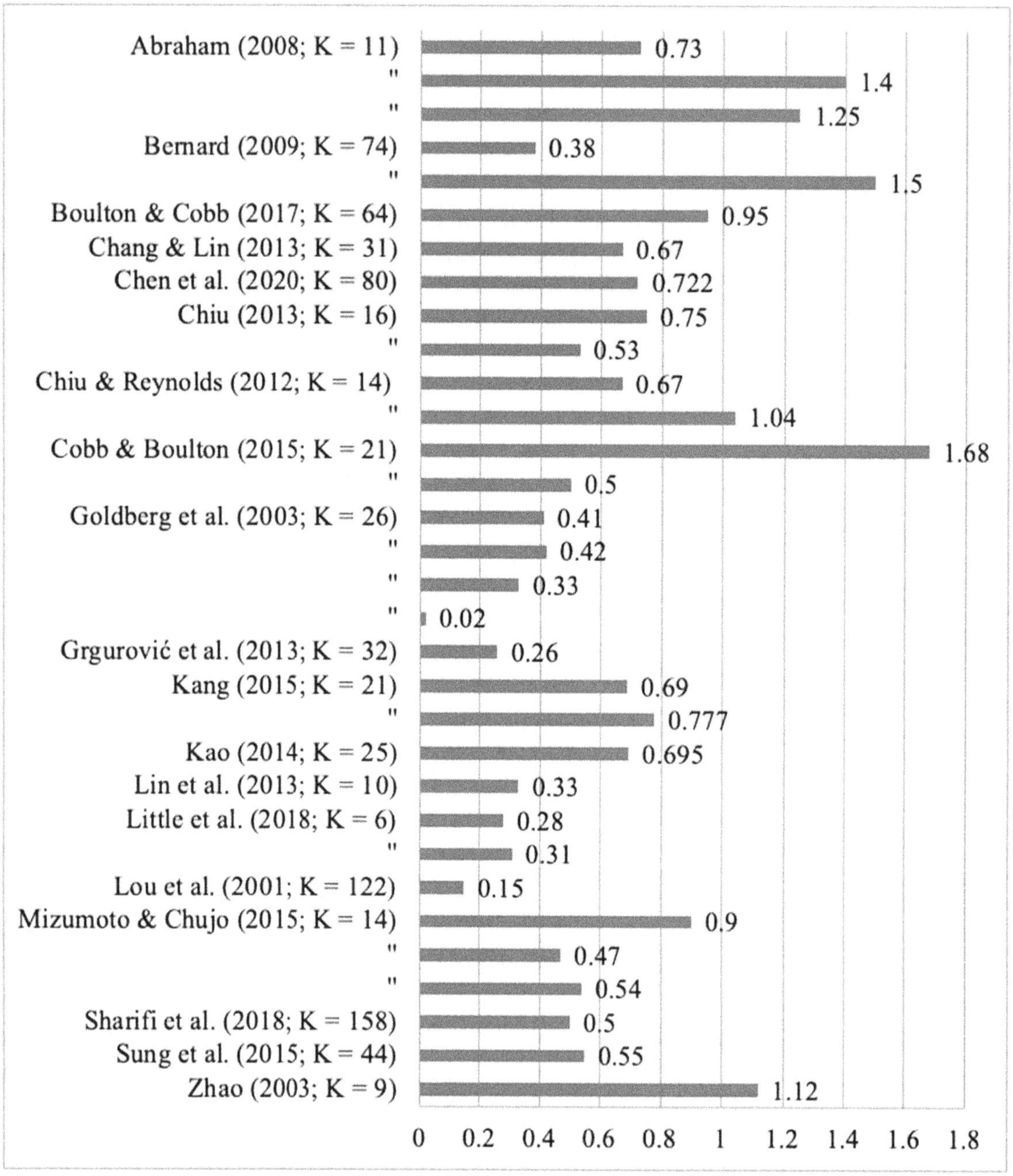

METHODS

Identifying Primary Studies

In attempting to identify relevant primary studies, two library-based databases including Educational Resources Information Center (ERIC) and Linguistics and Language Behavior Abstracts (LLBA) and one non-library-based database, Google Scholar were searched, using combinations of keywords (i.e., listening, technology, CALL, second language, and multiple media). Additionally, several journals related to CALL (i.e., Australasian Journal of Educational Technology, British Journal of Educational Technology, CALICO Journal, Computer Assisted Language Learning (CALL), Computer in Education, Computers in Human Behavior, European Journal of Engineering Education, Journal of Computer Assisted Learning, Journal of Educational Technology & Society, Language Learning & Technology, Learning, media and Technology, ReCALL, and Research in Learning Technology) were searched through their table of contents, using combinations of keywords (i.e., listening and second language). The last searching date was April 5th, 2020.

Exclusion and Inclusion Criteria

With reference to the exclusion/inclusion criteria followed in precious meta-analyses, and considering the purpose of the current meta-analysis, during the first phase, studies were excluded if:

1. The articles were not empirical studies, but summaries on different websites, introductions on one specific technology, or theoretical summaries
2. Dependent variables were not listening comprehension, but listening behavior
3. Independent variables were not related to the influence of technology, but related to variables such as text, gender, and context
4. There was no independent measurement on listening comprehension,

Lastly, during the second phase, studies that used a pre-post (within-group), control/comparison-experimental (between-group), or mixed designs were included for the meta-analysis.

The Coding Scheme and Inter-Coder Reliability

Articles were coded using a coding convention adapted from Plonsky and Oswald (2012) (see Appendix). Specifically, they were coded by identification, study context, design and treatment, measures, and results. After the data were coded by the first coder, a second coder was trained and coded for five random selected studies. Their agreement was around 90%.

Quantitative Synthesis Methods

The fixed effect size used in this section was Cohen's d, and the formula was:

$$ES = \frac{M_1 - M_2}{\sqrt{\dfrac{SD_1^2 \times (n_1 - 1) + SD_2^2 \times (n_2 - 1)}{n_1 + n_2 - 2}}}$$

(Lipsey & Wilson, 2001),where M_1 is the mean of group 1, M_2 is the mean of group 2, SD_1 is the standard deviation of group 1, SD_2 is the standard deviation of group 2, n_1 is the number of subjects in group 1, and n_2 is the number of subjects in group 2.

The detailed calculation procedures for the between-group contrasts are presented as follows. First, effect sizes based on experimental and control/comparison groups' performance on the posttests were calculated. For studies that included comparison groups, when calculating the overall effects of multiple media tools on L2 listening comprehension, group 1 was the group that received higher mean score, and group 2 was the group received lower mean score; however, when calculating the effects of specific multiple media tools on listening comprehension, group 1 was this specified multiple media tool, and group 2 was the other multiple media tool. According to previous discussion on the effective multiple media tools in listening comprehension, the specified multiple media tools and their comparison groups that had been selected in this meta-analysis study were video vs audio, self-regulated listening vs normal listening, slower speed vs faster speed, subtitles vs without subtitle, and L2 caption vs L1 subtitle.

Then, the average effect sizes were calculated and weighted based on individual sample sizes, and the formula was:

$$Weighted\ ES = \frac{ES_1 \times N_1 + ES_2 \times N_2 + ES_3 \times N_3 + ... + ES_k \times N_k}{N_1 + N_2 + N_3 + ... + N_k},$$

where ES_1 is the effect size of the first sample, N_1 is the total participants in the first sample, EF_2 is the effect size of the second sample, N_2 is the total participants in the second sample, and K is the total number of the samples. A 95% confidence interval around the weighted mean effect size was also calculated. If a confidence interval does not include zero, this suggests a statistical difference between the observed effect and null hypothesis of no effect (Norusis, 2008). To interpret the magnitude of the effects, this study used Cohen's (1988) criteria: small effect sizes are less or equal 0.2, medium are around 0.5, and large effect sizes are equal to or greater than 0.8. All of the above calculations were conducted through a pre-developed excel worksheet.

Additionally, moderator variable analysis was conducted by grouping, averaging, and weighting the effect size of individual samples according to the moderator variables. In order to aggregate the effects of technology on listening comprehension from different studies, some subcategories of moderators that include only one study were excluded in the final report. Four continuous moderator variables were also transformed into dichotomous categories, namely, age, instruction length, instruction duration, and technology training length.

RESULTS

Phase 1: The Research Synthesis

After the initial search, a total of 148 articles using technology to enhance second language learners' listening skills were identified. However, only 79 studies published from 1980 to April 2020 were selected following the exclusion criteria. As shown in Table 2, the numbers of the studies on multiple media tools in L2 listening increased chronologically, and 29 articles were published from 2014 to 2020. This section synthesizes the (a) multiple media tools that have been empirically analyzed, (b) the participants of the primary studies, (c) research methods that have been used, (d) how listening was operationalized, and (e) the reported effective multiple media tools among these 79 studies.

Table 2. Publication Years (K = 79)

Year	# of articles	%
1980-1985	1	1
1986-1990	1	1
1991-1995	1	1
1996-2000	4	5
2001-2005	6	8
2006-2010	18	23
2011- 2014	19	24
2014-2020	29	37

Multiple Media Tools That Have Been Utilized

Over that last 40 years, a growing number of empirical studies have analyzed the effectiveness of multiple media tools in L2 listening. Table 3 illustrates the multiple media tools that have been utilized in the 79 selected empirical studies. Most of the previous empirical studies compared the effectiveness among audio, video, and still image ($k = 21$; e.g., Ockey, 2007). Multiple media tools such as caption/subtitle/glosses ($k = 13$; e.g., Hayati & Mohmedi, 2011), self-regulated listening ($k = 9$; e.g., Roussel, 2011), personal digital assistant (PDA) ($k = 9$; e.g., Chen & Chang, 2011), online websites ($k = 7$; e.g., Pardo-Ballester, 2012), and podcasts ($k = 4$; e.g., Abdous, Camarena, & Facer, 2009) have also been studied by different researchers. There are other multiple media tools that have started to draw researchers' attentions: web-based course ($k = 2$; e.g., Chen, Belkada, & Okamoto, 2004), digital story ($k = 2$; e.g., Verdugo & Belmonte, 2007), text-to-speech conversion ($k = 2$; Sha, 2010), radio program ($k = 1$; Absalom & Rizzi, 2008), TV ($k = 1$; Fallahkhair, Pemberton, & Griffiths, 2007), Longman English Interactive program ($k = 1$; Cardenas-Claros & Gruba, 2013), and English for specific purpose (ESP) multiple media program ($k = 1$; Bani Salameh, Kabilan, & Bani-Salalmeh, 2011). There are other technology tools that have started to draw researchers' attention after 2015, such as various CALL tools (*k*

= 2; e.g., Bahari, 2019), game (k = 2; e.g., Hwang, Hsu, Lai, & Hsueh, 2017), WeChat (k = 1; Jia & Hew, 2019), and Second Life and Skype (k = 1; Levak & Son, 2017).

Table 3. Multiple Media Tools (K = 79)

Tools	# of studies	%
audio vs video vs still image	21	27
caption/subtitle/glosses	13	16
self-regulated listening/speed	9	11
PDA	9	11
online websites/program	7	9
podcasts	4	5
web-based course	2	3
digital story	2	3
various CALL tools	2	3
game	2	3
text-to-speech conversion	2	3
radio program	1	1
TV	1	1
Longman English Interactive program	1	1
ESP multiple media program	1	1
WeChat	1	1
Second Life and Skype	1	1

It is worth mentioning that, in the field of language education, technological devices that are portable or mobile are also known as m-learning or MALL. M-learning or MALL is incorporated with any form of teaching and learning that involve a mobile device, which includes mobile phones, tablet computers, MP3 players, PDAs, electronic dictionaries, and game players (Kukulska-Hulme & Shield, 2008). With regards to MALL, 28% of the articles employed mobile devices, and all of them were published after 2007 (k = 22; e.g., Nah, White, & Sussex, 2008). In that case, MALL seems to be a promising topic in this field.

Mobile devices have been widespread among college students over that last 15 years. According to a national report on college students' mobile device survey in 2013, 75% of the college students owned a smartphone, 40% of the college students owned a tablet, and more 50% of the college students intended to purchase a mobile device in the next six months (Pearson Student Mobile Device Survey 2013). Mobile device nowadays has become such a significant part of student life. The average number of hours that college students reported using their smartphones each day was nine hours, which was more than half of their daily waking hours (Roberts, Yaya, & Manolis, 2014). In light of college students' attitudes towards mobile devices, over 80% of the college students believe that mobile devices could transform the way they learn and make learning more fun (Pearson Student Mobile Device Survey 2013, 2013). In this regard, researchers and teachers started to develop various educational applications for mobile devices to promote teaching and learning. M-learning is currently considered as the next generation of e-learning and has the advantages of extending students' learning from inside the classroom to outside the classroom and allowing students to study whenever and whatever they want.

A meta-analysis, conducted by Wu et al., (2012), analyzed the results of 164 studies on m-learning. Some of the important findings are summarized as follows: First, in terms of the research purposes, evaluating the effects of m-learning was most common (58%), followed by designing a mobile system for learning (32%) and investigating the affective domain during mobile learning (5%). Second, 86% of the 164 studies reported positive research outcomes. Third, with regards to different mobile devices, mobile phones were most commonly used for m-learning (36.5%), followed by PDAs (30.96%), laptop computers (9.14%), iPods (4.06%), mp3/mp4 players (2.54%), podcasts (2.03%), and cameras (1.52%). Fourth, among different academic disciplines, applied sciences published the biggest amount of m-learning studies (29%), followed by humanities (20%), formal sciences (16%), social sciences (4%), and natural sciences (3%). Specifically, languages and linguistics courses were the most common focus (17.5%) when compared to other sub-disciplines. Additionally, Hwang and Tsai's (2011) meta-analysis on m-learning and ubiquitous learning reported the major contributing countries from 2001 to 2010 (K =154). In the first five years, US authors contributed the most publications (7), followed by UK authors (5) and Taiwanese authors (4). However, in the second 5 years, Taiwanese authors published 51 studies and became the number one contributing country.

In the field of language education, the first project of MALL was conducted by the Stanford Learning Lab with the aim of facilitating foreign language learners' listening and speaking skills (Brown, 2001). MALL has the advantages of extending students' learning from inside the classroom to outside the classroom and allowing students to study whenever and whatever they want.

According to a meta-analysis conducted by Duman, Orhon, and Gedik (2014), MALL research increased at the fastest pace around 2008. The most frequently used mobile devices in MALL were cell phones ($k = 29$) among 69 identified studies, followed by PDAs ($k = 17$) and portable music/video player ($k = 10$). Also, the results revealed that a large scope of language learning topics have been discussed in previous MALL studies, and vocabulary was most commonly addressed ($k = 28$). The number of listening and speaking/pronunciation in MALL studies were only eight and six. Kukulska-Hulme and Shield (2008) also pointed that even though cell phones were originally developed for oral communication, the MALL studies failed to make significant use of this capability. However, it is inspiring that listening, speaking, and pronunciation, along with learners' perceptions and attitudes, have become "permanent areas of focus" (Duman et al., 2014, p. 6) since 2008.

The Participants of the Primary Studies

Table 4 summarizes the background information of the participants in the primary studies. First, most of the participants were from the foreign language (FL) learning context ($k = 61$; e.g., Yang & Chang, 2014), and only19 studies were conducted in the second language (SL) environment (e.g., Sueyoshi, & Hardison, 2005). This result might be due to the accessibility of "authentic language" in different learning contexts. For instance, in the FL environment, it is harder for students to get access to and communicate with native speakers when compared to the SL environment. Students in the FL environment

Table 4. Summary of the Participants (K = 79)

		# of studies	%
SL/FL	SL	19	24
	FL	61	76
Learning Institution	university	61	77
	high school	2	3
	middle school	2	3
	elementary	6	8
	preschool	1	1
	others & missing data	7	9
First Language	Chinese	22	28
	English	11	14
	Japanese	6	8
	Spanish	5	6
	Korean	4	5
	mixed	13	16
	others & missing data	18	23
Target Language	English	67	85
	Spanish	3	4
	French	2	3
	mixed	2	3
	others & missing data	5	6

preferred using multiple media tools, which are globally accessible, to improve their L2 listening skill. Also, it is worth noting that one of the studies, Levak & Son (2017), was conducted both in the ESL and EFL environment. Meanwhile, the average age of the participants was 19.34-year-old. A majority of them was university students (k = 61; e.g., Cross, 2011), followed by elementary (k = 6; e.g., Suh, Kim, & Kim, 2010), high school students (k =2; e.g., Yasin, Mustafa, & Permatasari, 2017), middle school students (k =2; e.g., Leveridge, & Yang, 2013). The only study that included preschool students was published in 2016 (Tarantino).

Participants' first languages included Chinese (k = 22; e.g., Chen, 2011), English (k = 11; e.g., Winke, Gass, & Sydorenko, 2010), Japanese (k = 6; e.g., Kondo et al., 2012), Spanish (k = 5; e.g., Verdugo & Belmonte, 2007), Korean (k = 4; e.g., Bani-Salameh et al., 2011), etc. Most often, the target language was English (k = 67; e.g., Hui, Hu, Clark, Tam, & Milton, 2008). Therefore, we might be able to draw the conclusion that most of the multiple media tools nowadays have been designed for English language learners. However, since the library-based databases and journals in this meta-analysis were in English, it also made English as the most studied target language reasonable.

Research Methods That Were Most Commonly Used

As shown in Table 5, most of the studies used either quantitative research method ($k = 42$; e.g., Markham, Peter, & McCarthy, 2001) or mixed-methods ($k = 28$; e.g., Wagner, 2010). Qualitative research method was least frequently used ($k = 9$; e.g., Dehghani & Jowkar, 2005).

Table 5. Summary of the Research Methods (K = 79)

Research method	# of studies	%
quantitative	42	53
qualitative	9	11
mixed-methods	28	35

How Listening was Operationalized

In terms of operationalizing listening skills, most studies used listening comprehension tests to measure participants' listening skills. The different task types or outcome measures in these listening comprehension tests include multiple-choice questions (e.g., Shin, 1998), true-false questions (e.g., Gruba, 1993), written recall (e.g., Markham et al., 2001), and oral recall (e.g., Roussel, 2011). Meanwhile, questionnaires and interviews have also been utilized to measure learners' perceptions of their listening skills (e.g., Nah et al., 2008).

Table 6. Reported Effective of the Multiple Media Tools (K = 79)

				95% Confidence Interval	
	M (d)	K[a]	SE	Lower	Upper
overall effects[b]	.70	140	.01	.68	.72
effects of self-regulated listening & slower speed vs. normal listening & regular speed[c]	.75	14	.04	.67	.83
effects of video vs. audio[d]	.49	25	.03	.43	.55
effects of subtitles/captions vs. no subtitles/captions[e]	1.06	19	.03	1.01	1.11

Notes: [a] *Unique samples.* [b] *41 unique studies.* [c] *5 unique studies.* [d] *9 unique studies.* [e] *8 unique studies.*

Table 7. Weighted Effects of Multiple Media Tools on L2 Listening in Between-group Designs

Tools	Effective tools	References
audio vs video	video	Kim, 2015; Parry & Meredith, 1984; Shin, 1998; Wagner, 2010; Wagner 2013; Woottipong, 2014; Yang, 2014a
caption/subtitle/glosses	subtitle; target language captions	Chen, 2011; Hayati & Mohmedi, 2011; Hsu, Hwang, & Chan, 2014; Mirzaei, Meshgi, Akita, & Kawahara, 2017; Winke et al., 2010; Yang (2014b)
self-regulated listening/speed	self-regulated listening; slower speed	de la Fuente, 2014; East & King, 2012; McBride, 2011; Roussel, 2011; Zhao, 1997
PDA	PDA	Chang, Warden, Liang, & Chou, 2018; Hwang, Huang, Shadiev, & Chen, 2014; Kondo et al., 2012; Nah et al., 2008; Tarantino, 2016
online websites	online websites	Bower & Rutson-Griffiths, 2016; Lin, Winaitham, & Saitakham, 2008; Liu, Chen, & Hwang, 2018; Pardo-Ballester, 2012; Suh et al., 2010
podcasts	podcasts	Abdous, Camarena, & Facer, 2009; O'Bryan, & Hegelheimer, 2007; Şendağ, Gedik, & Toker, 2018

Reported Effective Multiple Media Tools

Most of the multiple media tools revealed effectiveness on learners' listening comprehension. In light of the well discussed multiple media tools, a summary of the results is demonstrated in Table 6. Here, the "well discussed multiple media tools" means the multiple media tools that have been empirically tested by at least two studies. Such multiple media tools included video, self-regulated listening, video with subtitle, video with target language captions, PDA, different online websites, and podcasts. As shown in the second column, the effective multiple media tools embrace video when compared to audio, self-regulated listening, slower speed, subtitle, target language captions, PDA, different online websites, and podcasts. The third column listed all the related studies.

Thus far, the phase I synthesis revealed several effective multiple media tools including video, self-regulated listening, video with subtitle, video with target language captions, PDA, and different online websites. Participants in these studies were mostly university students from the English foreign language (EFL) learning context. Most of the studies employed either quantitative or mixed research methods, and participants' listening skills were measured by listening comprehension tests, questionnaires, and interviews. Pedagogically, owing to the effectiveness of different multiple media listening tools that have been discovered by previous studies, language teachers are encouraged to employ these multiple media tools to their listening instruction.

PHASE II. THE QUANTITATIVE META-ANALYSIS

RQ1: What is the Overall Effectiveness of Multiple Media Tools in L2 Listening Instruction?

During this selection phase, 52 studies used either a pre-post (within-group), control/comparison-experimental (between-group), or mixed designs and were included. As discussed in Norris and Ortega (2006), one-group pre-post test designs tend to overestimate the effects when compared to the between-group design, and because the number of the pre-post (within-group) studies was relatively small (k = 13), only the between-group contrasts from the control/comparison-experimental (between-group) and mixed designs were utilized in this section. A total of 38 between-group (control/comparison-experimental) and mixed studies, which included 140 contrasts, were used in the final meta-analysis in this section.

As shown in Table 7, there was a medium-to-large overall effect of technology on listening comprehension with a confidence interval ranging from .68 to .72 (d = .70) in between-group designs, obtained from 41 unique studies and 140 unique samples with unweighted effect sizes ranging from 0 to 8.34.

Table 7 also includes the effectiveness of individual multiple media tools on L2 listening comprehension observed in current meta-analysis. The overall effects of self-regulated listening, video, and subtitles/captions remained the same from medium to large.

First, the effectiveness of self-regulated listening and slower speed over normal listening and faster speed on L2 listening comprehension in the between-group designs was drawn from three research studies published from 1997 to 2012 (East & King, 2012; Roussel, 2011; Zhao, 1997) and two newly identified studies (Botero, Restrepo, Zhu, & Questier, 2019; de la Fuente, 2014). The results revealed that the average weighted effect size was medium, d = .75 [.67, .83]. In Roussel's (2011) study, the researcher recorded listeners' (n = 29) movements of the mouse while listening to MP3-tracks on computers to track learners' self-regulated listening strategies. The study consisted of four experiments to testify the effects of self-regulated listening on listening comprehension across different listening conditions, learners' L2 (German) listening proficiencies, and speech tasks. Participants' listening comprehension was measured by their written recall of the listening discourses. As an indicator of L2 learners' metacognitive activity, the results showed that self-regulated listening resulted in a better understanding of various speech tasks compared to other listening conditions. In addition, learners with higher L2 listening proficiency levels tended to use more self-regulated strategies than less-skilled L2 listeners. Pedagogically, the researcher encouraged L2 teachers to use more individual listening devices with the aim of enhancing L2 learners' listening comprehension. Additionally, Zhao (1997) reported that English as a SL learners (n = 15) reacted positively to the use of computers to control speech rate, and those who were in self-controlled listening situations achieved better listening comprehension scores than in non-controlled situations. In this research, participants' listening comprehension was measured by having them choose a sentence that had the closest meaning to the sentence that they had heard. East and King (2012) also further conformed the effectiveness of using software to reduce the speed to enhance test takers' (n = 120) listening scores on the IELTS. This study compared the normal speed to three levels of reduced speech rate, 15%, 22.5%, and 30%. The results showed that reduced speech rate could facilitate test takers' listening comprehension, whereas when compared test takers' performance on the three speech rates, students who listened to the listening passages that were 22.5% slower than the normal speed outperformed the students who listened the listening passages with 15% or 30% reduced speech rates. Similarly, the results observed from de la Fuente (2014) revealed that self-regulated listening would enhance L2 learners' (n

= 20) bottom-up and top-down comprehension. However, a more recent study conducted in 2019 by Botero et al. reported Spanish learners who are learning French as a second language (n = 48) did not demonstrate better listening comprehension when using self-regulation and scaffolding.

Second, in terms of the effectiveness of video over audio on L2 listening comprehension, the observed average Cohen's d value in the between-group designs was .49[.45, .55]. A total of nine studies were involved in the analysis, and mixed results were revealed: seven studies discovered that video was better than audio to enhance L2 learners' listening comprehension (Becker & Sturm, 2017; Gruba, 1993; Shin, 1998; Suvorov, 2011; Wagner, 2013; Yang, 2014a; Yasin et al., 2017), one study showed an opposite result, audio was better than video (Coniam, 2001), and one study had a mixed result (Brett, 1997). Gruba's study was conducted in 1993. Ninety-one advanced English as SL learners completed a multiple-choice and true/false listening comprehension test. Students who watched video performed better than students who listened to the traditional audio tape player. Later on, results from different research have conformed this finding. On the other hand, in Coniam's (2001) study, the researcher reported that pre-service and in-service English language teachers in Hong Kong (n = 104) received higher scores on the audio version of the Hong Kong English Language Benchmark Test when compared to the video version of the test. Since in Brett's (1997) study the results from 12 listening comprehension and cloze tasks were reported individually, the effects of audio over video or vice versa were inconsistent. However, in most the tasks, students (n = 43) performed better while using video than audio.

Third, as shown in Tables 7, there were large effects of subtitles/captions on L2 listening comprehension when compared to no subtitles/captions in between-group designs (d = 1.06 [1.01, 1.11]). This comparison involved eight unique studies (Çakmak & Erçetin, 2018; Chen, 2011; Hayati & Mohmedi, 2005; Hsieh, 2019; Markham et al., 2001; Michael & Webb, 2017; Winke et al., 2010; Yang, 2014b), and most of them tested that subtitles/captions were more effective than no subtitles/captions even though the research methods varied. In Chen's (2011) study, the author applied a video-text synchronization system, and 83 English as FL students in Taiwan were asked to watch around 30 synchronized videos in one semester. Hayati and Mohmedi (2005) recruited 90 English as FL students in Iran to watch a documentary film over six weeks. Participants in Markham et al.'s (2001) and Winke et al.'s (2010) studies were Spanish as FL learners who watched videos with and without subtitles/captions. To measure participants listening comprehension, multiple-choice (Hayati & Mohmedi, 2005; Winke et al., 2010) and written summary (Markham et al., 2001; Chen, 2011) were utilized. However, two most recently published studies (Hsieh, 2019; Michael & Webb, 2017) observed different results, and both of them reported that there were no significant differences of listening comprehension between the group with and without captions.

Table 8. Weighted Between-group Effects by Subcategories

Moderator	Subcategory	Value	M(d)	K[a]	N[b]	SE	95% Confidence Interval Lower	Upper
Context	SL/FL	SL	.66	58	2316	.02	.62	.70
		FL	.71	82	7338	.01	.69	.74
	country	US	.74	40	2162	.02	.69	.78
		Taiwan	.80	24	1615	.02	.75	.85
	L1	English	.70	24	1354	.03	.65	.75
		Chinese	.71	34	2315	.02	.67	.75
		mixed	.70	52	1956	.02	.66	.75
	L2	English	.71	113	8179	.01	.68	.73
	age	< 15	.88	7	987	.03	.82	.94
		> 19	.90	42	2594	.02	.86	.93
	institution	university	.69	125	7181	.01	.67	.71
		elementary	.77	9	2045	.02	.73	.81
	class	classroom	.67	63	5185	.01	.64	.70
		laboratory	.74	77	4439	.02	.71	.77
Design	technology tool	self-regulated/speed	.75	14	581	.04	.67	.83
		audio/video/still image	.66	57	2397	.02	.62	.71
		subtitle/caption	.80	47	2893	.02	.77	.84
	MALL	yes	1.11	13	1050	.03	1.05	1.17
		no	.65	127	8574	.01	.63	.67
	delayed posttest	yes	.56	7	648	.04	.48	.63
		no	.71	133	8976	.01	.69	.73
	# of groups	2	.73	50	4333	.02	.70	.76
		3	.69	65	4157	.02	.66	.72
		4	.85	15	714	.04	.78	.93
		5	.36	10	420	.05	.26	.45
	random assignment	yes	.66	82	6211	.01	.63	.68
		no	1.00	10	1363	.03	.95	1.05
		not mentioned	.65	48	2050	.02	.60	.69
	instruction length	0 hours	.64	71	3482	.02	.60	.67
		0.5~4.5 hours	1.16	37	2005	.02	1.11	1.20
		5~60 hours	.76	18	1791	.02	.72	.81
	instruction duration	0 days	.64	71	3482	.02	.60	.67
		1~2 days	.96	38	1828	.02	.91	1.00
		10~175 days	.63	24	3796	.02	.60	.66
	technology training	yes	.76	33	2301	.02	.72	.80
		no/not reported	.69	107	7323	.01	.67	.71
	technology training length	0.3~1 hour	.46	19	1484	.03	.41	.51
		3~7 hours	.40	3	364	.05	.30	.50
Measures	listening comprehension test	closed-ended questions	.66	124	8531	.01	.64	.68
		open-ended questions	1.05	16	1093	.03	.99	1.11
	reliability reported	yes	.66	51	6071	.01	.63	.68
		no	.79	89	3553	.02	.75	.82

Notes: [a] *Unique samples.* [b] *# of participants.*

RQ2: How is the Effectiveness of Multiple Media Tools in L2 Listening Instruction Affected by Different Learning Contexts, Treatments, and Measures?

Different context, design, and measures in different studies could result in various results. Tables 8 presents the moderator effects of context, design, and measures in the between-group designs. A larger weighted effect size was obtained in the FL environment (*d* = .71 [.69, .74]) when compared to the SL environment (*d* = .66 [.62, .70]). The effects of multiple media tools on L2 listening comprehension were medium in the U.S. (*d* = .74 [.69, .78]) and Taiwan (*d* = .80 [.75, .85]). Additionally, the Cohen's *d* value observed for L2 English speakers was .71 [.67, .73], and larger effects were also observed in the laboratory (*d* = .74 [.71, .77]) when compared to the classroom environment (*d* = .67 [.64, .70]). Among different L1s, multiple media tools showed similar effects on L1 English speakers (*d* = .70 [.65, .75]), L1 Chinese speakers (*d* = .71 [.67, .75]), and mixed L1 participants (*d* = .70 [.66, .75]). Older participants (> 19 years old; *d* = .90 [.86, .93]) tended to receive similar influence from the multiple media tools as the younger participants (< 15 years old; *d* = .88 [.82, .94]), and there were more effects of multiple media tools on listening comprehension in the elementary schools (*d* = .77 [.73, .81]) than in the universities (*d* = .69 [.67, .71]).

In light of the research designs, subtitle/caption (*d* = .80 [.77, .84]) revealed a slightly higher effect on L2 listening comprehension than self-regulated listening/slow speed (*d* = .75 [.67, .83]) and audio/video/still image (*d* = .66 [.62, .71]). MALL related studies (*d* = 1.11 [1.05, 1.17]) had a larger effect than the other studies (*d* = .65 [.63, .67]). Additionally, studies that included technology training (*d* = .76 [.72, .80]) showed larger effects than the ones excluded or did not report them. On the other hand, 4 research groups (*d* = .85 [.78, 93]), instruction length from 0.5 to 0.45 hours (*d* = 1.16 [1.11, 1.20]), instruction duration from 1 to 2 days (*d* = .96 [.91, 1.00]), and technology training from 0.3 to 1 hour (*d* = .46 [.41, .51]) resulted in large weighted effect sizes. Studies that included delayed posttests (*d* = .56 [.48, .63]) and random assignment (*d* = .66 [.63, .68]) showed smaller effects than the ones excluded or did not report them.

In terms of the measures, when L2 listening comprehension was measured by open-ended questions, such as written and oral summary, larger weighted effect sizes were gained (*d* = 1.05 [.99, 1.11]), when compared to the closed-ended questions, such as multiple-choice and cloze questions (*d* = .66 [.64, .68]). However, listening comprehension tests that reported reliabilities revealed smaller effects (*d* = .66 [.63, .68]) than the ones did not report reliabilities (*d* = .79 [.75, .82]).

DISCUSSION

The overall effects of multiple media tools on listening comprehension were .70 in the between-group designs, which were almost identical to the results observed in Kang (2015; *d* = .69) and slightly exceeded the average weighted effect size in previous meta-analysis on CALL, .67. A total number of 29 articles were identified since 2014, and 17 of them were able to be added to calculate the overall weighted effect size. It is also interesting to discover the fact that, even after five years, the overall effects of multiple media tools on listening comprehension remained the same, medium-to-large, and match up with the effects of technology on learning other L2 skills. Technology tools should still be highly recommended to the L2 listening instruction. Owing to the nature of the meta-analyses, effectiveness of previous empirically tested multiple media tools has been statistically synthesized. The average Cohen's *d* values for each of the multiple media tools in the between-group designs were 1.06 when compared subtitle/captions to no subtitle/captions, .75 when self-regulated listening and slow speed compared to normal listening and regular speed, and .49 when video compared to audio, which were similar to the results achieved in Kang (2015). Applying these results to L2 instruction and assessment, on the aim of improving L2 learners' listening comprehension, subtitles/captions, as well as self-regulated listening and slow speed, should be considered by teachers and test developers. However, the effectiveness of video over audio was relatively small, which suggested L2 teachers and test developers to prevent the overestimations of video over audio in L2 listening. In that case, in previous empirical studies, showing there was a significant difference between video and audio was not sufficient enough and could easily mislead readers.

Specifically, self-regulated listening is highly recommended to the L2 listening instructors. In order to better understand the process of listening and teach listening more effectively, metacognition, which means "the ability of learners to control their thoughts and to regulate their own learning" (Vandergrift & Goh, 2012, p. 5), has been widely discussed. Researchers argued that self-regulated learning, which refers to learners' mastering of their learning process by controlling their own thoughts, feelings, and actions during the learning process, could develop learners' self-efficacy and motivate them towards greater success (Vandergrift & Goh, 2012). There are different ways to track learners' metacognitive listening strategies (e.g., think-aloud and recall). In recent years, due to the development of computer technology, researchers started to use technical functions, such as speech rate (Zhao, 1997), pause, rewind, and fast-forward (Roussel, 2011), to indicate learners' self-regulated listening strategies. The followings are two self-regulated listening activities proposed by the author.

First, according to the listening material development principles proposed by Morley (2001), the listening passage used in this activity could be downloaded from *VOA Learning English* website, and the topic should be related to the topics that they have studied in their Listening and Speaking class. The listening passage is always about 10 minutes long, and the teacher could cut them into small segments. Since one of the top-down processing goals for lower level students is to get the main idea of a passage (Peterson, 2010), after listening to each segment twice, students are asked to complete two main idea questions in the format of multiple choices. Additionally, students should be familiar with using the audio player. During the instruction, the teacher first presents the four types of listening process that have been discovered in Roussel's (2008) study: 1) global listen + detailed listen, 2) detailed listen + global listen, 3) global listen + global listen, and 4) detailed listen + detailed listen. Then, teacher asks students to guess which is the most effective way of listening based on their own experience. After this short discussion, teacher proposes the most efficient listening process, global listen + detailed listen, based on Roussel's (2008) results. With this in mind, teacher asks students to follow this listening process to listen to one listening segment through their own computer and answer the main idea questions. Once they finished, the teacher could check the answers with the whole class and ask individual student's feedback on using this listening strategy. In the end, during the practice time, the teacher could encourage students to try the other listening process which could help them to find out the advantages and disadvantages of each listening process and to testify which listening process is best suitable to their own learning style.

Moreover, oftentimes, teachers were advised to model the metacognitive strategies (Goh, 2013). However, since teachers and students are at very different proficiency levels, the teacher could use successful students to model the metacognitive process. For the preparation of this activity, teacher needs to find 3-4 short listening passages (1-2 minutes) which are appropriate for students' L2 level, and the content should also relate to the topics that they have covered in their listening class. According to Peterson (2001), one of the bottom-up processing goals for lower level students is to get the details from the text. In that case, the three multiple choice items for each listening passage are detail questions. Additionally, students should be familiar with using the audio player and the screen recording software (e.g., *CamStudio*). During the instruction, the teacher asks students to complete the detail questions after listening to the first audio twice through their own computer. Meanwhile, their listening process should be recorded. Then, the teacher checks the answers with the students. A student who answered them all correctly should send his/her screen recordings to the teacher and then use the think-aloud method to demonstrate his/her listening process to other students. During this process, the teacher could help student to recall the listening process (e.g., Why did you pause/rewind here?). After illustrating the successful student's listening process, a short discussion section should be included to conclude the strategies that have been utilized. In the end, this class ends with 10 minutes of practice on applying the strategies into their own listening process.

Regarding to the results of moderating effects, several moderator variables were worth discussing. First, in terms of the instruction and technology training length, a shorter instruction (0.5~4.5 hours) and technology training (0.3~1 hour) were sufficient, and no treatment or tedious instruction could result in reduced effects. Pedagogically, in attempting to improve L2 learners' listening comprehension, teachers and test developers should not expect students' automatic mastery of a new technology; instead, a sufficient instruction and technology training period should be provided.

Second, in terms of contextual moderators, according to Kang (2015), multiple media tools tended to have larger effects on older learners when compared to younger learners in the between-group designs. Specifically, participants who were older than 19 years old (d = .78) and college students (d = .74) benefited more from the multiple media tools than participants who were younger than 15 years old (d = .44) and from elementary schools (d = .45). Five years later, after recruiting more studies, the results showed that younger learners (d = .88) tended to receive similar influence from the multipole media tools as the older learners (d = .90), and even more effects were found in the elementary schools (d = .77) than in the universities (d = .69). Similarly, both meta-analyses discovered that, in the between-group designs, larger weighted effect sizes were observed in the laboratory settings than in the classroom settings. Therefore, the ideal context to employ multiple media tools in L2 listening was for both young and adult learners in laboratory settings.

Third, current study further confirmed that researchers used different instruments to measure listening comprehension. Often times, closed-ended questions such as multiple-choice were employed. However, the results showed that open-ended questions had a tendency of involving more authentic tasks and capturing more and large influences from the multiple media tools (d = .97 (Kang, 2015) & d = 1.05). Unlike the results observed in Kang (2015), reporting test reliabilities showed a tendency of producing less effects. Therefore, to ensure the quality of the research, both open-ended and closed-ended questions were highly recommended to be included in future studies when measuring listening comprehension. Reporting instrument reliabilities were essential, and researchers should always remember to include this critical information in their studies to warrant the reliability of the research instruments.

FUTURE RESEARCH DIRECTIONS

Another limitation that has been identified in this meta-analysis was the small number of within-group effects. Subcategory values under several moderator variables were also excluded owing to the small number of within-group effects. Future researchers are suggested to utilize equivalent pretest and posttest and employ within-group and mixed research designs.

Additionally, several research gaps have been identified based on previous empirical studies. First, in terms of participants, expanding the age range, recruiting students from institutions other than universities, and exploring the SL learning environment are highly needed. Second, even though some of the multiple media tools have been widely discussed, only a few of them have been empirically testified. Therefore, some other newly developed technology tools (e.g., podcasts, text-to-speech conversion, MALL, WeChat) are also worth analyzing in the future.

CONCLUSION

This chapter recruited the most recently published empirical studies and meta-analyzed available evidence on the effects of different multiple media tools on L2 listening comprehension. Unlike traditional synthesis approach that mainly qualitatively summarized previous studies, this chapter has applied the quantitative synthesis approach to precisely conclude the effectiveness of different multiple media tools. Multiple media tools such as video, self-regulated listening, subtitles/captions, PDA, and different online websites have been tested and recommended by different researchers based on the qualitative synthesis. The effects of video, self-regulated listening, and subtitles/captions were further conformed in the meta-analysis. Therefore, these findings should be widely applied in the L2 listening instructions. Additionally, moderator analysis helped L2 educators and test developers make decisions on applying different multiple media tools in different situations.

Yet, inevitably, a note of caution must be added when interpreting the results. First of all, since all of the primary studies included in this meta-analysis are published journal articles, and the published studies are more likely to report significant results, the actual effect size might be lower when adding more unpublished studies. According to Borenstein, Hedges, Higgins, & Rothstein, the only true test for publication bias is to compare effects in the published studies formally with effects in the unpublished studies (2011). Future researchers could perform a more comprehensive search by broadening the publication types by adding conference papers or emailing the authors of the retained studies to ask for unpublished manuscripts as Lin (2015) and Little et al. (2018) used in their meta-analyses. However, since the primary studies in this meta-analysis were not coded for conducting the statistical analyses for publication bias, in the future, methods such as the funnel plot, which plotted the effect size on the X axis and standard error on the Y axis, Rosenthal's Fail-safe N test, which computes how many missing studies would need to make the effect size non-significant, and the Duval and Tweedie's Trim and Fill analysis, which removes extreme effect sizes, should be included to gauge the impact of publication bias (Borenstein et al., 2011).

Meanwhile, search bias is also occurred in this meta-analysis since all the included studies are in English which made English the target language in 67 cases which was about 85% of the entire primary studies. In the future, more search could be done in different languages, such as those frequently identified first languages in this study, like Chines, Japanese, Spanish, and Korean.

REFERENCES

Abdous, M., Camarena, M., & Facer, B. (2009). MALL technology: Use of academic podcasting in the foreign language classroom. *ReCALL, 21*(1), 76–95. doi:10.1017/S0958344009000020

Abraham, L. B. (2008). Computer-mediated glosses in second language reading comprehension and vocabulary learning: A meta-analysis. *Computer Assisted Language Learning, 21*(3), 199–226. doi:10.1080/09588220802090246

Absalom, M., & Rizzi, A. (2008). Comparing the outcomes of online listening versus online text-based tasks in university level Italian L2 study. *ReCALL, 20*(1), 55–66. doi:10.1017/S0958344008000517

Allen, M., Mabry, E., Mattrey, M., Bourhis, J., Titsworth, S., & Burrell, N. (2004). Evaluating the effectiveness of distance learning: A comparison using meta-analysis. *Journal of Communication, 54*(3), 402–420. doi:10.1111/j.1460-2466.2004.tb02636.x

Alvarez-Marinelli, H., Blanco, M., Lara-Alecio, R., Irby, B. J., Tong, F., Stanley, K., & Fan, Y. (2016). Computer assisted English language learning in Costa Rican elementary schools: An experimental study. *Computer Assisted Language Learning, 29*(1), 103–126. doi:10.1080/09588221.2014.903977

Bahari, A. (2019). FonF practice model from theory to practice: CALL via focus on form approach and non-linear dynamic motivation to develop listening and speaking proficiency. *Computers & Education, 130*, 40–58. doi:10.1016/j.compedu.2018.11.009

Bani-Salameh, Z. A., Kabilan, M. K., & Bani-Salalmeh, L. (2011). Utilising multiple media ESP programme in enhancing flight attendants' safety knowledge and problem solving skills. *British Journal of Educational Technology, 42*(6), 1003–1015. doi:10.1111/j.1467-8535.2010.01124.x

Becker, S. R., & Sturm, J. L. (2017). Effects of audiovisual media on L2 listening comprehension: A preliminary study in French. *CALICO Journal, 34*, 147–177.

Bernard, R. M., Borokhovski, E., Schmid, R. F., & Tamim, R. M. (2018). Gauging the effectiveness of educational technology integration in education: what the best-quality meta-analyses tell us. *Learning, Design, and Technology: An International Compendium of Theory, Research, Practice, and Policy*, 1-25.

Blau, E. K. (1990). The effect of syntax, speed, and pauses on listening comprehension. *TESOL Quarterly, 24*(4), 746–753. doi:10.2307/3587129

Bonk, W. J. (2000). Second language lexical knowledge and listening comprehension. *International Journal of Listening, 14*(1), 14–31. doi:10.1080/10904018.2000.10499033

Borenstein, M., Hedges, L. V., Higgins, J. P., & Rothstein, H. R. (2011). Publication bias. In *Introduction to meta-analysis*. John Wiley & Sons.

Botero, G. G., Restrepo, M. A. B., Zhu, C., & Questier, F. (2019). Complementing in-class language learning with voluntary out-of-class MALL. Does training in self-regulation and scaffolding make a difference? *Computer Assisted Language Learning*, 1–27. doi:10.1080/09588221.2019.1650780

Boulton, A., & Cobb, T. (2017). Corpus use in language learning: A meta-analysis. *Language Learning, 67*(2), 348–393. doi:10.1111/lang.12224

Bower, J. V., & Rutson-Griffiths, A. (2016). The relationship between the use of spaced repetition software with a TOEIC word list and TOEIC score gains. *Computer Assisted Language Learning*, *29*(7), 1238–1248. doi:10.1080/09588221.2016.1222444

Brett, P. (1997). A comparative study of the effects of the use of multiple media on listening comprehension. *System*, *25*(1), 39–53. doi:10.1016/S0346-251X(96)00059-0

Brown, E. (Ed.). (2001). Mobile learning explorations at the Stanford Learning Lab. *Speaking of Computers*, *55*. Retrieved September 24, 2014, from http://cgi.stanford.edu/~dept-ctl/tomprof/posting.php?ID=289

Buck, G., & Tatsuoka, K. (1998). Application of the rule-space procedure to language testing: Examining attributes of a free response listening test. *Language Testing*, *15*(2), 119–157. doi:10.1177/026553229801500201

Burley-Allen, M. (1995). *Listening: The forgotten skill*. John Wiley & Sons, Inc.

Burston, J. (2014). Twenty years of MALL project implementation: A meta-analysis of learning outcomes. *ReCALL*, 1–17.

Çakmak, F., & Erçetin, G. (2018). Effects of gloss type on text recall and incidental vocabulary learning in mobile-assisted L2 listening. *ReCALL*, *30*(1), 24–47. doi:10.1017/S0958344017000155

Cardenas-Claros, M., & Gruba, P. (2013). Decoding the "CoDe": A framework for conceptualizing and designing help options in computer-based second language listening. *ReCALL*, *25*(2), 250–271. doi:10.1017/S0958344013000049

Cauldwell, R. T. (1996). Direct encounters with fast speech on CD-Audio to teach listening. *System*, *24*(4), 521–528. doi:10.1016/S0346-251X(96)00046-2

Cavus, N., & Ibrahim, D. (2017). Learning English using children's stories in mobile devices. *British Journal of Educational Technology*, *48*(2), 625–641. doi:10.1111/bjet.12427

Chang, C. C., Lei, H., & Tseng, J. S. (2011). Media presentation mode, English listening comprehension and cognitive load in ubiquitous learning environments: Modality effect or redundancy effect? *Australasian Journal of Educational Technology*, *27*(4), 633–654. doi:10.14742/ajet.942

Chang, C. C., Warden, C. A., Liang, C., & Chou, P. N. (2018). Performance, cognitive load, and behaviour of technology-assisted English listening learning: From CALL to MALL. *Journal of Computer Assisted Learning*, *34*(2), 105–114. doi:10.1111/jcal.12218

Chang, M. M., & Lin, M. C. (2013). Strategy-oriented web-based English instruction—A meta-analysis. *Australasian Journal of Educational Technology*, *29*(2), 203–216. doi:10.14742/ajet.67

Chen, H. (2011). Developing and evaluating SyntoLearn, a fully automatic video and transcript synchronization tool for EFL learners. *Computer Assisted Language Learning*, *24*(2), 117–130. doi:10.1080/09588221.2010.526947

Chen, I., & Chang, C. (2011). Content presentation modes in mobile language listening tasks: English proficiency as a moderator. *Computer Assisted Language Learning*, *24*(5), 451–470. doi:10.1080/09588221.2011.577749

Chen, J., Belkada, S., & Okamoto, T. (2004). How a web-based course facilitates acquisition of English for academic purposes. *Language Learning & Technology, 8*, 33–49.

Chen, Z., Chen, W., Jia, J., & An, H. (2020). The effects of using mobile devices on language learning: A meta-analysis. *Educational Technology Research and Development, 68*(4), 1769–1789. doi:10.100711423-020-09801-5

Cheon, J., & Grant, M. (2009). Web-based assessment tool for communication and active listening skill development. *TechTrends, 53*(6), 24–34. doi:10.100711528-009-0340-4

Chiu, Y., Kao, C., & Reynolds, B. (2012). The relative effectiveness of digital game-based learning types in English as a foreign language setting: A meta-analysis. *British Journal of Educational Technology, 43*(4), E104–E107. doi:10.1111/j.1467-8535.2012.01295.x

Chiu, Y. H. (2013). Computer-assisted second language vocabulary instruction: A meta-analysis. *British Journal of Educational Technology, 44*(2), E52–E56. doi:10.1111/j.1467-8535.2012.01342.x

Cobb, T., & Boulton, A. (2015). Classroom applications of corpus analysis. In D. Biber & R. Reppen (Eds.), *The Cambridge handbook of corpus linguistics* (pp. 478–497). Cambridge University Press. doi:10.1017/CBO9781139764377.027

Cohen, J. (1988). *Statistical power analysis for the behavioral sciences* (2nd ed.). Erlbaum.

Coniam, D. (2001). The use of audio or video comprehension as an assessment instrument in the certification of English language teachers: A case study. *System, 29*(1), 1–14. doi:10.1016/S0346-251X(00)00057-9

Conrad, L. (1985). Semantic versus syntactic cues in listening comprehension. *Studies in Second Language Acquisition, 7*(1), 59–69. doi:10.1017/S0272263100005155

Cross, J. (2011). Comprehending news videotexts: The influence of the visual content. *Language Learning & Technology, 15*, 44–68.

De la Fuente, M. J. (2014). Learners' attention to input during focus on form listening tasks: The role of mobile technology in the second language classroom. *Computer Assisted Language Learning, 27*(3), 261–276. doi:10.1080/09588221.2012.733710

Dehghani, A. P., & Jowkar, M. (2005). The impact of computer-video projector on EFL learners' listening comprehension. *Academic Research International, 3*, 106–112.

Duman, G., Orhon, G., & Gedik, N. (2014). Research trends in mobile assisted language learning from 2000 to 2012. *ReCALL, FirstView*, 1-20.

East, M., & King, C. (2012). L2 learners' engagement with high stakes listening tests: Does technology have a beneficial role to play? *CALICO Journal, 29*(2), 208–223. doi:10.11139/cj.29.2.208-223

Fallahkhair, S., Pemberton, L., & Griffiths, R. (2007). Development of a cross-platform ubiquitous language learning service via mobile phone and interactive television. *Journal of Computer Assisted Learning, 23*(4), 312–325. doi:10.1111/j.1365-2729.2007.00236.x

Flowerdew, J., & Miller, L. (2005). *Developing listening skills through technology. In Second Language Listening: Theory and Practice*. Cambridge University Press.

Freedle, R., & Kostin, I. (1999). Does the text matter in a multiple-choice test of comprehension? The case for the construct validity of TOEFL's minitalks. *Language Testing*, *16*(1), 2–32.

Goh, C. (2013). Second language listening comprehension: Process and pedagogy. In M. Celce-Murica (Ed.), *Teaching English as a second or foreign language* (pp. 72–89). Heinle & Heinle.

Goldberg, A., Russell, M., & Cook, A. (2003). The effect of computers on student writing: A meta-analysis of studies from 1992 to 2002. *The Journal of Technology, Learning, and Assessment*, *2*, 1–51.

Grgurović, M., Chapelle, C. A., & Shelley, M. C. (2013). A meta-analysis of effectiveness studies on computer technology-supported language learning. *ReCALL*, *25*(2), 165–198. doi:10.1017/S0958344013000013

Grgurović, M., & Hegelheimer, V. (2007). Help options and multiple media listening: Students' use of subtitles and the transcript. *Language Learning & Technology*, *11*, 45–66.

Griffiths, R. (1990). Facilitating listening comprehension through rate-control. *RELC Journal*, *21*(1), 55–65. doi:10.1177/003368829002100104

Griffiths, R. (1992). Speech rate and listening comprehension: Further evidence of the relationship. *TESOL Quarterly*, *26*(2), 385–390. doi:10.2307/3587015

Gruba, P. (1993). A comparison study of audio and video in language testing. *JALT Journal*, *15*, 85–88.

Gruba, P. (2006). Playing the videotext: A media literacy perspective on video-mediated L2 listening. *Language Learning & Technology*, *10*, 77–92.

Hayati, A., & Mohmedi, F. (2011). The effect of films with and without subtitles on listening comprehension of EFL learners. *British Journal of Educational Technology*, *42*(1), 181–192. doi:10.1111/j.1467-8535.2009.01004.x

Herron, C., York, H., Corrie, C., & Cole, S. P. (2006). A comparison study of the effects of a story-based video instructional package versus a text-based instructional package in the intermediate-level foreign language classroom. *CALICO Journal*, *23*(2), 281–307. doi:10.1558/cj.v23i2.281-307

Hochart, J. (1998). Improving listening and speaking skills in English through the use of authoring systems. *ReCALL*, *10*(2), 18–24. doi:10.1017/S0958344000003700

Hsieh, Y. (2019). Effects of video captioning on EFL vocabulary learning and listening comprehension. *Computer Assisted Language Learning*, 1–23.

Hsu, C. K., Hwang, G. J., & Chan, C. K. (2014). An automatic caption filtering and partial hiding approach to improving the English listening comprehension of EFL students. *Journal of Educational Technology & Society*, *17*, 270–283.

Hsu, C. K., Hwang, G. J., Chang, Y. T., & Chang, C. K. (2013). Effects of Video Caption Modes on English Listening Comprehension and Vocabulary Acquisition Using Handheld Devices. *Journal of Educational Technology & Society*, *16*, 403–414.

Hui, W. W., Hu, P. H., Clark, T. K., Tam, K. Y., & Milton, J. J. (2008). Technology-assisted learning: A longitudinal field study of knowledge category, learning effectiveness and satisfaction in language learning. *Journal of Computer Assisted Learning*, *24*(3), 245–259. doi:10.1111/j.1365-2729.2007.00257.x

Hwang, G. J., Hsu, T. C., Lai, C. L., & Hsueh, C. J. (2017). Interaction of problem-based gaming and learning anxiety in language students' English listening performance and progressive behavioral patterns. *Computers & Education*, *106*, 26–42. doi:10.1016/j.compedu.2016.11.010

Hwang, W. Y., Huang, Y. M., Shadiev, R., Wu, S. Y., & Chen, S. L. (2014). Effects of using mobile devices on English listening diversity and speaking for EFL elementary students. *Australasian Journal of Educational Technology*, *30*(5), 503–516. doi:10.14742/ajet.237

Hwang, W. Y., Shih, T. K., Ma, Z. H., Shadiev, R., & Chen, S. Y. (2016). Evaluating listening and speaking skills in a mobile game-based learning environment with situational contexts. *Computer Assisted Language Learning*, *29*(4), 639–657. doi:10.1080/09588221.2015.1016438

Hwu, F. (2004). On the applicability of the input-enhancement hypothesis and input processing theory in multiple media CALL: The case of Spanish preterite and imperfect instruction in an input application. *CALICO Journal*, *21*(2), 317–338. doi:10.1558/cj.v21i2.317-338

Jakobsdóttir, S., & Hooper, S. (1995). Computer-assisted foreign language learning: Effects of text, context, and gender on listening comprehension and motivation. *Educational Technology Research and Development*, *43*(4), 43–59. doi:10.1007/BF02300490

Jensen, E. D., & Vinther, T. (2003). Exact repetition as input enhancement in second language acquisition. *Language Learning*, *53*(3), 373–428. doi:10.1111/1467-9922.00230

Jia, C., & Hew, K. F. T. (2019). Supporting lower-level processes in EFL listening: The effect on learners' listening proficiency of a dictation program supported by a mobile instant messaging app. *Computer Assisted Language Learning*, 1–28. doi:10.1080/09588221.2019.1671462

Kang, T. (2015). The effectiveness of multiple media tools in L2 listening: A meta-analysis. In V. Turel (Ed.), *Intelligent design of interactive multimedia listening software* (pp. 255–284). IGI Global. doi:10.4018/978-1-4666-8499-7.ch010

Kao, C. W. (2014). The effects of digital game-based learning task in English as a foreign language contexts: A meta-analysis. *Education Journal*, *42*, 113–141.

Kim, H. S. (2015). Using Authentic Videos to Improve EFL Students' Listening Comprehension. *International Journal of Contents*, *11*(4), 15–24. doi:10.5392/IJoC.2015.11.4.015

Kondo, M., Ishikawa, Y., Smith, C., Skamoto, K., Shimomura, H., & Wada, N. (2012). Mobile assisted language learning in university EFL courses in Japan: Developing attitudes and skills for self-regulated learning. *ReCALL*, *24*(2), 169–187. doi:10.1017/S0958344012000055

Kukulska-Hulme, A., & Shield, L. (2008). An overview of mobile assisted language learning: From content delivery to supported collaboration and interaction. *ReCALL*, *20*(3), 271–289. doi:10.1017/S0958344008000335

Levak, N., & Son, J. B. (2017). Facilitating second language learners' listening comprehension with Second Life and Skype. *ReCALL, 29*(2), 200–218. doi:10.1017/S0958344016000215

Leveridge, A., & Yang, J. (2013). Testing learner reliance on caption supports in second language listening comprehension multiple media environments. *ReCALL, 25*(2), 199–214. doi:10.1017/S0958344013000074

Lin, H. (2015). A meta-synthesis of empirical research on the effectiveness of computer-mediated communication (CMC) in SLA. *Language Learning & Technology, 19*, 85–117.

Lin, S., Winaitham, W., & Saitakham, K. (2008). *The use of websites for practicing listening skills of undergraduate students. A case study at Suranaree University of Technology*. Online Submission.

Lin, W. C., Huang, H.-T., & Liou, H.-C. (2013). The effects of text-based SCMC on SLA: A metaanalysis. *Language Learning & Technology, 17*, 123–142.

Lipsey, M. W., & Wilson, D. B. (2001). *Practical meta-analysis*. Sage.

Little, C. W., Clark, J. C., Tani, N. E., & Connor, C. M. (2018). Improving writing skills through technology-based instruction: A meta-analysis. *Review of Education, 6*(2), 183–201. doi:10.1002/rev3.3114

Liu, G. Z., Chen, J. Y., & Hwang, G. J. (2018). Mobile-based collaborative learning in the fitness center: A case study on the development of English listening comprehension with a context-aware application. *British Journal of Educational Technology, 49*(2), 305–320. doi:10.1111/bjet.12581

Liu, T. Y. (2009). A context-aware ubiquitous learning environment for language listening and speaking. *Journal of Computer Assisted Learning, 25*(6), 515–527. doi:10.1111/j.1365-2729.2009.00329.x

Long, D. R. (1990). What you don't know can't help you. *Studies in Second Language Acquisition, 12*(1), 65–80. doi:10.1017/S0272263100008743

Lou, Y., Abrami, P. C., & d'Apollonia, S. (2001). Small group and individual learning with technology: A meta-analysis. *Review of Educational Research, 71*(3), 449–521. doi:10.3102/00346543071003449

Macaro, E., Vanderplank, R., & Graham, S. (2005). A systematic review of the role of prior knowledge in unidirectional listening comprehension. In *Research Evidence in Education Library*. EPPI-Centre, Social Science Research Unit, Institute of Education, University of London.

Markham, P. L., Peter, L. A., & McCarthy, T. J. (2001). The effects of native language vs. target language captions on foreign language students' DVD video comprehension. *Foreign Language Annals, 34*(5), 439–445. doi:10.1111/j.1944-9720.2001.tb02083.x

Matthew, G. (2020). The effect of adding same-language subtitles to recorded lectures for non-native, English speakers in e-learning environments. *Research in Learning Technology, 28*(0), 1–16. doi:10.25304/rlt.v28.2340

Matthews, J., Cheng, J., & O'Toole, J. M. (2015). Computer-mediated input, output and feedback in the development of L2 word recognition from speech. *ReCALL, 27*(3), 321–339. doi:10.1017/S0958344014000421

McBride, K. (2011). The effect of rate of speech and distributed practice on the development of listening comprehension. *Computer Assisted Language Learning, 24*(2), 131–154. doi:10.1080/09588221.2010.528777

Mecartty, F. H. (2000). Lexical and grammatical knowledge in reading and listening comprehension by foreign language learners of Spanish. *Applied Language Learning, 11*, 323–348.

Meinardi, M. (2009). Speed bumps for authentic listening material. *ReCALL, 21*(3), 302–318. doi:10.1017/S0958344009990048

Michael, P. H., & Webb, S. (2017). The effects of captions on EFL learners' comprehension of English-language television programs. *CALICO Journal, 34*, 20–38.

Mirzaei, M. S., Meshgi, K., Akita, Y., & Kawahara, T. (2017). Partial and synchronized captioning: A new tool to assist learners in developing second language listening skill. *ReCALL, 29*(2), 178–199. doi:10.1017/S0958344017000039

Mizumoto, A., & Chujo, K. (2015). A meta-analysis of data-driven learning approach in the Japanese EFL classroom. *English Corpus Studies, 22*, 1–18.

Morley, J. (2001). Aural comprehension instruction: Principles and practice. In M. Celce-Murcia (Ed.), *Teaching English as a second or foreign language* (pp. 69–86). Heinle & Heinle.

Nah, K. C. (2011). Optimising the use of wireless application protocol (WAP) sites for listening activities in a Korean English as a foreign language (EFL) context. *Computer Assisted Language Learning, 24*(2), 103–116. doi:10.1080/09588221.2010.526946

Nah, K. C., White, P., & Sussex, R. (2008). The potential of using a mobile phone to access the Internet for learning EFL listening skills within a Korean context. *ReCALL, 20*(3), 331–347. doi:10.1017/S0958344008000633

Nation, I. S. P., & Newton, J. (2009). *Teaching ESL/EFL listening and speaking*. Routledge.

Norris, J., & Ortega, L. (2006). The value and practice of research synthesis for language learning. In J. Norris & L. Ortega (Eds.), *Synthesizing research on language learning and teaching* (pp. 3–52). Benjamins. doi:10.1075/lllt.13.04nor

Norusis, M. J. (2008). *SPSS 16.0 guide to data analysis*. Prentice Hall.

O'Bryan, A., & Hegelheimer, V. (2007). Integrating CALL into the classroom: The role of podcasting in an ESL listening strategies course. *ReCALL, 19*(2), 162–180. doi:10.1017/S0958344007000523

Ockey, G. J. (2007). Construct implications of including still image or video in computer-based listening tests. *Language Testing, 24*(4), 517–537. doi:10.1177/0265532207080771

Pardo-Ballester, C. (2012). CALL evaluation: Students' perception and use of LoMasTv. *CALICO Journal, 29*(3), 532–547. doi:10.11139/cj.29.3.532-547

Parry, T. S., & Meredith, R. A. (1984). Videotape vs. Audiotape for Listening Comprehension Tests: An Experiment. *OMLTA Journal*, 47-53.

Pearson Student Mobile Device Survey 2013. (2013). Retrieved September 19, 2014, from https://www.pearsoned.com/wp-content/uploads/Pearson-Student-Mobile-Device-Survey-2013-National-Report-on-College-Students-public-release.pdf

Peterson, M. (2010). Computerized games and simulations in computer-assisted language learning: A meta-analysis of research. *Simulation & Gaming*, *41*(1), 72–93. doi:10.1177/1046878109355684

Peterson, P. W. (2001). Skills and strategies for proficient listening. In M. Celce-Murica (Ed.), *Teaching English as a second or foreign language* (pp. 87–100). Heinle & Heinle.

Plonsky, L., & Oswald, F. L. (2012). How to do a meta-analysis. In A. Mackey & S. M. Gass (Eds.), *Research methods in second language acquisition: A practical guide* (pp. 275–295). Basil Blackwell. doi:10.1002/9781444347340.ch14

Roberts, J. A., Yaya, L. H. P., & Manolis, C. (2014). The invisible addiction: Cell-phone activities and addiction among male and female college students. *Journal of Behavioral Addictions*, *3*(4), 1–12. doi:10.1556/JBA.3.2014.015 PMID:25595966

Rost, M. (2011). *Teaching and researching Listening*. Longman.

Roussel, S. (2011). A computer assisted method to track listening strategies in second language learning. *ReCALL*, *23*(2), 98–116. doi:10.1017/S0958344011000036

Şendağ, S., Gedik, N., & Toker, S. (2018). Impact of repetitive listening, listening-aid and podcast length on EFL podcast listening. *Computers & Education*, *125*, 273–283. doi:10.1016/j.compedu.2018.06.019

Sha, G. (2010). Using TTS voices to develop audio materials for listening comprehension: A digital approach. *British Journal of Educational Technology*, *41*(4), 632–641. doi:10.1111/j.1467-8535.2009.01025.x

Sharifi, M., Rostami AbuSaeedi, A., Jafarigohar, M., & Zandi, B. (2018). Retrospect and prospect of computer assisted English language learning: A meta-analysis of the empirical literature. *Computer Assisted Language Learning*, *31*(4), 413–436. doi:10.1080/09588221.2017.1412325

Shin, D. (1998). Using video-taped lectures for testing academic language. *International Journal of Listening*, *12*, 56–79. doi:10.1080/10904018.1998.10499019

Sueyoshi, A., & Hardison, D. M. (2005). The role of gestures and facial cues in second language listening comprehension. *Language Learning*, *55*(4), 661–699. doi:10.1111/j.0023-8333.2005.00320.x

Suh, S. S., Kim, S. W., & Kim, N. J. (2010). Effectiveness of MMORPG-based instruction in elementary English education in Korea. *Journal of Computer Assisted Learning*, *26*(5), 370–378. doi:10.1111/j.1365-2729.2010.00353.x

Sung, Y. T., Chang, K. E., & Yang, J. M. (2015). How effective are mobile devices for language learning? A meta-analysis. *Educational Research Review*, *16*, 68–84. doi:10.1016/j.edurev.2015.09.001

Suvorov, R. (2011). The effects of context visuals on L2 listening comprehension. *University of Cambridge ESOL Examinations Research Notes*, *45*, 2–8.

Tarantino, J. (2016). Examining the effects of independent MALL on vocabulary recall and listening comprehension: An exploratory case study of preschool children. *CALICO Journal*, *33*, 260–277.

Taylor, A. M. (2009). CALL-based versus paper-based glosses: Is there a difference in reading comprehension? *CALICO Journal*, *27*(1), 147–160. doi:10.11139/cj.27.1.147-160

Trofimovich, P., & Isaacs, T. (2012). Disentangling accent from comprehensibility. *Bilingualism: Language and Cognition*, *15*(4), 905–916. doi:10.1017/S1366728912000168

Tsai, S. C. (2015). Implementing courseware as the primary mode of task-based ESP instruction: A case study of EFL students. *Computer Assisted Language Learning*, *28*(2), 171–186. doi:10.1080/09588221.2013.818554

Vandergrift, L., & Goh, C. C. M. (2012). Listening in multiple media environments. In *Teaching and Learning Second Language Listening* (pp. 217–238). Routledge, Taylor & Francis Group.

Verdugo, D. R., & Belmonte, I. A. (2007). Using digital stories to improve listening comprehension with Spanish young learners of English. *Language Learning & Technology*, *11*, 87–101.

Wagner, E. (2007). Are they watching? Test-taker viewing behavior during an L2 video listening test. *Language Learning & Technology*, *11*, 67–86.

Wagner, E. (2010). The effect of the use of video texts on ESL listening test-taker performance. *Language Testing*, *27*(4), 493–513. doi:10.1177/0265532209355668

Wagner, E. (2013). An investigation of how the channel of input and access to test questions affect l2 listening test performance. *Language Assessment Quarterly*, *10*(2), 178–195. doi:10.1080/15434303.2013.769552

Weinberg, A., Knoerr, H., & Vandergrift, L. (2011). Creating podcasts for academic listening in French: Student perceptions of enjoyment and usefulness. *CALICO Journal*, *28*(3), 588–605. doi:10.11139/cj.28.3.588-605

Wilberschied, L., & Berman, P. M. (2004). Effect of using photos from authentic video as advance organizers on listening comprehension in an FLES Chinese class. *Foreign Language Annals*, *37*(4), 534–543. doi:10.1111/j.1944-9720.2004.tb02420.x

Winke, P., Gass, S., & Sydorenko, T. (2010). The effects of captioning videos used for foreign language listening activities. *Language Learning & Technology*, *14*, 65–86.

Woottipong, K. (2014). Effect of using video materials in the teaching of listening skills for university students. *International Journal of Linguistics*, *6*(4), 200–212. doi:10.5296/ijl.v6i4.5870

Wu, W. H., Wu, Y. C., Chen, C. Y., Kao, H. Y., Lin, C. H., & Huang, S. H. (2012). Review of trends from mobile learning studies: A meta-analysis. *Computers & Education*, *59*(2), 817–827. doi:10.1016/j.compedu.2012.03.016

Yang, H. Y. (2014a). Does multimedia support individual differences?–EFL learners' listening comprehension and cognitive load. *Australasian Journal of Educational Technology*, *30*(6), 699–713. doi:10.14742/ajet.639

Yang, H. Y. (2014b). The effects of advance organizers and subtitles on EFL learners' listening comprehension skills. *CALICO Journal*, *31*(3), 345–373. doi:10.11139/cj.31.3.345-373

Yang, J., & Chang, P. (2014). Captions and reduced forms instruction: The impact on EFL students' listening comprehension. *ReCALL, 26*(1), 44–61. doi:10.1017/S0958344013000219

Yasin, B., Mustafa, F., & Permatasari, R. (2017). How Much Videos Win over Audios in Listening Instruction for EFL Learners. *Turkish Online Journal of Educational Technology-TOJET, 17*, 92–100.

Yun, J. (2011). The effects of hypertext glosses on L2 vocabulary acquisition: A meta-analysis. *Computer Assisted Language Learning, 24*(1), 39–58. doi:10.1080/09588221.2010.523285

Zhao, Y. (1997). The effects of listeners' control of speech rate on second language comprehension. *Applied Linguistics, 18*(1), 49–68. doi:10.1093/applin/18.1.49

Zhao, Y. (2003). Recent developments in technology and language learning: A literature review and meta-analysis. *CALICO Journal, 21*(1), 7–27. doi:10.1558/cj.v21i1.7-27

KEY TERMS AND DEFINITIONS

Between-Group Design: In experimental research, participants are divided into two or more groups, each group completes a different task, and the comparisons are made between different groups.

Computer-Assisted Language Learning (CALL): Any form of linguistic teaching or study that involves computer technology.

Cohen's *d*: An effect size used to indicate the standardized difference between two means.

Listening Comprehension: Second language learners' receptive oral skill, or to what extent a listening passage has been understood.

Meta-Analysis: A systematic technique for reviewing, analyzing, and summarizing quantitative research studies on specific topics or questions.

Mobile-Assisted Language Learning (MALL): Any form of linguistic teaching or study that involves a portable device (e.g., mobile phones, tablet computers, and MP3 players).

Within-Group Design: In experimental research, all the participants complete two or more different tasks, and the comparisons were made between their performances on these different tasks.

APPENDIX. CODING CONVENTIONS

Moderator	Value	Definition
Identification		
ID	1, 2, 3...	study's identification #
authors	as it is	study's authors
year	as it is ( )	study's publish year
journal	as it is	the journal study published
title	as it is	study's title
Study Context		
SL/FL	SL=0; FL=1	language learning situation
country	as it is	the place that the study conducted
L1	English=0; Chinese=1; Spanish=2; French=3; Italian=4; Korean=5; Mixed=9....	participants' first language
L2 (target language)	English=0; Chinese=1; Spanish=2; French=3; Italian=4; Mixed=9....	participants' target language
L2 level	low=0; intermediate=1; high=2	target language proficiency
age	in years/range of years	participants' age
institution	university=0; high school=1; middle school=2; elementary=3	the institution that participants come from
classroom/laboratory	classroom=0; laboratory=1	where the study conduced
Design and treatment		
tool	as it is	the multiple media tool that used in this study on the purpose of improving listening skills
MALL	no=0; yes=1	the multiple media tool in this study is a type of mobile assisted language learning
delayed posttest	no=0; yes=1	include a delayed-posttest
# of delayed posttest	#	total # of delayed posttest
intervals between treatment and delayed posttest	in days (if there is more than one delayed posttest, separate it using ;)	intervals between treatment and delayed posttest
# of groups	#	total groups
random assignment or random assignment by group	no=0; yes=1	the participants and the group were randomly assigned
experimental group	as it is	name of the treatment
# of comparison/ control group	#	total # of comparison/ control group
comparison/control group	no=0; yes=as it is (if there is more than one group, separate it using ;)	treatment of the comparison/compare group
sample experimental	#	# of participants in the experimental group; if there is only one group in the study, put the total # of participants
sample comparison/control	# (if there is more than one group, separate it using ;)	# of participants in the comparison/control group
assessment/instruction	assessment=0; instruction=1	whether the study is for assessment or instruction purpose
instruction length	in hours	actual treatment length
instruction duration	in days	treatment duration
technology training	no=0; yes=1	include technology training during/before the treatment
length of technology training	in hours	length of technology training
Measures		
listening comprehension tests	as it is: DV (e.g., oral summary, multiple-choices, cloze) (if there is more than one instrument, separate it using ;)	tests used to measure listening comprehension
reliability	alpha, inter-rater	instrument reliability
Results		
outcome 1	as it is	briefly state the first finding
ES 1	Cohen's d	the effect size of outcome 1
outcome 2	as it is	briefly state the second finding
ES 2	Cohen's d	the effect size of outcome 2
outcome 3	as it is	briefly state the third finding
ES 3	Cohen's d	the effect size of outcome 3

Note: missing data=999

Chapter 2
An Instructional Design Model for Design and Development of Adaptive Hypermedia Listening Environments

Vehbi Turel
Bingol University, Turkey

ABSTRACT

The design and development of adaptive hypermedia (AH) listening environments consist of different stages. Each stage in design and development of AH listening applications is vital in the design and development process so that cost-effective AH listening software can be designed and developed in a true sense. Pedagogically, epistemologically, and technically, a wide range of design principles and guidelines have to be heeded at each design and development stage so that effective and efficient AH listening environments can be designed and developed.

INTRODUCTION

Adaptive hypermedia (AH) is the combination and instantaneous delivery of a wide range of different digital learning elements such as audio, video, text, pictures, graphics and so forth on the same digital platform that (a) are provided through links to retrieve information on the same digital environment, (b) are interactive and navigational, which provide ease of use and user control, and (c) are personalised (Brusilovsky, 1996; Shute & Zapata-Rivera, 2012; Turel, 2015a; 2018). In other words, when we –designers and developers of AH learning environments- speak of AH, mostly one thing comes to mind. It is the use of optimum combinations and instantaneous delivery of digital video, audio, texts, visuals (i.e. pictures, images, photographs, graphics, tables, figures and so on), animations, hyperlinks, instructions, tasks, activities, feedback, help, glossary, captions etc. on the same digital platform, which are totally computerised and under computer as well as learners' control. This digital platform also enables learners to make preferences. These preferences, their individual needs and learning goals are recorded,

DOI: 10.4018/978-1-7998-7876-6.ch002

and then AH listening application "uses them throughout interaction with the learners in order to meet their personal needs so that they can learn better" (Turel, 2018, p. 2357; see also Brusilovsky, 2007; Brusilovsky 2012; Brusilovsky, Eklund & Schwarz, 1998; Brusilovsky & Millán, 2007).

The design and development of AH listening environments consist of different stages (e.g. Turel 2015a). These stages are: (1) Needs analysis (i.e. feasibility), (2) Setting-up (a) team(s) of experts, (3) Design, (4) Development (i.e. programming), (5) Testing and (6) Evaluation (Figure 1). Each design and development stage of AH listening environments is vital in the design and development process of cost effective AH listening applications. A wide range of design principles and guidelines have to be borne in mind at each design and development stage of AH learning systems so that effective learning environments can be designed and developed in a true sense.

To achieve creating cost-effective AH learning environments, a very wide range of instructional design models, which aim to provide pedagogically and epistemologically effective and efficient design principles and guideless for creating instructional activities for different purposes, have been suggested, prescribed as well as accepted (e.g. Branch & Kopcha, 2013, Şimşek, 2016). While some of these instructional design models are already tested (and accepted), the others have yet to be.

In this chapter, an instructional design model for design and development of AH listening environments is suggested. Each stage of the suggested instructional design model (i.e. AHLE Model – Adaptive Hypermedia Listening Environments Model) is explained thoroughly and the pedagogical and epistemological principles and guidelines that need to be taken into heed at each stage are explained clearly. Let us now look at these stages one by one and clearly indicate what needs to be taken into heed at every single stage.

THE DESIGN AND DEVELOPMENT STAGES OF HYPERMEDIA LISTENING ENVIRONMENTS

1. Needs Analysis

At the needs analysis (i.e. feasibility) stage, there are certain questions that need to be answered by the design and development team(s) of AH listening environments. These questions are: (a) who the target learners and what their needs are, (b) what the scope (i.e. contents) of the AH listening application is, and (c) what the output of the AH listening application is.

a. Target Learners

Who are they and what and why do they want to learn? The design and development team of the AH listening environments need to define or know the target learners (i.e. their education, age, interests, computing skills, abilities, needs and so on) and the target objectives. As emphasised by Turel & McKenna (2013, p.191), a learner-based approach (i.e. 'a needs-based approach') needs to be followed (Bath & Klein Wohl, 2011, pp. 88-98). This approach focuses on target earners' needs. What this means is that: 'an understanding of why the LLs [language learners] wish to learn a language, what purposes this learning will serve, how they best learn a language, which learning strategies they favour, and when and where they prefer this learning to take place' (Watts, 1997, p. 3). These are valuable aspects and are considered the essential central features of an instructional design model (Hoven, 1999, p. 92; Brown, 1997, p. 6; see Turel & McKenna, 2013, p.190-91 for details).

Figure 1. The design and development stages of AH listening environments

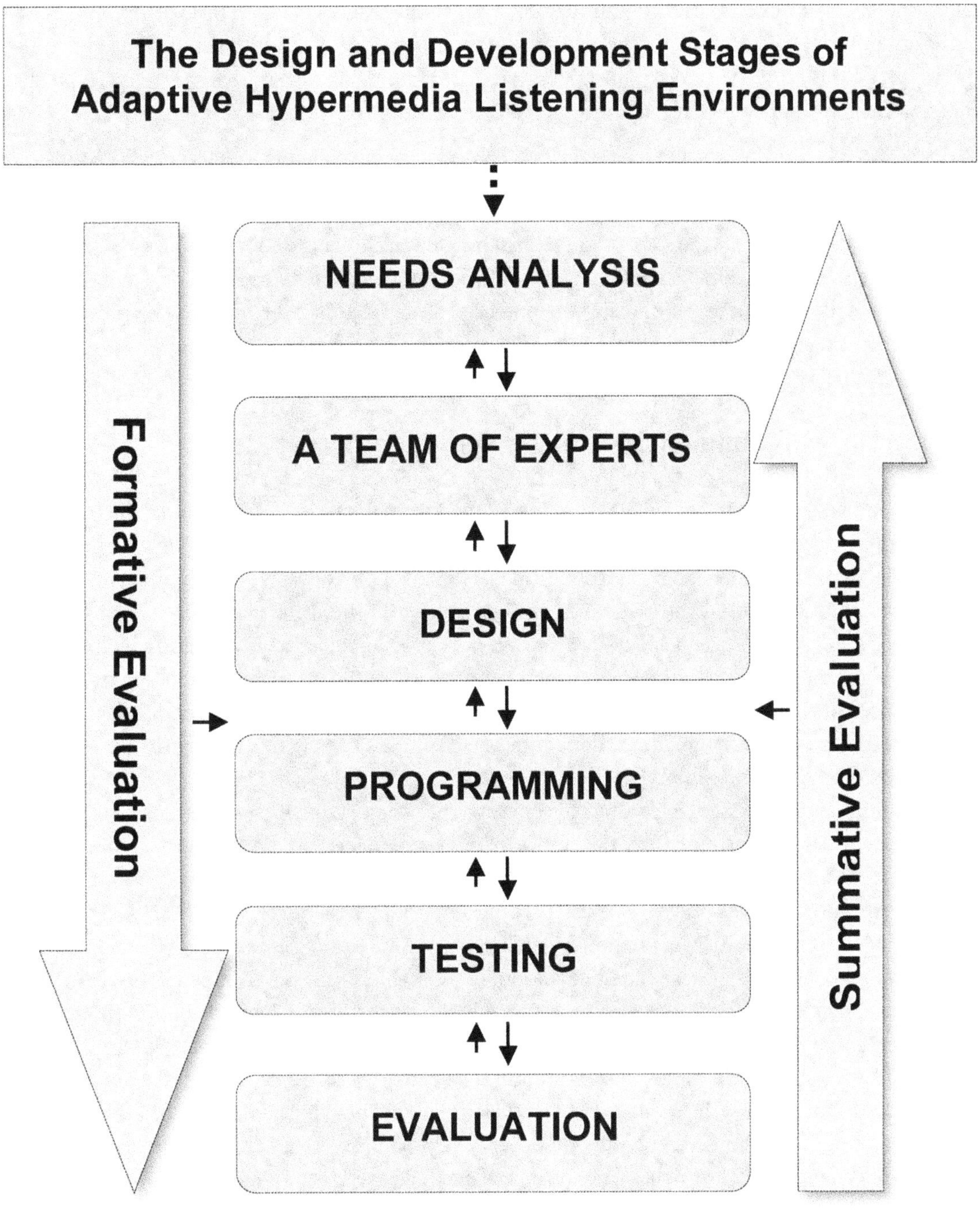

b. Scope

The design and development team of the AH listening environments need to know what the contents of the potential AH listening environment are, what has to be included in the AH listening environment and what is optional. In other words, the team need to define the learning objectives, the needs of the target learners or the target(s) that need to be achieved through a wide range of instructional activities for different purposes. The scope can be determined by different authorities such as those in charge of the pertinent positions at the ministry of education, local education authorities, head teachers, subject teachers, managers, and so on depending on which part of the world we are working.

c. Output

The design and development team(s) of the AH listening environments need to know what the potential AH listening environment aims to achieve. In other words, what knowledge, skills, and abilities the target learners will gain as a result of using the potential target AH listening environment, which totally depends on why target learners want to learn a second/foreign language and 'what purposes this learning will serve' (Watts, 1997, p. 3) as emphasised above.

2. Setting Up (a) Team(s) of Specialists and Experts

Having decided what needs to be designed and developed from the points of view of (a) the target learners' needs, (b) the scope to be taught, and the output to be gained, accordingly, (a) team(s) of pertinent experts need(s) to be set up. Under normal conditions, from the point of view of designing and developing AH listening application (a) team(s) of experts might consist of different specialists such as: computer programmers, subject experts and specialists (i.e. subject specialist language /listening lecturers, tutors, instructors, teachers), authors, graphic designers, animators, audio engineers, video engineers, musicians, film directors/specialists, voice actors, artists, photographers, native speakers, target learners, AH designers, pedagogues, and many more (depending on the needs clearly outlined at the needs analysis / feasibility stage, as shown in Figure 2.

To be able to create pedagogically and psychologically sophisticated and cost effective AH listening environments, the involvement of the above-mentioned specialists and experts is necessary and vital (Nicholson & Ngai, 1996). To this end, it is, for instance, said that target learners' involvement, "produces more useable and effective" AH learning environments (Nikolova, 2002, p. 112; see also Kennedy & McNaught, 1997; MacGregor, 1993; Eraut, 1988), In the same way, the lack of specialist programmers in the process of design and development of AH learning environments hinders the use of the maximum potential of the programming tools, which can result in a lack of the minimum requirements. For example, in one case, the design and development work came to a point where the learning environment developers had to ask a "programming specialist to take over ... the development work" (Grob & Wolff, 2001, p. 249). In another case, many learners did not continue studying with the learning environment due to the technical problems the learners encountered (Lyall & McNamara, 2000). Such digital learning environments, which are not designed and developed by teams of experts, are even called 'shovelware' (Clifford, 1998; Le Mon, 1988, Cagiltay & Göktaş, 2016). The assumptions underlying such shortages are the lack of setting up teams of pedagogically and technically needed experts for the design and development of AH learning environments. The importance of inclusion of all needed teams of specialists and experts during the design and development process is also emphasized in other studies (e.g. Sayin & Seferoglu, 2015).

Figure 2. A wide range of specialists and experts that might be essential for real sense and cost effective design and development of AH listening environments (Turel & McKenna, 2013, p. 189)

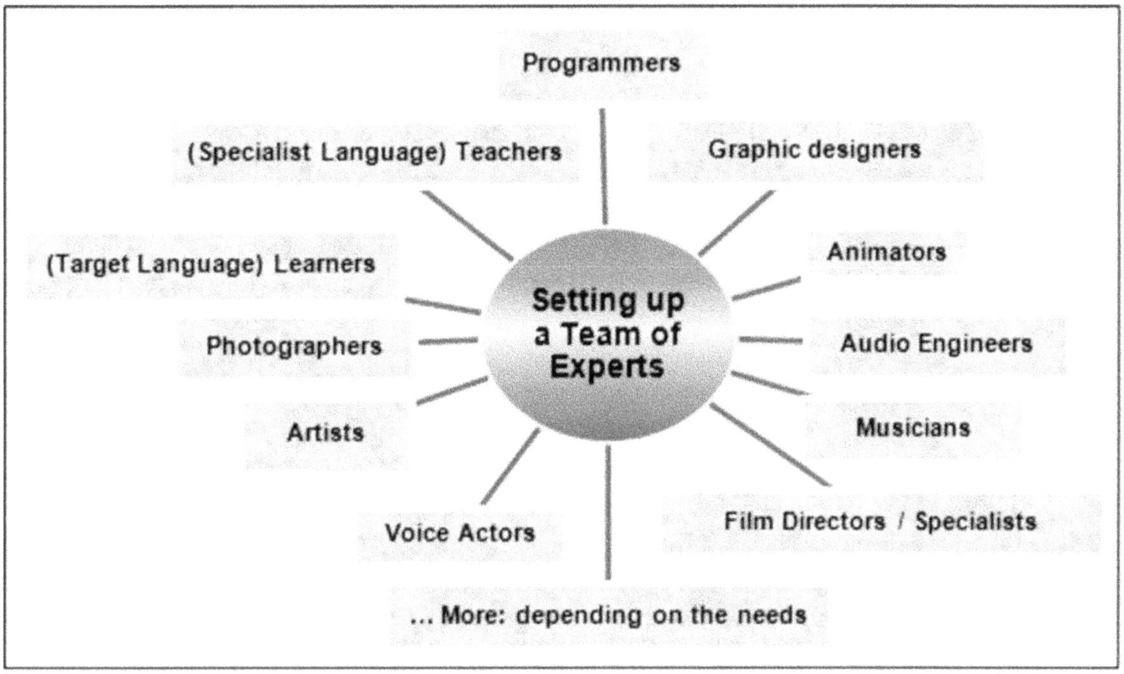

3. Design

From the point of view of designing and developing AH listening environments, there are many different design issues that need to be taken heed of throughout the design and development process at design stage. These are:

a. General Considerations

General considerations such as 'design approaches'/'navigation', 'simplicity', 'consistency', 'clarity', 'brevity', 'interactivity', flexibility', 'motivation', 'icons' and 'colour' have to be taken into account and the pertinent hypothesis, theories, models, principles and guidelines have to be taken into account. For example, element properties, precise positioning, general layout/designing screen layouts, interconnections between different screens/pages, user friendly interface, simplicity, clarity, consistency, dynamic functionality and interactivity, hyperlinking, one storyboard per screen, screen size and colour, button dimensions and consistent positions, caption font and colour, responsiveness, compatibility, and early evaluations are only some of the issues that need to be taken into heed so that AH listening environments can be designed effectively and efficiently (see for details Turel & McKenna 2013).

b. Specific Considerations

Throughout the design and development process of AH listening environments, all specific considerations such as the priority and design of AH learning elements (i.e. instructions, text, visuals, media types - audio, video, audio + animation, talking + supplementary contextual visuals, audio + supplementary contextual visuals -, visuals, cultural differences, unfamiliar items, glossary, tasks/activities, feedback, self-assessment tests etc.) have to be taken into heed (see for details Turel & McKenna, 2015a; Turel & McKenna, 2015b; Turel, 2014; Turel & Kılıç 2014; Turel, 2014b, 2014c;). For example, the design and development team should use any media type only when these media types are really effective and essential (Turel, 2014d). Similarly, for instance, clear visual messages as well as interactive and personalised feedback have to be provided (Turel, 2018, 2015a, 2012, 2014e). It is de facto that usually learning environments designed with the target learners in mind. Having said that, in the same way, target learners' involvement is still efficient and effective (Eraut, 1988; Kennedy & McNaught, 1997; MacGregor, 1993; Nikolova, 2002). Therefore, designers and developers of AH listening environments should not ignore such pedagogical aspects, as well.

b. Designing according to the Needs Outlined at the Needs Analysis Stage

The design and development team(s) need(s) to remember the initial needs clearly outlined and indicated at the feasibility stage, and design all learning elements accordingly. They need to chunk the needs into related areas and decide on sub-levels of the target learners' needs. The design and development teams need to make sure that they go down to the "lowest" level of detail. Without enough skills, efforts and cooperation, no one or no teams can ever achieve any target in a true sense.

c. Working and Communicating with all Members of the Team(s)

The design and development team(s) need to make sure that they work and communicate with all of the team members (as well as the clients) whenever needed regarding task allocation, useful common standards, systematic means of reference, task and resource allocations, the design and priority of the optimum elements, and their evaluation.

To make sure that the deadlines are met, the design and development team(s) need(s) to plan their time as (a) teams of specialists and experts. The teams have to be realistic and stick to their schedule. To achieve this, they have to systemise their resources and documents at all stages. The teams also need to indicate roughly what should be completed when. Appropriate Gantt chart that indicates periods of time for each design, evaluation, development, re-evaluation, incremental development, testing, and production tasks can be helpful in terms of meeting the deadlines.

d. Everything has to be Detailed

The design and development teams need to make sure that everything about the design and development of the AH listening application is detailed such as screen number and name, storyboard date, names of the allocated team members for designing the particular parts or learning elements of the AH listening environment.

e. Incremental Evaluation and Development

The design and development team(s) should not forget that only efficient, pedagogically and psychologically well thought storyboards can produce a better prototype, which is more likely to result in a sophisticated development. Such a result can only be achieved through 'well and thoroughly team thinking', 'design', 'evaluation', and 're-design'/'incremental development'. As a result, the final objective, - which is to help the target learners to acquire new knowledge, new skills and abilities (and improve their existing ones)-, is achieved. This can only happen if the learning conditions emphasised in Figure 3 are thoroughly provided (see for details Turel, 2015b, 2018).

f. Right Tool for the Right Element

The design and development team(s) need(s) to make sure that they use powerful design tools at this stage (as well as at the following stages). All elements on every single storyboard must be clear and comprehensive, and they need to be related to other storyboards so that everything can be crystal clear for developers or programmers at the development (i.e. programming) stage.

4. Programming

At this stage, a team of professional computer programmers of the design and development team implement what was decided and designed at the previous stage – the design stage -. What needs to be taken into heed at the programming stage is that in a true sense computer programming specialists have to be in charge of programming the AH listening environments. As emphasised above, in one case, the design and development work came to a point where the learning application developers had to ask a 'programming specialist to take over ... the development work' (Grob & Wolff, 2001, p. 249). Otherwise, technical problems related to programming will be unavoidable. In such situations, many learners will quit studying with the learning environment due to the technical problems. This is what happened in the past. The target learners stopped studying with the learning environments due to the technical problems they encountered (Lyall & McNamara, 2000). Therefore, the inclusion of all needed teams of specialists and experts during the every single stage of the design and development process of AH listening environments should never be underestimated. After implementing and completing the production tasks (i.e. programming stage), a prototype (i.e. an AH listening application), which is a part of the AH listening environment life cycle, is created.

5. Testing

The created prototype (i.e. the designed and developed AH listening application by teams of specialists and experts) is technically tested and formatively evaluated on different required digital devices, operating systems, Internet browsers, screens and so on. Similarly, the prototype is technically and pedagogically tested and formatively evaluated with a similar group of the target language learners. The sample group of the target learners study with the prototype AH listening environment for long-enough period of time. They study and complete all parts of the prototype as they are supposed to. After this, all necessary data are collected and analysed. Following all of these, the technical as well as pedagogical weaknesses and

Figure 3. The stages of listening acquiring process or the conditions that enable listening acquisition (Turel, 2015, p. XXVI)

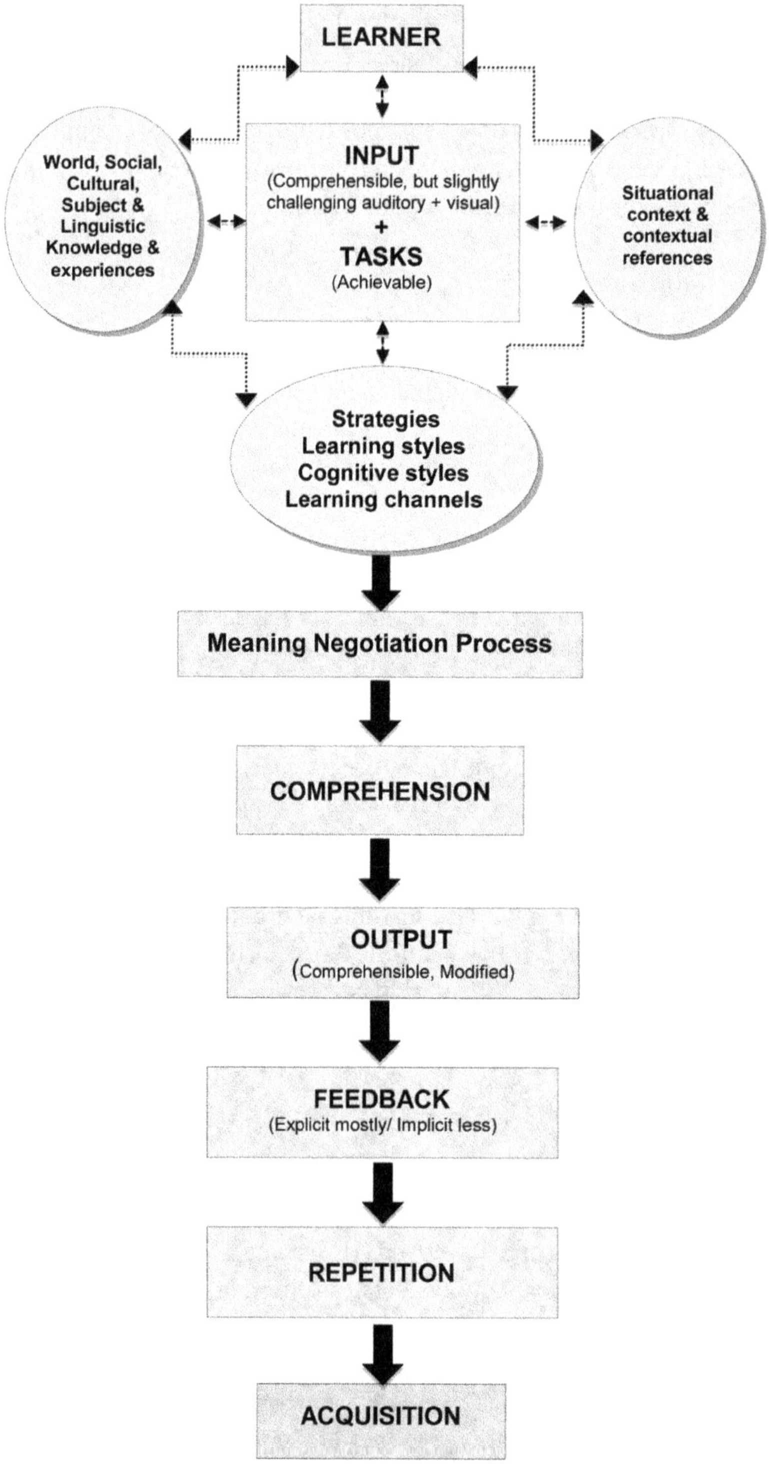

faults related to the prototype are diagnosed and improved, and then the final version of the prototype is completed, as emphasised in Figure 4.

Figure 4. The prototype incremental development process

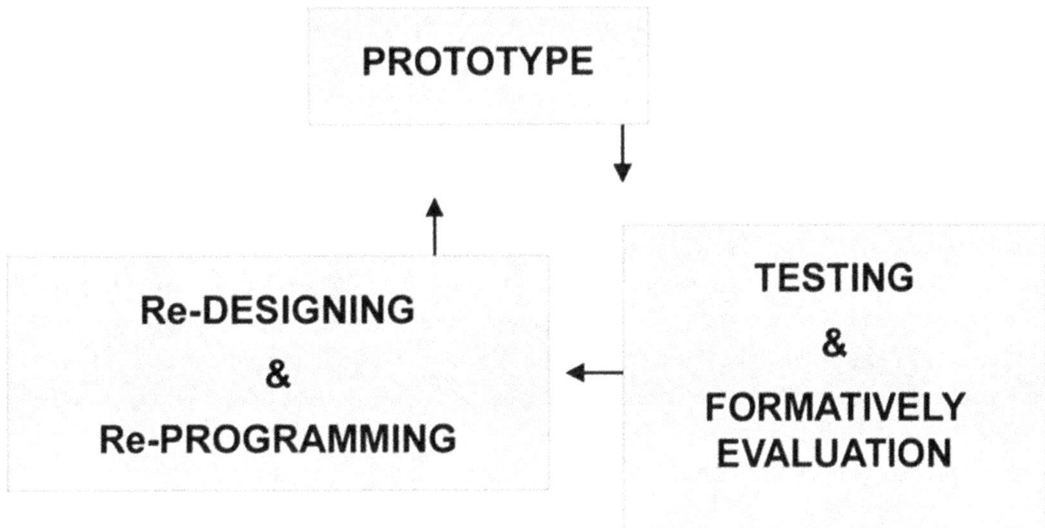

The design and development team(s) should implement formative evaluation of the (prototype) AH listening environment in the early life cycle (i.e. at testing stage) both technically and pedagogically so that the formative evaluation can help the design process (i.e. design – testing and formatively evaluation – re-designing and re-programming) by working closely with the sample group of the target learners. After getting an acceptable number of the target learners' opinions regarding different aspects of the prototype, the prototype AH listening environment can be improved accordingly. As a result, the design and development teams have less work left to do after the 'testing' and 'evaluation' stages.

6. Evaluation

At this stage, summative evaluation, which is not testing and takes longer, is required. Summative evaluation is done to make sure that AH listening environments are effective, efficient, and satisfying in terms of usability (i.e. learner satisfaction in terms of ease of use), and learnability (i.e. acquiring the targeted knowledge, skills and abilities). This can be measured through qualitative and quantitative data by conducting Likert scale questionnaires and interviews. It might take at least a few months and require a big sample of the target students to evaluate an AH listening application. Furthermore, think aloud, incident diaries, observations, and log data evaluation techniques can also be made use of so that a proper evaluation of AH listening environments can be conducted in a true sense.

To be able to effectively carry out evaluation, the design and development teams need to identify and decide in terms of which aspects (i.e. usability and learnability) they want to evaluate AH listening environments. Similarly, the teams have to conduct evaluations with groups of similar target learners under similar conditions. The groups of the target learners have to use the AH listening application (prototype) for enough period of time and under right conditions. After that, all necessary data for evaluation need to be collected in appropriate ways and times, and analysed accurately so that we can have a technically and pedagogically clear evaluative picture of the AH listening application prototype.

The design and development team should not forget that summative evaluation is an essential part of the AH listening environment design and development process. Therefore, as it is the case with all other stages, evaluation stage of the design and development process also requires a team of required experts who do proper planning, carefully plan their approach, put enough effort, time and apply sophisticated evaluation methods and techniques; collect accurate and reliable evaluation data and analyse correctly.

Having conducted summative evaluation (i.e. conducted right techniques under appropriate conditions and at right times, collected accurate and reliable data), and analysed the gathered data, according to the results obtained, if needed, new incremental and final improvements are done to the designed and developed AH listening application prototype. After all these, the final product (i.e. engineer product) is produced. Then the AH listening environment is used by the real target learners and further evaluations are conducted to find out whether the final version of the AH listening application has achieved the aimed targets or not (i.e. enabling the target learners to gain new knowledge, acquire new and efficient skills and abilities as well as increase their existing ones). Accordingly, if needed, either the existing AH listening environment is further improved or the new or additional ones are designed and developed.

FUTURE RESEARCH DIRECTIONS

Further research might try to find out whether implementing the above emphasized and illustrated principles and guidelines of instructional design model for design and development of adaptive hypermedia listening environments (i.e. AHLE Model) enable the designers and developers of AH listening environments to create cost effective AH listening applications in a true sense or not.

CONCLUSION

As can be seen from the above, the design and development of AH listening environments have their own life cycle process, and this process consists of six different design and development stages. For "true" cost effective AH listening environments, a proper allocation of teams of specialists and experts, physical and mental effort, intelligence and time is needed at every single of these stages so that technically, pedagogically and epistemologically cost effective AH listening applications can be designed and developed in a true sense.

RERERENCES

Branch, R. M., & Kopcha, T. J. (2013). Instructional Design Models. In Handbook of Research on Educational Communications and Technology (pp. 77-87). DOI: doi:10.1007/978-1-4614-3185-5_7

Brusilovsky, P. (1996). Methods and techniques of adaptive hypermedia. *User Modeling and User-Adapted Interaction*, *6*(2-3), 87–129. doi:10.1007/ BF00143964

Brusilovsky, P. (2007). Adaptive Navigation Support. In P. Brusilovski, A. Kobsa, & W. Nejdl (Eds.), *The Adaptive Web: Methods and Strategies of Web Personalization* (Vol. 4321, pp. 263–290). Springer-Verlag.

Brusilovsky, P. (2012). Adaptive Hypermedia for Education and Training. In P. J. Durlach & A. M. Lesgold (Eds.), *Adaptive Technologies for Training and Education* (pp. 46–68). CUP.

Brusilovsky, P., Eklund, J., & Schwarz, E. (1998). Web-based education for all: A tool for development adaptive courseware. *Computer Networks and ISDN Systems*, *30*(1), 291–300.

Brusilovsky, P., & Millán, E. (2007). User models for adaptive hypermedia and adaptive educational systems. In P. Brusilovski, A. Kobsa, & W. Nejdl (Eds.), *The Adaptive Web: Methods and Strategies of Web Personalization* (Vol. 4321, pp. 3–53). Springer-Verlag.

Çağıltay, K., & Göktaş, Y. (Eds.). (2016). Öğretim Teknolojilerinin Temelleri: Teoriler, Araştırmalar, Eğilimler. Pegem Akademi.

Clifford, R. (1998). Mirror, mirror, on the wall: Reflections on computer assisted language learning. *CALICO Journal*, *16*(1), 1–11.

Eraut, M. (1988). What has happened to learning design. In H. Mathias, N. Rushby, & R. Budgett (Eds.), *Aspects of Educational Technology XXI: Designing New Systems and Technologies for Learning*. Kogan Page.

Kennedy, D. M., & McNaught, C. (1997). Design elements for interactive multimedia. *Australian Journal of Educational Technology*, *13*(1), 1–22.

Le Mon, R. (1988). Cultural differences affecting the international transfer of computerised language learning. *System*, *16*(1), 37–40. doi:10.1016/0346- 251X(88)90008-5

MacGregor, R. C. (1993). Are we recognising the organisational impact on educational soft- ware design? *Australian Journal of Educational Technology*, *9*(1), 59–68.

Nicholson, A. Y. W., & Ngai, J. Y. K. (1996). Managing the development and production of interactive multimedia courseware in education. *Australian Journal of Educational Technology*, *12*(1), 35–45.

Nikolova, O. R. (2002). Effects of students' participation in authoring of multimedia materials on student acquisition of vocabulary. *Language Learning & Technology*, *6*(1), 100–122.

Sayın, Z., & Seferoğlu, S. S. (2015). Çevrim-içi öğrenmeye yeni bir bakış açısı: Kitlesel açık çevrimiçi dersler. In B. Akkoyunlu, A. İşman, & H. F. Odabaşı (Eds.), *Eğitim teknolojileri okumaları* (pp. 525–538). TOJET-The Turkish Online Journal of Educational Technology.

Shute, V. J., & Zapata-Rivera, D. (2012). Adaptive educational systems. In J. D. Paula & M. L. Alan (Eds.), *Adaptive technologies for training and education* (pp. 7–27). Cambridge University Press., doi:10.1017/ CBO9781139049580.004

Şimşek, A. (2016). Öğretim tasarımı ve modelleri. In K. Çağıltay & Y. Göktaş (Eds.), Öğretim Teknolojilerinin Temelleri: Teoriler, Araştırmalar, Eğilimler (pp. 105-122). Pegem Akademi.

Türel, V. (2012). Design of feedback in interactive multimedia language learning environments. *Linguistik Online, 54*(4), 35-49. Retrieved June 1, 2021, http://www.linguistik-online.de/54_12/tuerel.pdf

Turel, V. (2014). The use and design of supplementary visuals for the enhancement of listening skills in hypermedia. In C. M. Akrivopoulou & N. Garipidis (Eds.), *Human rights and the impact of ICT in the public sphere: Participation, democracy, and political autonomy* (pp. 268–291). IGI-Global. doi:10.4018/978-1-4666-6248-3.ch016

Turel, V. (2014b). Factors affecting listening. In C. M. Akrivopoulou & N. Garipidis (Eds.), *Human rights and the impact of ICT in the public sphere: Participation, democracy, and political autonomy* (pp. 310–331). IGI-Global. doi:10.4018/978-1-4666-6248-3.ch018

Turel, V. (2014c). Design of cultural differences in hypermedia environments. *International Journal of Human Rights and Constitutional Studies, 2*(2), 150–170. doi:10.1504/IJHRCS.2014.062766

Turel, V. (2014d). Priority of listening materials for autonomous intermediate language learners. In C. M. Akrivopoulou & N. Garipidis (Eds.), *Human rights and the impact of ICT in the public sphere: Participation, democracy, and political autonomy* (pp. 292–309). IGI-Global. doi:10.4018/978-1-4666-6248-3.ch017

Turel, V. (2014e). The use and design of supplementary visuals for the enhancement of listening skills in hypermedia. In C. M. Akrivopoulou & N. Garipidis (Eds.), *Human rights and the impact of ICT in the public sphere: Participation, democracy, and political autonomy* (pp. 268–291). IGI-Global. doi:10.4018/978-1-4666-6248-3.ch016

Turel, V. (2015a). Hypermedia and its' role in learning. In M. Khosrow-Pour (Ed.), *Encyclopaedia of information science and technology* (3rd ed., pp. 2495–2505). IGI-Global. doi:10.4018/978-1-4666-5888-2.ch243

Turel, V. (2015b). Introduction: Listening, multimedia and optimum design. In V. Turel (Ed.), *Intelligent Design of Interactive Multimedia Listening Software* (pp. xx–xxxiiv). IGI-Global.

Turel, V. (2018). Adaptive Hypermedia in Education. In M. Khosrow-Pour (Ed.), Encyclopaedia of information science and technology (4th ed.). (Vols. 1–10; pp. 2357–2368). IGI-Global. doi:10.4018/978-1-5225-2255-3

Turel, V., & Kılıç, E. (2014). The inclusion and design of cultural differences in Interactive multimedia environments. In C. M. Akrivopoulou & N. Garipidis (Eds.), *Human rights and the impact of ICT in the public sphere: Participation, democracy, and political autonomy* (pp. 245–267). IGI-Global. doi:10.4018/978-1-4666-6248-3.ch015

Turel, V., & McKenna, P. (2013). Design of language learning software. In *Computer-assisted foreign language teaching and learning: Technological advances* (pp. 188–209). IGI Global. doi:10.4018/978-1-4666-2821-2

Turel, V., & McKenna, P. (2015a). Design of multimedia listening software: Unfamiliar items, glossary, aural texts and speech rate. In V. Turel (Ed.), *Intelligent design of interactive multimedia listening software* (pp. 170–197). IGI-Global. doi:10.4018/978-1-4666-8499-7.ch007

Turel, V., & McKenna, P. (2015b). Design of multimedia listening software: Instructions, tasks, texts and self-assessment tests. In V. Turel (Ed.), *Intelligent design of interactive multimedia listening software* (pp. 142–169). IGI-Global. doi:10.4018/978-1-4666-8499-7.ch006

ADDITIONAL READING

Bailey, C., Hall, W., Millard, D. E., & Weal, M. J. (2007). Adaptive hypermedia through contextual- ized open hypermedia structures. *Journal ACM Transactions on Information Systems*, *25*(4), 1–36.

Baltova, I. (1999). Multisensory language teaching in a multidimensional curriculum: The use of authentic bimodal video in core French. *Canadian Modern Language Review*, *56*(1), 31–48. doi:10.3138/cmlr.56.1.31

Branch, R. M., & Kopcha, T. J. (2014). Instructional design models. In *Handbook of research on educational communications and technology* (pp. 77–87). Springer New York. doi:10.1007/978-1-4614-3185-5_7

Branden, K. V. (2006). *Task-based language education: From theory to practice*. CUP., doi:10.1017/CBO9780511667282

Cristea, A. (2005). Authoring of adaptive hypermedia. *Journal of Educational Technology & Society*, *8*(3), 6–8.

Driscoll, M., & Carliner, S. (2005). *Advanced Web-Based Training: Adapting Real World Strategies in Your Online Learning*. Pfeiffer.

Grgurović, M., & Hegelheimer, V. (2007). Help options and multimedia: Students' use of subtitles and the transcript. *Language Learning & Technology*, *11*(1), 45–66.

Guillroy, H. G. (1998). The effects of keyword captions to authentic French video on learner comprehension. *CALICO Journal*, *15*(1-3), 89–108. doi:10.1558/cj.v15i1-3.89-108

Gustafson, K. L., & Branch, R. M. (2002). What is instructional design? *Trends and issues in instructional design and technology*, 16-25.

Nitta, R., & Nakatsuhara, F. (2014). A multifaceted approach to investigating pre-task planning effects on paired oral test performance. *Language Testing*, *31*(2), 147–175. doi:10.1177/0265532213514401

Plass, J. L., Chun, D. M., Mayer, R. E., & Leutner, D. (2003). Cognitive load in reading a foreign language text with multimedia aids and the influence of verbal and spatial abilities. *Computers in Human Behavior*, *19*(2), 221–243. doi:10.1016/S0747-5632(02)00015-8

Pujolà, J. T. (2002). CALLing for help: Researching language learning strategies using help facilities in a web-based multimedia program. *ReCALL*, *14*(2), 235–262. doi:10.1017/S0958344002000423

Shapiro, A. M. (2008). Hypermedia design as learner scaffolding. *Educational Technology Research and Development*, *56*(1), 29–44. doi:10.100711423-007-9063-4

Tochon, F. V., Karaman, A. C., & Ökten, C. E. (2014). Online instructional personal environment for deep language learning. *International Online Journal of Education & Teaching*, *1*(2), 71–100. Retrieved April 28, 2021, from https://www.iojet.org/index.php/IOJET/article/view/32/52

Tschirner, E. (Ed.). (2012). *Aligning frameworks of reference in language testing: The ACTFL proficiency guidelines and the common European framework of reference for languages*. Stauffenburg Verlag.

Turel, V. (2010). Advanced Turkish. *ReCALL, 22*(3), 396-401. doi:10.1017/S0958344010000224

Turel, V. (2011a). Learners' attitudes to repetitious exposure in multimedia listening software. *The EUROCALL Review, 19*(September), 57-83, Retrieved December 12, 2020, from http://www.eurocall-languages.org/wordpress/wp-content/uploads/2014/01/review19.pdf

Turel, V. (2011b). Beginning Kurmanji Kurdish. *CALICO Journal*, *28*(3), 766–776. doi:10.11139/cj.28.3.766-776

KEY TERMS AND DEFINITIONS

Adaptive Hypermedia: It refers to the pedagogically effective and efficient combination and instantaneous delivery of a wide range of digital elements (i.e., input, texts, sound, audio, visuals, animation, video, feedback, instructions, glossaries, self-assessment tests, etc.) on the same digital platform that adapts the learning contents to the needs of different learners who are diverse in terms of abilities, levels, interests, backgrounds, and other characteristics. In other words, it is the hypermedia environment that provides 'personalized learning'.

Hyperlinks: It refers to a word, button, image, icon and so on in a digital environment on which target users can click to navigate to any part of the digital environment.

Hypermedia: Interactive multimedia is also called hypermedia.

Hypertext: It is a type of text retrieval system that enables target users to access instantaneously particular media types in certain locations or files in digital applications, webpages, or other digital environments.

Instructional Design Model: From the point of view of designing and developing AH listening environments, an instructional design model is a set of principles and guidelines that enable designers and developers of adaptive hypermedia listening environments to design and develop pedagogically cost-effective listening applications so that target instructional goals can be achieved.

Interactive Multimedia: The pedagogically and instructionally effective and efficient combination and instantaneous delivery of a wide range of digital elements on the same digital platform which have links between elements in the form of hypertexts, buttons, hotspots, or hyperlinks to create an interactive application in which target users can navigate in the way and as they need to.

Listening: It is a process of giving conscious and purposeful attention, following, and understanding input/messages so that meaning can be derived out from that input/message.

Multimedia: It is the effective and efficient combination and instantaneous delivery of digital learning elements on the same digital platform.

Chapter 3
How Best to Use Audio/ Subtitle Combinations in the Use of Films for Nonnative Learners of English

Buğra Zengin
Namık Kemal University, Turkey

ABSTRACT

Use of movies and TV shows for foreign language learning purposes pervades informal settings where viewers enjoy the multiplicity of soundtracks and of subtitles. However, this is not an integral part of formal language education policies. Hoping to inform policy making processes, this study aimed to investigate the most efficient use of audio/subtitle combinations for specific purposes. As a result, reversed subtitling was found to be effective for vocabulary learning whereas, generally, bimodal subtitling was considered to benefit accent-related problems. The participants self-reported preferrence for non-subtitled version in case of listening comprehension in general, which was due to the subtitle effect in the case of students at lower levels. Considering that mostly interlingual subtitled content is the only version in most platforms, these findings make it necessary to prefer a platform presenting a variety of alternative combinations in terms of soundtrack and subtitles.

INTRODUCTION

In our daily lives, it is highly likely to come across someone saying s/he has watched films (movies or TV shows) in response to the question inquiring how s/he learned his/her foreign language. This generalization can be seen in many research studies in terms of vocabulary (Ashcroft, Garner, & Hadingham, 2018; Vahdami & Azaraki, 2015; Zengin & Aker, 2016), both structure and vocabulary (Danan, 1992), Caimi & Mariotti, 2015), captions (Başaran & Dilber, 2013, Danan, 2004, Vanderplan, 2016), listenin skills (Danan, 2004), awareness as to what soundtrack/subtitling options to prefer (Mitterer & McQueen, 2009), as well as the use of film scripts (Zengin, 2019). Not only video but also the use of audio-only

DOI: 10.4018/978-1-7998-7876-6.ch003

texts and video-only texts have also been studied (Basal, Gulizer, & Demir, 2015), which demonstrated a significantly higher success for the former. Given the pervasive preference for using movies as tools, this has implications for the need to develop ways of coping with possible difficulties in their use of these tools. Another pervasive practice is the potential of mobile applications for foreign language learning. For instance, the study of Basal, Yilmaz, Tanrıverdi and Sari (2016, p. 55) "contributed to the literature in that using mobile applications in vocabulary teaching can be a more effective tool when compared to traditional paper-based activities" and highlighted the teachers' need for know-how in terms of evaluating "the usefulness of these applications for vocabulary teaching". For instance an Android app developed by Benjamin Bisinger, subtitle viewer is a good example for real-time displaying of movie subtitles.

It seems the language education curriculums have not recognized the importance of above-mentioned resources. It is possible to see a similar view in terms of materials development. Most ELT materials do not include the content of movies and TV shows. The reality within the classroom should be fed with the rich variety of sources that pervade out-of-class settings. Although the informal use of films for language learning is pervasive, formal education does not tap into this wealth of resources. It seems the solutions can be achieved with the collaborations between stakeholders in the education and broadcasting sectors. This paper is an attempt to inform the stakeholders in both sectors for the educational potential of the audio/subtitle combinations in viewing movies and TV shows.

BACKGROUND

The study of Ashcroft, Garner and Hadingham (2018) suggests that as a resource for meaning-focused audio-visual input movie watching can lead to incidental vocabulary gains indicating the potential of movies to boost language learners' vocabulary acquisition rates. Confirming the benefits of captions in their study, Başaran and Dilber (2013) recommend learners to study the captions of the English TV shows before watching them to aid their comprehension and enhance their motivation. They state that repeated studies of captions lower the learners' affective filter. They also add that memory works most effectively with the stimulation of both auditory and visual registers. Vahdami and Azaraki (2015) share a similar view in terms of memory stating that L2 learners' vocabulary comprehension is enhanced with simultaneous exposure to text, sound, and picture. They argue that even inexperienced language learners' vocabulary is promoted with authentic English captioned videos.

Danan (2004) believes that whether it is captioning or interlingual subtitling, listening comprehension skills are improved. Language learning is facilitated with captioning in that it aids the learners' zvisualization of what they hear especially considering the kind of input that is beyond their linguistic ability. In subtitling, the increase in language comprehension is reached with the cognitive benefit of deeper processing. However, Danan (2004) stresses the learners' need for development through training of active viewing strategies to use captioned and subtitled material efficiently.

Movies are invaluable and rich in terms of exposure to different accents and authentic conversations (Goctu, 2017). Presenting "colloquial English in real-life contexts rather than artificial situations" movies provide exposure "to different native speaker voices, stress, accents, and dialects" (King, 2002; as cited in Goctu, 2017, p. 121). Mitterer and McQueen (2009) suggest that it is now time to exploit the possibilities of multiple audio channels and multiple types of subtitles with the advent of digital television broadcasting.

Mitterer and McQueen (2009) highlight the use of the same language subtitles in terms of accents. Experimenting with the combination of French soundtrack and English subtitles (reversed subtitling), Danan (1992) found out that it was effective not only for the vocabulary but also the structure of the target language. Zengin and Akcr's (2016) study concludes with the beneficial impact on vocabulary learning of viewers with films Turkish dubbed-English subtitled. Telling from their experience, Caimi and Mariotti (2015, p. 235) concede that students at the beginning or lower-intermediate level working in small groups may see the benefit of reversed subtitles. However, they also mention about their students self-reporting that if they lost concentration, they avoided reading L2 subtitles and just listened to the L1 soundtrack."

In a course designed for non-language university students to practice English in the context of their subject matter through L2 intralingual subtitled audiovisual aids, the emerging outcomes were the students' speeding up of the reading abilities and increased self-confidence in target language use (Caimi & Mariotti, 2015). Intralingual subtitling "turned out to be an opportunity for foreign language learning, especially for students with at least intermediate competence in the L2" (Caimi & Mariotti, 2015, p. 235).

Broadcasting its content with a variety of audio-subtitle/caption combinations, Netflix is an ideal platform that non-native learners of English can benefit from. However, the out-of-class and in-class settings are not integrated enough to make the most of the possibilities of the technology. Vanderplan (2016, p. 5) states that whereas "informal language learners may already be making good use of captioned viewing, the potential for teachers and learners in drawing on captioned TV and films in the classroom remains largely unfulfilled."

METHODOLOGY

Pre-intermediate to intermediate Turkish students of the English philology department at a state university in Turkey were asked to choose one out of four audio/subtitle combinations for different scenarios focusing on different aspects of learning English having experimented with different audio-subtitle combinations. The study was participated by 84 university students of the English language and literature department of a state university. Thirty students were from the preparatory class and fifty-four students were first year students. Five imaginary scenarios were presented to the participants who were then asked to choose their preference for an audio/subtitle combination in case they lived the presented scenarios. The multiple choice combination question included non-subtitled English audio, bimodal/intralingual/same language subtitles (English audio/English subtitle), interlingual/standard subtitles (English audio/Turkish subtitle), reversed subtitles (Turkish dubbed/English subtitles). Besides, there was one five-point Likert scale item requiring the participant to report the actual frequency of use of the preferred combination in case of the above mentioned scenarios. Netflix was preferred to watch movies and TV shows to experience content rich in accent, jargons and colloqualisms since it was the only streaming platform to include the widest variety of audio/subtitle options. Among others, the variety has also included viewing Turkish content (movies and TV shows) as English dubbed.

Procedure

As part of the activities in the basic English course in the prep class of the English language and literature, language education materials were constructed out of the scripts of film and TV shows (TV series). The activities consisted of detecting chunks and jargons in the original language (i.e., English) script files and their translation equivalents in the files of translations of these scripts and compiling these chunks and jargons with their contexts as well as their translations. The learners were trained on how to select what to compile out of these scripts. The instructor demonstrated how to do the activity to each learner providing one-on-one guidance and instruction with the instructor doing the job himself to provide a model to do the tasks. Even if the time spent with each learner was not long, this was necessary. Some students found some chunks to be easy and showed tendencies to pass them as unnecessary but they did not have the productive knowledge of the word in question. Some general knowledge was presented. For instance, they were asked to check all phrasal verbs, or they were asked to include in their lists anything to do with the word "get" since it is very frequent in conversational English. The above-mentioned activities had been started by the first-year students (who participated in the study as well) of the same department in their preparation class year and continued these as first-year students. Besides, as a new source of material, the use of Netflix for language learning was started with prep year students in the basic English lessons and with first-year studies in the advanced English lessons. Netflix facilitated the activities with the multiplicity of audio/subtitle combinations.

As part of benefiting from the movies and TV shows (series) for language education, education materials were produced focusing on chunks, jargon, and accents. Chunks and jargons in the scripts were compiled together with their translation equivalents both in the dubbing and subtitles. In time, a set of criteria were developed to include two samples from five genres (politics, military, police/crime, law, and medicine) five samples from the general category of conversational chunks. The learners were also required to list two samples from the content of Turkish movies and TV shows and the English translation of the Turkish content. Since the direction of translation is from Turkish (the original language of the Turkish production) to English, provided samples which are more likely to reflect a more naturalistic and authentic use of the Turkish language and the possibility of seeing their translations. Besides, since it is likely to see some differences between the dubbed translation and the subbed translation, three samples were required from each type of translation. All the samples were recorded with chronological information (minute and second of the utterances). Whether the learners did the activities meeting the criteria set was checked with the teacher having one-on-one sessions with each learner, who was provided with individualized feedback, which resulted in the correction of the errors in the learners' assignments.

1. Participants

Eighty-four university students (30 from preparatory class, 54 freshmen) participated in the study. Out of these 84 participants, those delivering the assignments number 56 (25 from the prep class, 31 freshmen). Forty-eight assignments were delivered in a printed format (22 prep, 26 freshmen), eight assignments were delivered online (3 prep, 5 freshmen).

Twenty-eight participants' (5 from the prep class and 23 freshmen) assignments were incomplete. Although some students did not meet all the criteria of the assignment, they were observed to do their utmost effort. Since many activities were carried out in groups or as pair work, peers more able or more advanced led others less able or advanced resulting in total quality.

Some of the difficulties experienced were payment security concerns regarding the costs of Netflix's streaming plans, as well as problems related to laptops and the Internet (problems with affordability and technicalities). Besides, the disproportion of the number of computers in the class lab and that of the students was obvious, however, the guidance provided was of the kind that helped them extend their in-class practices and continue them in out-of-class settings.

The participants' level was within the pre-intermediate – intermediate range, therefore their level was suitable for this study. According to Yeldham (2018) watching texts providing access to the phonological, lexico-grammatical, and pragmatic features of the authentic language is suitable for levels approximately intermediate or slightly above. For lower levels, the texts can be simplified for pedagogical purposes to suit them. Bianci and Ciabattoni (2018, p. 87) stated that when proficiency level was not very high, text aids did not favor language-in-use acquisition "but text aids in general and subtitles in particular gradually acquired greater relevance when the proficiency level rose" (Bianci & Ciabattoni, 2008, p. 87).

2. Instrument

The students were presented with imaginative five scenarios and asked which audio-subtitle combinations they would prefer if they experienced the situation described in that scenario. There were four choices regarding audio-subtitle combinations: English audio – no subtitle, English audio – English captioned, English audio – Turkish subtitled (subbed), Turkish dubbed – English subbed. There was also one 5-point Likert scale item asking what frequency they would actually apply their audio-subtitle preferences for the above-mentioned scenarios.

Scenarios

1. When you have difficulty with listening comprehension due to jargons what combination do you use if your purpose is to learn examples of specialized language (jargons) when watching a movie or a TV show (series)
2. If your purpose is to have accent familiarity, what audio-subtitle combination do you prefer?
3. You are having difficulty with comprehending the accent and you want to develop your listening skill accents, what combination do you prefer?
4. When you have difficulty with listening comprehension due to accent, what combination do you use?
5. What combination do you use if you want to develop your listening skill in general?

Audio-Subtitle Combination Varieties

1. English language – no subtitles
2. English language – Turkish subtitles
3. English language – English subtitles
4. Turkish language – English subtitles

The choices are expressed in ways to facilitate the participants' comprehension of the combinations. For instance, in the relevant literature, the combination 'C' comes across as bimodal subtitles, subtitles in the same language as the audio, foreign language audio with foreign language subtitles/captions, intralingual subtitles, and captions. It is also necessary to mention the two different kinds of captions:

The closed captions (CC) and open captions (OC). In the case of the closed captioning, there is an option to turn it off or on.

RESULTS

The following table shows percentages of the participants who delivered their assignments (the 3rd column), those who did not deliver (the 2nd column), and the total number (the 1st column). The letters A, B, C, and D show the choices of audio/subtitle combinations. There are also five rows each demonstrating the scenarios in the five-item scenario list above (numbers in percentages).

Table 1. Participants' preferences for audio/subtitle combinations

	1st column				2nd column				3rd column			
	Prep+1st year total				Prep+1st year (did not deliver)				Prep+1st year (did not deliver)			
	A	B	C	D	A	B	C	D	A	B	C	D
1	0.00	35.71	33.33	30.95	0.00	32.14	42.86	25.00	0.00	37.50	28.57	33.93
2	32.14	21.43	46.43	0.00	28.57	21.43	50.00	0.00	33.93	21.43	44.64	0.00
3	2.38	21.43	75.00	1.19	7.14	10.71	78.57	3.57	0.00	26.79	73.21	0.00
4	0.00	27.38	46.43	26.19	0.00	39.29	32.14	28.57	0.00	21.43	53.57	25.00
5	54.76	9.52	35.71	0.00	64.29	14.29	21.43	0.00	50.00	7.14	42.86	0.00

Table 2. The 5-point likert scale frequency the first-year students used their preferred audio/subtitle combinations

	Audio/subtitle combination:	never	rarely	sometimes	often	always
A	English language – no subtitles	X	7	5	5	2
B	English language – Turkish subtitles	X	4	5	6	4
C	English language – English subtitles	X	2	8	4	5
D	Turkish language – English subtitles	11	4	4	X	X

Table 3. The 5-point likert scale frequency the preparation-year students used their preferred audio/ subtitle combinations

	Audio/subtitle combination:	never	rarely	sometimes	often	always
A	English language – no subtitles	7	15	4	4	X
B	English language – Turkish subtitles	1	3	5	10	11
C	English language – English subtitles	X	1	8	11	10
D	Turkish language – English subtitles	11	15	3	1	X

DISCUSSION AND CONCLUSION

In the scenario in which both the problem area and the learning focus is jargons, it is worth noticing to see the preference for the Turkish dubbed-English subbed combination to be about one-third of the participants. Some results overlap with the findings of the previous studies. With regard to the learners' preference for the Turkish dubbed-English subbed version when the extra focus was the learning of the targeted vocabulary, there is convergence on the findings of this study and those of a study by Zengin and Aker (2016) whose semi-structured interviews with English philology students demonstrated their use of Turkish audio of an originally English movie (or TV show) combined with the English caption/ subtitle at the beginning of their learning process though this was self-reported to decline and evolve to English-only preferences in the final stage. The study graphic (Zengin & Aker, 2016, p. 491) demonstrated their view on the Turkish dubbed-English subbed strategy with 30% actually utilizing the combination and 41% realizing and /or acknowledging its potential. The rest (29%) neither used nor accepted its potential. The study also underlined that the learners' preferences were affected by those likely to accompany them watching the film.

In the previous studies on the use of backseat TV systems for learning English, one questionnaire item specifically inquired if the respondents prefer Turkish dubbed English subbed combination. In Zengin and Çubukçu's study (2013) the participants' levels were at a point within the beginning-pre intermediate range, the mean value for the preference for the Turkish dubbed-English subbed combination was 2.1756 (Zengin & Çubukçu, 2013, p. 1572). A similar study (Zengin, Doğan, & Çubukçu, 2015, p. 292) but participated by ELT students demonstrated a similar mean value of 2.2449 (number (n): 8 frequency (f): %16.3 n: 4 f: %8.2 n: 37 f: %75.5).

In contrast to the claim that defends monolingualism for the learners' progress, the number of participants watching Turkish dubbed/English subbed was considerably high (in Zengin & Aker, 2016) or at a level which should not be underestimated given the studies in question mainly focused on the backseat TV systems of the buses (in Zengin & Çubukçu, 2013; Zengin, Zengin, Doğan, & Çubukçu, 2015). The reason for the difference can also be attributed to the individual differences, or differences with regard to the learning culture, or availability/ownership of technology with audio/subtitle options for the participants. Valid for many contexts/settings, the influence of monolingualism is prevalent, pragmatic use of translation is confused with outdated practices of the grammar-translation method. Supporting the learners' own languages and translation in the learning of English, Hall and Cook (2012, 2013) demonstrated a significantly high number of quality evidence for the use of the students' own languages and translation. Translation made for movies and TV shows was one aspect of the questionnaire in Zengin's (2019) study, which focused on the use of the English language scripts and their Turkish translation script files in teaching English.

The use of reversed subtitles for vocabulary gains is also supported by a very sophisticated study conducted in Iran. In their study aiming to investigate Iranian EFL learners watching English language movies in three different conditions (reversed subtitled, standard subtitled - namely English soundtrack with Persian subtitles, and non-subtitled), Fazilatfa, Ghorbani and Samavarchi (2011) concluded that the reversed subtitled mode was the most superior to the other two modes in terms of incidental vocabulary acquisition. In other words, those hearing the Persian translations of the English words read in the subtitle (dialogues in Persian with subtitles in English) outperformed those in the other groups. Those in the standard subtitled condition (dialogues in English with subtitles in Persian) outperformed those in the non-subtitled condition.

For the reliability of data, all three groups were tested with three different test types (word-form recognition, meaning, multiple-choice, meaning translation). The different exam performances between groups were found to be statistically significant, except for the form recognition tests. Besides the listening test administered ensured the participants' homogeneity (Fazilarfa, Ghorbani & Samavarchi, 2011).

The topic of the second, third, and fourth questions are what combination is to be preferred when faced with problems with the accent. Mainly the tendency for the second question is to prefer the language of both the audio and the caption/subtitle as English. Although these questions overlap in terms of accent difficulty, there is an emphasis on the comprehension of what is listened to especially in the third question, this is likely to be the reason for some respondents to choose English audio – Turkish subtitle combination. Some respondents might be more sensitive to the comprehension part reflected in the different wordings of the question, which might be explained with perceptual selectivity. In fact there are even those choosing Turkish dubbed English subbed version, which demonstrates the need for comprehension of what is listened to.

The findings confirm the conclusions of other studies on bimodal subtitling. According to Bird and Williams (2002, p. 19) same-language subtitling benefits second language learning with its "significant facilitatory impact on long-term implicit and explicit learning of spoken word forms". The intra-lingual subtitles "neatly divide the boundaries between words in the speech stream and are unaffected by accent variations and sound degradation which can adversely affect auditory comprehension" (Vanderplank, 1993: as cited in Bird & Williams, 2002, p. 509).

Listeners' difficulties with understanding unfamiliar regional accents, which are magnified in the case of a foreign language can be facilitated with same-language subtitles through the lexically-guided retuning mechanism. Speech learning can be boosted and language comprehension can be facilitated even in the case of foreign language words and the unfamiliar regional accents they are pronounced in. In short, bimodal subtitles have been shown to facilitate nonnative listeners' perceptual learning of accents that are unfamiliar to them (Mitterer & McQueen, 2009).

Conducting a study adopting an experimental research design with participation of English language prep classes of a state university, Şevik (2017) investigated the effects of bimodal subtitles on the prep class students' listening comprehension. Outperforming the control group, the experimental group demonstrated that bimodal subtitling has positive effects on listening comprehension and that caption reliance did not have any negative effects for the participants.

As for the question of what combination the respondents use if they want to improve their listening skill in general, almost two-thirds of the first-year students preferred English audio with no subtitles, and the English audio – English subtitle ranked second with a bit over one-quarter of the respondents preferring this combination (almost one third in the group completing their assignments). On the other hand, as for the English preparatory class students there was a different picture with the English audio – English subbed/captioned combination favored by a bit over half of the preparatory class students. Those using English audio with no subtitles/captions constituted a bit one-third of the sample.

The reason for the high rate of viewing with no subtitles may be explained with what Shepherd, Simonian and Trussler (2017) call the "subtitle effect". In their experimental study, it is possible to explain the number of students who perceive same-language subtitles to be as less necessary or not necessary at all with the "subtitle effect". Likewise, as pointed out by Shepherd et al. (2017), through the alternate access to the lexical information in the audio provided by bimodal subtitles, the facilitation of perceptual learning of unfamiliar accents is made possible with bimodal subtitles, through which viewers quickly adjust their sound-phoneme mappings. However, this can also be perceived as less necessary or not necessary at all by the viewers, who have actually been the beneficiaries of this. Bimodal subtitles "provide viewers with an alternate way of accessing the same lexical information that is present (albeit not necessarily readily available) in the audio" (Sheperd et al., 2017, p. 357). However, in the light of findings reached in their experiments, Shepherd et al. (2017,p. 357) warn against the potential of such an illusion and stress that "perceptions of bimodal subtitles as unnecessary must be interpreted with caution".

There are also some factors leading to differences between interlingual and intralingual subtitles. For instance, it is possible for some colloquialism or colloquial expression may be missubtitled. There is also the fact that euphemisation of offensive language (e.g. swearing) in the original version, on the other hand, more direct translation may be seen in the subtitled version. After all, something being heard spreads but being read is something occurring rather individually. It is my observation I got from my interactions with students that slangs play a large part in their evaluations between equivalences between languages although it is not significant for language learning for the most part if the aim is general language skills.

To conclude, this study demonstrated that in general, the effective strategy for accent related problems is bimodal subtitling. However, there is a tendency towards standard subtitles when the respondents focused on the listening comprehension part of the question of whether the question includes accent factor or not. As for vocabulary learning, the effective strategy is reversed subtitling. Watching movies without subtitles is generally aimed but the responses demonstrate "subtitle effect". It is also possible to explain some responses as results of social desirability. Some people might perceive the use of subtitles or captions to be a symptom of a problem. The students should be provided with strategy training to make the most of the multiplicity of audio-subtitle/caption combinations.

FUTURE RESEARCH DIRECTIONS

It should also be said that companies like Netflix should invest more heavily in language learning materials development. Language learning with Netflix extension is an excellent tool for learners of English for speakers of languages other than English. The extension features a whole script on the side bar and dual subtitles (simultaneous on-screen bimodal and interlingual subtitling). However, the design lacks bilingual dictionary for multiword units. The company should get the academics' recommendations for the upgrade of the product. There is also another extension that makes up for the need for social distancing, Teleparty (formerly, Netflix Party), which "synchronizes video playback and adds group chat to Netflix, Disney, Hulu and HBO" ([information about Teleparty], n.d.).

REFERENCES

Ashcroft, R. J., Garner, J., & Hadingham, O. (2018). Incidental vocabulary learning through watching movies. *Australian Journal of Applied Linguistics*, *1*(3), 135–147. doi:10.29140/ajal.v1n3.89

Başal, A., Gülizer, K., & Demir, I. (2015). Use of Video and Audio Texts in EFL Listening Test. *Journal of Education and Training Studies*, *3*(6), 83–89. doi:10.11114/jets.v3i6.1001

Basal, A., Yilmaz, S., Tanrıverdi, A., & Sari, L. (2016). Effectiveness of mobile applications in vocabulary teaching. *Contemporary Educational Teaching*, *7*(1), 47–59. doi:10.30935/cedtech/6162

Basaran, S., & Dilber, E. (2013). Effects of captioned TV shows on language learners motivation and perception. *Respectus Philologicus*, *23*(28), 83–96. doi:10.15388/RESPECTUS.2013.23.28.7

Bird, S. A., & Williams, J. N. (2002). The effect of bimodal input on implicit and explicit memory: An investigation into the benefits of within-language subtitling. *Applied Psycholinguistics*, *23*(4), 509–533. doi:10.1017/S0142716402004022

Caimi, A., & Mariotti, C. (2015). Beyond the book: The use of subtitled audiovisual material to promote content and language integrated learning in higher education. In J. Díaz-Cintas & J. Neves (Eds.), *Audiovisual translation: Taking stock* (pp. 230–243). Cambridge Scholars Publishing.

Charles, T. J. (2017). *The role of captioned video in developing speech segmentation for learners of English as a second language* (PhD thesis). University of York.

Danan, M. (1992). Reversed subtitling and dual coding theory: New directions for foreign language instruction. *Language Learning*, *42*(4), 497–527. doi:10.1111/j.1467-1770.1992.tb01042.x

Danan, M. (2004). Captioning and subtitling: Undervalued language learning strategies. *Meta*, *49*(1), 67–77. doi:10.7202/009021ar

Etemadi, A. (2012). Effects of bimodal subtitling of English movies on content comprehension and vocabulary recognition. *International Journal of English Linguistics*, *2*(1), 239–248. doi:10.5539/ijel.v2n1p239

Goctu, R. (2017). Using movies in EFL classrooms. *European Journal of Language and Literature Studies*, *3*(2), 121–124. doi:10.26417/ejls.v8i1.p121-124

Hall, G., & Cook, G. (2012). Own-language use in language teaching and learning. *Language Teaching*, *45*(3), 271–308. doi:10.1017/S0261444812000067

Hall, G., & Cook, G. (2013). *Own-language use in ELT: Exploring global practices and attitudes*. British Council.

Hoogendyk, F. W. C., & Arifin, Z. & Novita, D. (2014). Increasing students' vocabulary by using bimodal subtitling in short movie. *Jurnal Pendidikan dan Pembelajaran (Universitas Tanjungpura)*, *3*(11), 1-14.

Jaber, R. A., & Dollar, Y. K. (2019). A meta-analysis of the effect of bimodal subtitling on vocabulary learning among adult EFL learners. *Focus on ELT Journal*, *1*(1), 43–52. doi:10.14744/felt.2019.00001

Kanellopoulou, C. (2019). Film subtitles as a successful vocabulary learning tool. *Open Journal of Modern Linguistics*, *9*(02), 145–152. doi:10.4236/ojml.2019.92014

Köksal, D. (2005). *Çeviri eğitimi: Kuram ve uygulama*. Nobel.

Mardani, M., & Najmabadi, A. (2016). The Effect of movies with different types of subtitles on incidental English vocabulary learning of Iranian High School EFL learners. *Iranian Journal of Research in English Language Teaching (= Research in English Language Pedagogy)*, *4*(1), 52-63.

Mitterer, H., & McQueen, J. M. (2009). Foreign subtitles help but native-language subtitles harm foreign speech perception. *PLoS One*, *4*(11), e7785. Advance online publication. doi:10.1371/journal.pone.0007785 PMID:19918371

Information about Teleparty (n.d.). http://www.netflixpart.com

Şevik, M. (2017). Does bimodal subtitling effect listening comprehension in EFL contexts. In D. Köksal (Ed.), *Researching ELT: Classroom methodology and beyond* (pp. 55–64). Peter Lang International Academic Publisher.

Shepherd, M. A., Simonian, J. R., & Trussler, L. C. (2017). The Subtitle effect: How subtitle-facilitated perceptual learning can make subtitles (seem) less necessary. *Communication Research Reports*, *34*(4), 350–358. doi:10.1080/08824096.2017.1356712

Vahdany, F., & Azaraki, A. M. (2015). The effect of watching captioned movies on Iranian learners' vocabulary learning at pre-intermediate level. *Educational Technology*, *4*, 30619–30633.

Vanderplank, R. (2016). *Captioned media in foreign language learning and teaching: Subtitles for the deaf and hard-of-hearing as tools for language learning*. Palgrave MacMillan.

Yeldham, M. (2018). Viewing L2 captioned videos: What's in it for the listener? *Computer Assisted Language Learning*, *31*(4), 367–389.

Zengin, B. (2019). Use of film scripts and their translations in teaching English as a foreign language. *International Online Journal of Education & Teaching*, *6*(4), 944–958. https://iojet.org/index.php/IOJET/article/view/718

Zengin, B., & Çubukçu, F. (2013). English language learners' attitudes to backseat TV systems and subtitled movies. *Turkish Studies – International Periodical for the Languages, Literature and History of Turkish or Turkic*, *8*(8), 1561–1573.

Zengin, B., Doğan, D., & Çubukçu, F. (2015). English major students' attitudes towards movies and series as language resources. In V. Türel (Ed.), *Intelligent design of interactive multimedia listening software* (pp. 276–295).

Zengin, B., & Yetkin Aker, D. (2016). Education policy and the effect of subtitle options in learning process: A case of Turkish students as English language learners. *Humanitas. A Journal of Social Sciences*, *4*(7), 485–498.

ADDITIONAL READING

Danan, M. (1992). Reversed subtitling and dual coding theory. New directions for foreign language instruction. *Language Learning*, *42*(4), 497–527. doi:10.1111/j.1467-1770.1992.tb01042.x

Danan, M. (2016). Enhancing listening with captions and transcripts: Exploring learner differences. *Applied Language Learning*, *26*(2), 1–24.

Danan, M. (2016). Enhancing listening with captions and transcripts: Exploring learner differences. *Applied Language Learning*, *26*(2), 1–24.

Dooley, M. W. (2013). How movie subtitles influence EFL learners' listening comprehension. *Journal of Applied English*, *6*, 51–66.

Gilakjani, A. P., & Sabouri, N. B. (2016). Learners' listening comprehension difficulties in English language learning: A literature review. *English Language Teaching*, *9*(6), 123–133. doi:10.5539/elt.v9n6p123

Hosogoshi, K. (2016). Effects of captions and subtitles on the listening process. Insights from EFL learners' listening strategies. *The JALT CALL Journal*, *12*(3), 153–178. doi:10.29140/jaltcall.v12n3.206

Ismaili, M. (2013). The effectiveness of using movies in the EFL classroom–a study conducted at South East European University. *Academic Journal of Interdisciplinary Studies*, *2*(4), 121. doi:10.5901/ajis.2012.v2n4p121

Montero Perez, M., Van Den Noortgate, W., & Desmet, P. (2013). Captioned video for L2 listening and vocabulary learning. A meta-analysis. *System*, *41*(3), 720–739. doi:10.1016/j.system.2013.07.013

Murshidi, G. A. (2020). Effectiveness of movies in teaching and learning English as a foreign language at universities in UAE. *Psychology and Education*, *57*(6), 442–445.

Shang-Ikeda, S. (1994). Using captions increases comprehension. *JALT Journal*, *16*(1), 83–89.

Shea, P. (2000). Leveling the playing field. A Study of captioned interactive video for second language learning. *Journal of Educational Computing Research*, *22*(3), 243–263. doi:10.2190/3NEY-BNT0-FB28-VBWY

Sydorenko, T. (2010). Modality of input and vocabulary acquisition. *Language Learning & Technology*, *14*(2), 50–73.

Thammineni, H. B. (2016). Movies supplement English classroom to be effective in improving students' listening and speaking skills–a review. *International Journal of Studies in English Language and Literature*, *4*, 35–37.

KEY TERMS AND DEFINITIONS

Interlingual/Standard Subtitling: The kind of subtitling in which the soundtrack/audio language and the subtitle language are different. For instance, the soundtrack language is English and the subtitle language is Turkish for viewers whose native or first language (L1) is Turkish.

Intralingual/Bimodal Subtitling: The kind of subtitling in which both soundtrack and the subtitle are the same language (in this case subtitles are captions).

Reversed Subtitling: The kind of subtitling in which the soundtrack/audio language and the subtitle language are different, but in this case the former is the L1 and the latter is L2. For instance, the soundtrack language is Turkish and the subtitle language is English for viewers whose L1 is Turkish.

APPENDIX

Scenarios

1. When you have difficulty with listening comprehension due to jargons what combination do you use if your purpose is to learn examples of zspecialized language (jargons) when watching a movie or a TV show (series)
2. If your purpose is to have accent familiarity, what audio-subtitle combination do you prefer?
3. You are having difficulty with comprehending the accent and you want to develop your listening skill accents, what combination do you prefer?
4. When you have difficulty with listening comprehension due to accent, what combination do you use?
5. What combination do you use if you want to develop your listening skill in general?

Audio-Subtitle Combination Varieties

1. English language – no subtitles
2. English language – Turkish subtitles
3. English language – English subtitles
4. Turkish language – English subtitles

The choices are expressed in ways to facilitate the participants' comprehension of the combinations. For instance, in the relevant literature, the combination C is come across as bimodal subtitles, subtitles in the same language as the audio, foreign language audio with foreign language subtitles/captions, intralingual subtitles, and captions. It is also necessary to mention the two different kinds of captions: The closed captions (CC) and the open captions. In the case of the closed captioning, there is an option to turn it off or on.

The following table shows percentages of the participants who delivered their assignments (the 3rd column), those who did not deliver (the 2nd column) and the total number (the 1st column). The letters A, B, C and D show the choices of audio/subtitle combinations. There are also five rows each demonstrating the scenarios in the five-item scenario list above (numbers in percentages alongside with lower case a/b/c/d: e.g. 21a = twenty-one students who have chosen a).

Table 4. The frequency and percentage information for the first year participants

54 students 23 students 31 students
1st year students (total) 1st year students (did not deliver) 1st year students (delivered)
(Number in frequencies)
a b c d a b c d a b c d
0a 19b 19c 16d 0a 7b 11c 5d 0a 12b 8c 11d
21a 10b 23c 0d 5a 6b 12c 0d 16a 4b 11c 0d
2a 12b 39c 1d 2a 3b 17c 1d 0a 9b 22c 0d
0a 18b 22c 14d 0a 10b 6c 7d 0a 8b 16c 7d
35a 5b 14c 0d 15a 3b 5c 0d 20a 2b 9c 0d
1st year students (total) 1st year students (did not deliver) 1st year students (delivered) (numbers in percentages)
a b c d a b c d a b c d
0.00 35.19 35.19 29.63 0.00 30.43 47.83 21.74 0.00 38.71 25.81 35.48
38.88 18.52 42.59 0.00 21.74 26.09 52.17 0.00 51.61 12.90 35.48 0.00
3.70 22.22 72.22 1.85 8.70 13.04 73.91 4.35 0.00 29.03 70.97 0.00
0.00 33.33 40.74 25.93 0.00 43.48 26.09 30.43 0.00 25.81 51.61 22.58
64.81 9.26 25.93 0.00 65.22 13.04 21.74 0.00 64.52 6.45 29.03 0.00

Table 5. The frequency and percentage information for the prep class participants

30 students 5 students 25 students
Prep class (total) Prep class (did not deliver) Prep class (delivered)
(Numbers in frequencies)
a b c d a b c d a b c d
0a 11b 9c 10d 0a 2b 1c 2d 0a 9b 8c 8d
6a 8b 16c 0d 3a 0b 2c 0d 3a 8b 14c 0d
0a 6b 24c 0d 0a 0b 5c 0d 0a 6b 19c 0d
0a 5b 17c 8d 0a 1b 3c 1d 0a 4b 14c 7d
11a 3b 16c 0d 3a 1b 1c 0d 8a 2b 15c 0d
The above-mentioned information in percentages:
a b c d a b c d a b c d
0.00 36.66 30.00 33.33 0.00 40.00 20.00 40.00 0.00 36.00 32.00 32.00
20.00 26.66 53.33 0.00 60.00 0.00 40.00 0.00 12.00 32.00 56.00 0.00
0.00 20.00 80.00 0.00 0.00 0.00 100.00 0.00 0.00 24.00 76.00 0.00
0.00 16.66 56.66 26.66 0.00 20.00 60.00 20.00 0.00 16.00 56.00 28.00
36.66 10.00 53.33 0.00 60.00 20.00 20.00 0.00 32.00 8.00 60.00 0.00

Table 6. The frequency and percentage information for the first year and prep class participants

Prep+1st year (total) Prep+1st year (did not deliver) Prep +1st year (delivered)
(Numbers in frequencies)
a b c d a b c d a b c d
0a 30b 28c 26d 0a 9b 12c 7d 0a 21b 16c 19d
27a 18b 39c 0d 8a 6b 14c 0d 19a 12b 25c 0d
2a 18b 63c 1d 2a 3b 22c 1d 0a 15b 41c 0d
0a 23b 39c 22d 0a 11d 9c 8d 0a 12b 30c 14d
46a 8b 30c 0d 18a 4b 6c 0d 28a 4b 24c 0d
Prep+1st year (total) Prep+1st year (did not deliver) Prep +1st year (delivered)
The above mentioned information in percentages:
a b c d a b c d a b c d
0.00 35.71 33.33 30.95 0.00 32.14 42.86 25.00 0.00 37.50 28.57 33.93
32.14 21.43 46.43 0.00 28.57 21.43 50.00 0.00 3.93 21.43 44.64 0.00
2.38 21.43 75.00 1.19 7.14 10.71 78.57 3.57 0.00 26.79 73.21 0.00
0.00 27.38 46.43 26.19 0.00 39.29 32.14 28.57 0.00 21.43 53.57 25.00
54.76 9.52 35.71 0.00 64.29 14.29 21.43 0.00 50.00 7.14 42.86 0.00

Table 7. Participants' frequencies of use of each audio/subtitle combination (Those who delivered assignments)

Frequency of Prep Class studens (who delivered their assignments)
Never rarely sometimes often always
6 14 3 2 0
1 2 4 7 11
0 1 8 9 7
9 12 3 1 0
Frequency of First year students (who delivered their assignment)
Never rarely sometimes often always
3 10 9 3 2
0 4 5 12 6
0 1 11 9 6
15 5 6 1 0
Frequency of prep class students plus first year students in total (who delivered their assignments)
Never rarely sometimes often always
9 24 12 5 2
1 6 9 19 17
0 2 19 18 13
24 17 9 2 0

Table 8. Those students who didnot deliver their assignments

Frequency of Prep Class studens (who DID NOT deliver their assignments)
Never rarely sometimes often always
1 1 1 2 0
0 1 1 3 0
0 0 0 2 3
2 3 0 0 0
Frequency of First year students (who DID NOT deliver their assignment)
Never rarely sometimes often always
1 9 10 5 2
0 5 6 9 7
0 4 11 5 7
12 8 5 2 0
Frequency of prep class students plus first year students in total (who DID NOT deliver their assignments)
Never rarely sometimes often always
2 10 11 7 2
0 6 7 12 7
0 4 11 7 10
14 11 5 2 0

Chapter 4
Quality Indicators for Designing English Learning Podcasts

Tingting Kang
Lafayette College, USA

ABSTRACT

Podcasts, one type of mobile-assisted language learning, have been promoted as promising tools in second language (L2) learning. Yet, quality indicators of podcast programs have not been generalized. Applying previous empirical results on predictors of listening comprehension, the purpose of this chapter was to identify English language learning podcast programs in iTunes that were preferred by language learners and teachers and build up a list of the indicators that can be considered for selecting and designing high-quality English language learning podcasts. Participants were 26 intermediate English as second language (ESL) learners at a southwestern American university and two of their listening and speaking teachers. The results revealed that the ESL students and their teachers both preferred podcasts that had interesting topics, natural speed, and clear pronunciation.

INTRODUCTION

Podcast is a portmanteau which combines the "pod" from iPod, an MP3 player produced by Apple Inc., with "broadcast." It was coined in 2004 (Matthew, 2008) and praised as "the word of the year for 2005" by the New Oxford American Dictionary (Balas, 2005). Initially, podcasts were limited to iPod users who uploaded and downloaded audio files through *iTunes*. Nowadays, podcasts are not only available for Apple products but also other personal music players, e.g., MP3 players, cell phones, laptops, and PDAs (personal digital assistants). People can publish their own audio or even video podcasts on the Internet via different social networking websites, such as *iTunes*, *MySpace*, *Facebook*, and *YouTube*. Matthew (2008) explained the detailed phases of the podcasting process, which included recording, digitizing, editing, encoding, and uploading. After uploading the podcasts, people all over the world could download them on their mobile or PC devices and listen repeatedly (McMinn, 2008). Such software as *iTunes* also allows users to subscribe their favorite podcasts to receive automatic updates.

DOI: 10.4018/978-1-7998-7876-6.ch004

There are a large number of language learning podcasts that are now available in *iTunes*. Different language teachers and researchers recommended language learners to subscribe to these existing podcasts to improve their language skills as discussed in one of the recently published meta-analysis studies (Indahsari, 2020). However, one obstacle that the language teachers and researchers often face is the selection of the most suitable podcasts. Designing appropriate and high-quality podcasts is also the primary concern for the podcast developers. Therefore, the purpose of this chapter was to identify English language learning podcast programs in *iTunes* that were preferred by language learners and teachers and build up a list of the indicators that can be considered for selecting and designing high-quality English language learning podcasts.

LITERATURE REVIEW

MALL

M-learning is incorporated with any form of teaching and learning that involve a mobile device, which includes mobile phones, tablet computers, MP3 players, PDAs, electronic dictionaries, and game players (Kukulska-Hulme & Shield, 2008). Compared to personal computers, mobile devices had the advantages of being less expensive and easier to carry (Chinnery, 2006). Owing to the widespread ownership of mobile technologies, m-learning is one of the key current trends of applying technology to education, and it is also considered as the next generation of e-learning (Sharples, 2000).

In a meta-analysis, Wu, Wu, Chen, Kao, Lin, and Huang (2012) analyzed the results of 164 studies on m-learning. Some of the important findings are summarized as follows: first, in terms of the research purposes, evaluating the effects of m-learning was most common (58%), followed by designing a mobile system for learning (32%) and investigating the affective domain during mobile learning (5%). Second, 86% of the 164 studies reported positive research outcomes. Third, with regards to different mobile devices, mobile phones were most commonly used for m-learning (36.5%), followed by PDAs (30.96%), laptop computers (9.14%), iPods (4.06%), mp3/mp4 players (2.54%), podcasts (2.03%), and cameras (1.52%). Fourth, among different academic disciplines, applied sciences published the biggest amount of m-learning studies (29%), followed by humanities (20%), formal sciences (16%), social sciences (4%), and natural sciences (3%). Specifically, languages and linguistics courses were the most common focus (17.5%) when compared to other sub-disciplines. Additionally, Hwang and Tsai's (2011) meta-analysis on m-learning and ubiquitous learning (i.e., learning anywhere at anytime) reported the major contributing countries from 2001 to 2010 (K =154). In the first five years, US authors contributed the most publications (7), followed by UK authors (5) and Taiwanese authors (4). However, in the second 5 years, Taiwanese authors published 51 studies and became the number one contributing country.

In the field of language education, m-learning is called mobile-assisted language learning (MALL). The first project of MALL was conducted by the Stanford Learning Lab with the aim of facilitating foreign language learners' listening and speaking skills (Regan, Mabogunje, Nash & Licata, 2000). MALL has the advantages of extending students' learning from inside the classroom to outside the classroom and allowing students to study whenever and whatever they want.

According to a meta-analysis conducted by Duman, Orhon, and Gedik (2014), MALL research increased at fast pace around 2008. The most frequently used mobile devices in MALL were cell phones (29) among 69 identified studies, followed by PDAs (17) and portable music/video player (10). Also, the results revealed that a large scope of language learning topics have been discussed in previous MALL studies, and vocabulary was most commonly addressed (28). The number of listening and speaking/pronunciation in MALL studies were only eight and six. However, listening, speaking, and pronunciation, along with learners' perceptions and attitudes, have become "permanent areas of focus" (p. 6) since 2008. For instance, in 2012, Kondo et al. investigated whether using a mobile phone as an assisted language learning tool would enhance EFL college students' (n = 88) listening and reading skills and, in the long run, build up their self-study and self-regulated learning strategies. According to the statistical results, learners who used mobile phones to practice their English had significantly higher reading scores than the other group who studied English through regular textbooks. Meanwhile, the MALL group reported their beliefs in improvements of their language skills and interests in continuing to learn with the mobile device. However, delayed posttest provided valuable evidence showing how easily learners (n = 42) would abandon this MALL activity when the teachers departed from the mobile assisted learning process.

Podcasts

In the area of education, podcasts have been utilized in various teaching subjects, such as political science (e.g., Matthew, 2008), music (e.g., Polishook, 2005; Cooper, Dale, & Spencer, 2009), and language (e.g., Alfa, 2020; Nikolou & Darra, 2018; O'Bryan & Hegelheimer, 2007). Middlebury College, Duke University, and the University of Wisconsin were some of the pioneering schools that used iPods and podcasts to enhance the effectiveness of foreign language learning (McMinn, 2008). Based on different developers, Sze (2007) categorized podcasts to two types: "radio podcasts" derived from existing radio programs and "independent podcasts" produced by individuals and organizations (p. 117). In the following section, podcasts in language education will be presented by two categories, namely, receptive and productive podcasts.

Receptive podcasts: The first of the two types is the existing podcast programs that can be downloaded by students to their personal electronic devices. Basically, the podcasts for English language learning comprise the following types: "comprehensive, whole lessons, vocabulary, idioms, conversations, jokes, songs, phonetics, pronunciation, stories, and listening comprehension" (Sze, 2007, p. 118).

In O'Bryan and Hegelheimer's (2007) study, they illustrated some ESL podcasts for language learning, such as ESL Pod, English Idioms and Slang, and Learn a Song Podcast. They have also argued for integrating podcasts into language curriculum using a smooth and invisible way but not replacing the traditional teaching methods. Accordingly, they introduced a CALL integration sample curriculum. In the end, the instructor's and students' (n = 6) attitudes towards this new course were analyzed through a reflective journal and structured interviews and surveys, and the results showed positive attitudes from both instructor and students.

Other research (Abdous, Camarena, & Facer., 2009; Tai, 2012) tended to discover learners' language development from a holistic point of view, which included listening, speaking, reading, and writing four linguistic skills. Abdous et al. (2009) recruited 113 participants and compared the impact of the supplemental use of podcasting and the integrated use of podcasting in foreign language (i.e., German, French, Japanese, and Spanish) classrooms. According to the data collected from a survey at the end of a semester, 93% of the students downloaded the podcasts for assignments or projects, and only a few students used iPods or MP3 to listen to the course materials. In general, a great number of students reported positive effects of podcasts on their academic learning, and their L2 listening skills were mostly improved. In Tai's (2011) study, the author proposed applying mobile devices to task-based language learning with the aim of providing meaningful interaction, effective negotiation, authentic input, and fluent, accurate and complex output. Thirty-five sixth grade Taiwanese students were asked to complete a problem-solving task. A survey was also distributed to analyze their attitudes towards this MALL task. The results indicated that there was a significant improvement of learners' language skills, and they all showed positive attitudes towards this activity. However, their overall language proficiency was only measured by a set of written based grammar and sentence structure tests.

However, there are no empirical studies showing the exact amount of listening and instructional time that makes students become successful podcast listeners. Most often, researchers suggested integrating podcast activities into the entire course curricula (e.g., Copley, 2007; Herringtion & Kervin, 2007). In Abdous et al.'s (2009) and O'Bryan and Hegelheimer's (2007) studies, the podcast activities went through the whole semester, but owing to students had the freedom to listen to the podcasts after class, the podcast listening time may vary. Researchers, therefore, did not trace the exact amount of time that students spent on listening to the podcasts outside the classroom.

Productive podcasts: The second type of the educational podcasts is teacher- or student-created podcasts. Teachers could video or audio record their courses and make them accessible to their students or people around the world. By doing so, students who were slow note takers, missed classes, or failed to understand certain difficult topics could go over the course podcasts after classes as many times as they wanted (Matthew, 2008). From teachers' perspectives, podcasted lectures could assist their teachings and provide opportunities for them to reflect on their own teachings (Card, Polin, Parra, Rhoads, & Sartori, 2006; Miller, 2006). As Matthew (2008) suggested, there were several factors that contributed to the successful use of podcasted lectures, such as students' technical skills, course levels, and academic settings. For instance, students with higher technical background, such as computer science students, would have fewer difficulties manipulating the technical functions than students majored in political science. Also, upper-level courses and larger schools tended to be more suitable for podcasted lectures. Therefore, podcasted lectures were mostly used in the courses that contained a large number of students.

Student-created podcasts have been proposed as a great learning method to improve language learners' speaking skills. As discussed in Sze's (2007) article, student-created podcasts had the advantages of (a) enhancing learners' motivations by knowing their podcasts would be listened to by other students or even people around the world, (b) improving learners' pronunciation though practice and rehearsal, (c) developing learners' collaboration skills when working in groups, (d) improving learners' accuracy, and (e) providing more language practice opportunities for the learners who had fewer chances to practice in a large classroom setting or who were less confident to speak in front of other students. However, most of these assumptions have not been tested empirically. In 2008, Lord analyzed Spanish language learners' (n = 19) pronunciation and pronunciation attitudes improvement after making their own podcasts. During the treatment, participants created six podcasts (e.g., tongue twisters, short readings, and personal reflection on their own pronunciation) in groups of three or four over a semester. At the same time, they were required to give comments on each other's podcasts. In this study, the pronunciation construct was operationalized as the average score given by three raters on a 5-point scale of pronunciation ability. The results showed that both pronunciation abilities and attitudes towards pronunciation have significantly improved when comparing their performance before and after the treatment. Another empirical study on the topic of student-created podcasts was conducted by Ducate and Lomicka (2009). They compared German and French language learners' (n = 22) improvement on pronunciation after making eight student-created podcasts (i.e., five scripted podcasts and three extemporaneous podcasts) over one semester. The subconstructs of pronunciation were accentedness and comprehensibility, scored by two raters in each language on two separate 5-point scales. After comparing participants' pre- and post-test results, no significant improvement on pronunciation was found.

Factors in Listening Passages Influence Learners' Listening Comprehension

There is not a single definition of L2 listening. Wolvin and Coakley (1985) defined listening as "the process of receiving, attending to, and assigning meaning to aural stimuli" (p. 74). According to Rubin (1995), listening refers to "an active process in which listeners select and interpret information which comes from auditory and visual clues in order to define what is going on and what the speakers are trying to express" (p. 7). Listening is an important skill that appears at "the early stages of L2 learning" (Nation & Newton, 2009, p. 37), enables learners to receive language input, and facilitates the emergence of other language skills (Vandergrift & Goh, 2012). However, it has also been described as "the least understood and most overlooked of the four skills (Listening, Speaking, Reading, and Writing)" (Nation & Newton, 2009, p. 37). The continuous speech-flow made the listening comprehension process harder to analyze than reading comprehension. For instance, in the real spoken language, listeners could not slow it down or break it down into manageable chunks. Also, in most of the L2 listening class, although pre-listening activities were provided, once the audio or video began, students are expected to complete the listening task without any help along the way. The strategies on how to manage the listening input were seldom taught (Vandergrift & Goh, 2012). There are different ways to track learners' listening strategies (e.g., think-aloud and recall). In recent years, due to the development of computer technology, researchers started to use technical functions, such as speech rate (Zhao, 1997), pause, rewind, and fast-forward (Roussel, 2011), to indicate learners' self-regulated listening strategies.

In light of the possible factors in listening passages that influence listening success, Buck (2001) and Vandergrift and Goh (2012) have provided very comprehensive summarizations. Specifically, the statistically significant predictors of listening comprehension in listening passages that have been empirically tested are vocabulary, syntax/grammar, topics, speech rate, pronunciation/intonation, gender, and multiple media tools (Kang, 2015).

In conclusion, podcasts, along with MALL, have been promoted as promising tools in L2 learning, whereas quality indicators of such existing podcast programs in *iTunes* have not been generalized. Applying previous empirical results on predictors of listening comprehension in listening passages, the purpose of this research was to identify some effective English language learning podcast programs in iTunes and build up a list of the quality indicators for program developers and subscribers. Specifically, this study was conducted to explore the following two research questions: (a) From learners' perspectives, what are their favorite podcasts in iTunes, and what factors influence their preferences? and (b) From teachers' perspectives, what are their favorite podcasts in iTunes, and what factors influence their preferences?

METHOD

Participants

Students who attended this study were 26 ESL learners at a southwestern American university. They were pre-college undergraduate students, from ages16 to 32, and most of them are from Brazil, China, and Saudi Arabia. There are five levels in this ESL program. Level 1 is the lowest, and Level 6 is the highest. Level 4 is considered the intermediate level at this institution. To be assigned in this level, students have to meet at least one of the three criteria: a score between 45 to 56 out of 120 on the placement and the exit tests at the beginning and end of each semester, an iBT TOFEL score between 45 to 56, or the completion each course in Level 3 in the previous semester. In this study, there are two sections in Level 4: Section A ($n = 12$) and Section B ($n = 14$). Each week, learners from both sections take 24 instruction hours including six hours of Listening and Speaking class, six hours of Reading and Writing class, six hours of Content-based Instruction class, two hours of Writers' Workshop class, two hours of Readers' Workshop class, and two hours of Computer-Assisted Language Learning (CALL) class. This study was conducted during their CALL classes.

The two Level 4 Listening and Speaking teachers also participated in this study. Teacher A, from America, was a part-time instructor at this ESL program. Teacher B, from China, was a doctoral student majoring in Applied Linguistics and working as teaching assistant at this ESL program.

Procedures

Prior to learners' evaluations on different podcasts, the author selected six podcasts from the *iTunes* Store. This selection process is as follows. First, the author searched the keywords "language course English" and "language learning English" in *iTunes* Store. During this searching process, podcast programs that had less than five episodes or that had episodes shorter than two minutes in length were excluded. Second, the top 10 most popular and highly rated podcast programs were selected based on the following criteria. In *iTunes* Store, users could provide their feedback by rating the podcasts with a maximum score of five. The criteria used for selecting the best and most famous podcasts were ratings higher than four, and the number of raters larger than 47. Last, podcasts that had British accents or bad sound quality were excluded. Thus far, the podcast quality indicators that have been considered during this preliminary process were the length of each episode, the number of the episodes, users' rating scores, the number of the ratings, accents, and sound quality. Then, each of the six remaining podcasts was evaluated by four or five students, and the two teachers provided their perceptions on the six selected podcasts.

Instruments

Overall, three instruments were developed in this study: (a) the podcasts coding scheme, (b) the questionnaire for the students, and (c) the questionnaire for the teachers.

Podcast coding scheme: Each English language learning podcast was coded by title, year of the first episode published, a brief description of the content, range of the episode length in minute, video/audio, number of the episodes, gender of the host, accent, rating, and number of the raters. Table 1 presents all the coding schemes and their detailed descriptions.

Questionnaires for students: This questionnaire took about 15 minutes to complete. First, students followed the instructions on the second page to download the required episode. Then, they listened to two minutes of the episode and completed the two questions on the first page. In the first question, students gave an overall score of the podcast from 6 (highest) to 1 (lowest). In the second question, they checked reasons for why they liked the podcast. As shown in Table 2, the items in this question were created based on the predictors of listening comprehension in listening passages that have been identified in previous empirical studies. However, since the six initially selected podcasts did not include any video podcasts, the multimedia predictor was excluded.

Table 1. Podcasts Coding Schemes and Descriptions

Coding Schemes	Descriptions
title	title of the podcast
1st episode	the year of the first episode published
content	a brief description of the content
length (min)	the range of the episode length in minute
video/audio	"video" = video only; "audio" = audio only; "video > audio" = more video than audio; "video < audio" = more audio than video
# of episodes	total number of episodes posted in *iTunes*
host gender	the gender of the host(s)
accent	"UK" = British English accent; "US" = American English accent
rating/5	in *iTunes*, the rating of the podcast, out of 5
# of raters	in *iTunes*, the number of the raters that rated the podcast

Table 2. Predictors and Corresponding Questions

Predictors	Questions
vocabulary	a. The vocabulary words are easy to understand.
syntax/grammar	b. The grammar structures are simple.
prior knowledge/topic	c. The topics are interesting. & d. Students are familiar with the topic(s).
# of topics	e. It includes a lot of topics. & f. It only focuses on a few topics.
speed	g. The speed is slow. & h. The speed is natural.
pronunciation/intonation	i. The speaker(s) pronounce words clearly. & j. The speaker(s)' use appropriate intonation.
gender	k. The speaker(s) should be female. & l. The speaker(s) should be male.

Questionnaires for teachers: To complete this questionnaire, the author met with each teacher for about 25 minutes. After listening to two minutes for one episode in each of the six podcasts, teachers ranked their most favorite podcast, from 1 to 6, and checked for the factors that they have considered. It is worth mentioning that the teachers and the students listened to the same episode. Also, the factors provided in the second section of this questionnaire were identical to the factors in the students' questionnaire (see Table 2). At the end, teachers briefly explained their responses, including the advantages and disadvantages of each podcast, to the author.

RESULTS

The process of selecting the six initial podcast programs was presented in Figure 1. First, excluding the podcast programs that had less than five episodes or that had episodes shorter than two minutes in length, a total of 34 English language learning podcasts were identified and coded based on the coding scheme. Then, the top 10 most popular and highly rated podcasts were selected. Last, four podcasts that had British accents and bad sound quality were taken off from the top 10 podcasts list. Therefore, six podcasts were selected for students and teachers to evaluate, and one typical episode from each podcast was chosen (see Table 3). Students' and teachers' responses to the questionnaires will be discussed in the following section.

Figure 1. The process of selecting six initial podcasts

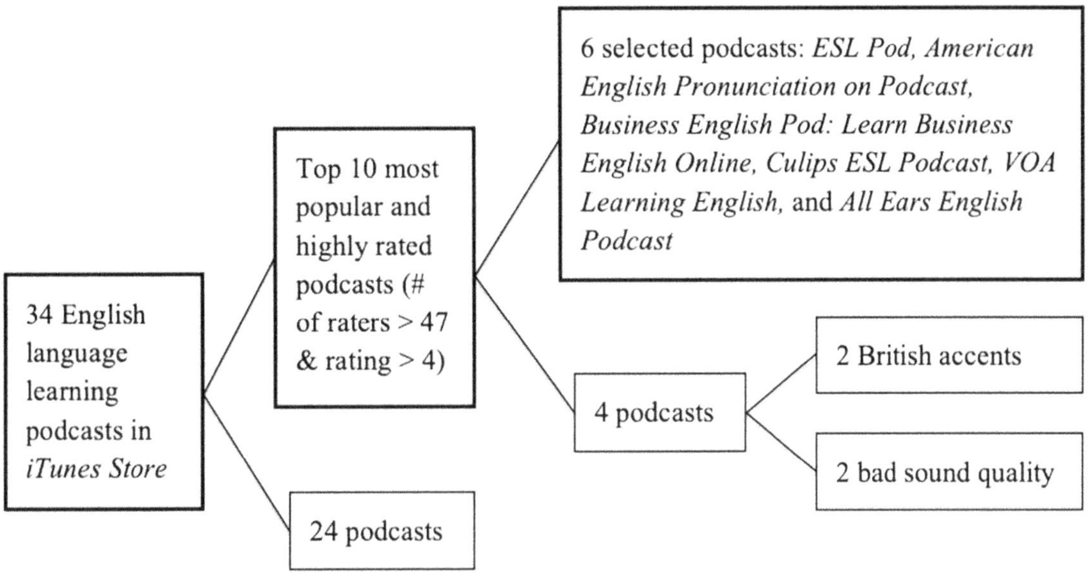

Table 3. Six podcasts and the episode used in this study

Podcast Program	**Episode**
ESL Pod	*Podcast 8 – Asking clarification questions*
American English Pronunciation on Podcast	*168: Stress pronunciation patterns in 3-syllable words*
Business English Pod: Learn Business English Online	*BEP 227 – Telephoning: Making sales calls (Part 1)*
Culips ESL Podcast	*Working around the clock*
VOA Learning English	*Learning English broadcast – March 26, 2014*
All Ears English Podcast	*AEE episode 3: When should you study English?*

Students' Responses

As shown in Table 4, each podcast was evaluated by four or five students, and the mean scores and standard deviations are listed in the last two columns. The top-three scored podcasts were *VOA Learning English* ($M = 5.25$; $SD = .96$), *All Ears English Podcast* ($M = 5.00$; $SD = .82$), and *American English Pronunciation on Podcast* ($M = 4.60$; $SD = .89$).

Additionally, the students' selections of their preference criteria are presented in Table 5. The items that more than half of the students ($n > 13$) selected as quality indicators of podcasts were (a) a. the vocabulary words in the podcast are easy to understand ($n = 22$), (b) i. the speakers pronounce words clearly ($n = 19$), (c) c. the topics are interesting ($n = 19$), (d) h. the speed is natural ($n = 18$), and (e) b. the grammar structures are simple ($n = 14$). Meanwhile, students did not show a specific preference on the gender of the speakers (k. I like female speakers ($n = 11$). l. I like male speakers ($n = 10$).). In summary, vocabulary, pronunciation, topics, speed, and grammar structures were the top five quality indicators that students considered for their preference of different podcasts.

Table 4. Students' Evaluation on Different Podcast Programs (N = 26)

Podcast Program	***n***	***M***	***SD***
ESL Pod	5	3.80	1.10
American English Pronunciation on Podcast	5	4.60	.89
Business English Pod: Learn Business English Online	4	4.50	1.00
Culips ESL Podcast	4	4.25	.96
VOA Learning English	4	5.25	.96
All Ears English Podcast	4	5.00	.82

Table 5. Students' Podcast Preference Criteria (N = 26)

Items	***n***	**Rank**
a. The vocabulary words are very easy to understand.	22	1
b. The grammar structures are simple.	14	4
c. The topics are interesting.	19	2
d. I am familiar with the topic(s).	10	7
e. It includes a lot of topics.	6	10
f. It only focuses on a few topics.	9	8
g. The speed is slow.	8	9
h. The speed is natural.	18	3
i. The speaker(s) pronounce words clearly	19	2
j. I like the speaker(s)' intonation.	13	5
k. I like female speakers.	11	6
l. I like male speakers.	10	7

Teachers' Responses

As shown in Table 6, *Culips ESL Podcast* and *VOA Learning English* were the two podcasts that both teachers favored. The specific advantages of these two podcasts mentioned by the two teachers are "natural sounding" (Teacher A) and "appropriate speed and topic" (Teacher B) for *Culips ESL Podcast* and "authentic news" (Teacher B) for *VOA Learning English*. Also, both of them disfavored *American English Pronunciation on Podcast* and *Business English Pod: Learn Business English Online*. For the *American English Pronunciation on Podcast*, Teacher A stated that "face-to-face instruction is better than using podcasts to teach pronunciation," and for the *Business English Pod: Learn Business English Online*, both teachers thought the topic might be suitable for students majoring in business but not as suitable for their Level 4 students, many of whom would not major in business. Teachers' preferences of the *ESL Pod* and *All Ears English Podcast* showed larger discrepancies. Teacher A liked *All Ears English Podcast* because it "provided learners different learning strategies," whereas Teacher B disliked it because "it was too casual." For *ESL Pod*, Teacher B thought "the topics were suitable for the academic purpose," but Teacher A disliked it because "the speed was too slow."

In addition, Table 7 summarizes the factors that influenced the teachers' preferences on different podcasts. The items that both teachers have checked in the questionnaire were (a) c. the topics are interesting, (b) h. the speed is natural, and (c) i. the speaker(s) pronounce words clearly. Therefore, topics, speed, and pronunciation were the three quality indicators with which both teachers were concerned.

Table 6. Teachers' Evaluation on Different Podcast Programs

Podcast Program	Teacher A Rank	Teacher B Rank
ESL Pod	4	1
American English Pronunciation on Podcast	5	4
Business English Pod: Learn Business English Online	6	5
Culips ESL Podcast	1	3
VOA Learning English	2	2
All Ears English Podcast	3	6

Table 7. Teachers' Podcast Preference Criteria

Items	Teacher A	Teacher B
a. The vocabulary words are very easy to understand.		√
c. The topics are interesting.	√	√
d. Students are familiar with the topic(s).	√	
f. It only focuses on a few topics.	√	
h. The speed is natural.	√	√
i. The speaker(s) pronounce words clearly	√	√

DISCUSSION

The results of this study provided empirical evidence on students with intermediate English language proficiency and their teachers' preferences of different language podcasts and the podcast quality indicators with which they were concerned. Students liked the *VOA Learning English*, *All Ears English Podcast*, and *American English Pronunciation on Podcast* programs. They preferred the podcast episodes that have understandable vocabulary words, interesting topics, natural speed, and simple grammar structures. Factors such as the gender of the speakers and the number of topics did not influence students' preferences on different podcast programs. In terms of the two teachers, *Culips ESL Podcast* and *VOA Learning English* were their two favorite podcasts; interesting topics, natural speed, and clear pronunciation were the top-3 factors that they considered when selecting appropriate podcasts for their students. In summary, the podcast programs that both students and teachers preferred were *VOA Learning English*, and both groups preferred podcasts that had interesting topics, natural speed, and clear pronunciation.

However, it is worth noting that there were great discrepancies between learners' and teachers' responses and between the two teachers' responses. For example, *All Ears English Podcast* and *American English Pronunciation on Podcast* were rated as the second and third favorite podcast by the students, whereas these two podcasts were disliked by the two teachers. *Culips ESL Podcast* was ranked very high by the two teachers but not favored among the students. In terms of the quality indicators, students preferred the podcast programs that had simple vocabulary words and grammatical structures, but these two indicators were not considered an essential factor that affected teachers' preferences on different podcast programs. Therefore, when selecting and designing the favorite podcast programs for students, the teachers and developers should not assume students' preferences since the preferences among various groups can be very different. Meanwhile, even between the two teachers, their preferences of *ESL Pod* and *All Ears English Podcast* differed, which indicates that teachers with different backgrounds and understanding of the appropriate podcast programs for their students varied from person to person.

FUTURE RESEARCH DIRECTIONS

MALL and using podcasts in language education are two promising topics in the field of linguistics. More empirical studies on different MALL programs could help educators and program developers summarize the success or drawbacks in the existing language learning programs to help them better select and develop effective and appropriate programs for language learners. Owing to the complexity of the students' backgrounds and learning contexts, future researchers could expand on this study by including students from different proficiency levels, age groups, or learning environments to make the understanding of the effectiveness of certain MALL programs more comprehensive.

CONCLUSION

To conclude, the factors that students considered when selecting their favorite podcast programs included the length of each episode, the number of the episodes, users' rating scores, the number of the ratings, accents, sound quality, topics, and the speed. Pedagogically, in attempting to choose the appropriate podcast programs, teachers should take students' preferences and concerns into consideration. Certain quality indicators, such as the length of each episode, speech rate, sound quality, and the podcast topics should be carefully considered when designing podcast programs. Further, depending on different learning contexts and student backgrounds, the selection and development of the podcast programs might need to be altered accordingly, which still requires more future empirical analyses.

REFERENCES

Abdous, M., Camarena, M., & Facer, B. (2009). MALL technology: Use of academic podcasting in the foreign language classroom. *ReCALL, 21*(1), 76–95. doi:10.1017/S0958344009000020

Alfa, R. R. (2020). Using podcast as authentic materials to develop students' speaking skill. *Journal of English Language Teaching*, *4*, 65–74.

Balas, J. L. (2005). Blogging is so last year: Now podcasting is hot. *Computers in Libraries*, *10*, 29–32.

Buck, G. (2001). *Assessing Listening*. Cambridge University Press. doi:10.1017/CBO9780511732959

Card, D., Polin, L., Parra, J., Rhoads, J. B., & Sartori, T. (2006). Can you hear me now: The return of voice to distance learning. Presented at the *Fifth International Conference on Web-based education of the International Association of Science and Technology for Development*, Puerto Vallarta, Mexico.

Chinnery, G. M. (2006). Emerging technologies going to the MALL: Mobile assisted language learning. *Language Learning & Technology*, *10*, 9–16.

Cooper, S., Dale, C., & Spencer, S. (2009). A tutor in your back pocket: Reflections on the use of iPods and podcasting in an undergraduate popular music programme. *British Journal of Music Education*, *26*(1), 85–97. doi:10.1017/S0265051708008280

Copley, J. (2007). Audio and video podcasts for lectures for campus-based students production and evaluation of student use. *Innovations in Education and Teaching International*, *44*(4), 387–399. doi:10.1080/14703290701602805

Ducate, L., & Lomicka, L. (2009). Podcasting: An effective tool for honing language students' pronunciation? *Language Learning & Technology*, *13*, 66–86.

Duman, G., Orhon, G., & Gedik, N. (2014). Research trends in mobile assisted language learning from 2000 to 2012. *ReCALL, FirstView*, 1-20.

Herrington, J., & Kervin, L. (2007). Authentic learning supported by technology: Ten suggestions and cases of integration in classroom. *Educational Media International*, *44*(3), 219–236. doi:10.1080/09523980701491666

Hwang, G. J., & Tsai, C. C. (2011). Research trends in mobile and ubiquitous learning: A review of publications in selected journals from 2001 to 2010. *British Journal of Educational Technology*, *42*(4), E65–E70. doi:10.1111/j.1467-8535.2011.01183.x

Indahsari, D. (2020). Using podcast for EFL students in language learning. *JEES*, *5*(2), 103–108. doi:10.21070/jees.v5i2.767

Kang, T. (2015). The effectiveness of multiple media tools in L2 listening: A meta-analysis. In V. Turel (Ed.), *Intelligent design of interactive multimedia listening software* (pp. 255–284). IGI Global. doi:10.4018/978-1-4666-8499-7.ch010

Kondo, M., Ishikawa, Y., Smith, C., Skamoto, K., Shimomura, H., & Wada, N. (2012). Mobile assisted language learning in university EFL courses in Japan: Developing attitudes and skills for self-regulated learning. *ReCALL, 24*(2), 169–187. doi:10.1017/S0958344012000055

Kukulska-Hulme, A., & Shield, L. (2008). An overview of mobile assisted language learning. *ReCALL, 20*(3), 271–289. doi:10.1017/S0958344008000335

Lord, G. (2008). Podcasting communities and second language pronunciation. *Foreign Language Annals, 41*(2), 364–379. doi:10.1111/j.1944-9720.2008.tb03297.x

Matthew, R. (2008). Adventures in podcasting. *PS, Political Science & Politics, 41*(03), 585–593. doi:10.1017/S1049096508080797

McMinn, S. W. J. (2008). *Podcasting possibilities: Increasing time and motivation in the language learning classroom*. Retrieved November 24, 2012, from http://www.eife-l.org/publications/proceedings/ilf08/contributions/improving-quality-of-learning-with-technologies/McMinn.pdf/view

Miller, D. B. (2006). Podcasting at the University of Connecticut: Enhancing the educational experience. *Campus Technology*. Retrieved November 24, 2012, from https://campustechnology.com/

Nation, I. S. P., & Newton, J. (2009). *Teaching ESL/EFL listening and speaking*. Routledge.

Nikolou, S., & Darra, M. (2018). The use and development of podcasting as a technological tool in secondary education in Greece: A case study. *International Education Studies, 11*(11), 109–109. doi:10.5539/ies.v11n11p109

O'Bryan, A., & Hegelheimer, V. (2007). Integrating CALL into the classroom: The role of podcasting in an ESL listening strategies course. *ReCALL, 19*(2), 162–180. doi:10.1017/S0958344007000523

Polishook, M. (2005). Handheld composing: Reconceptualizing artistic practice with PDAs. In A. Kukulska-Hulme & J. Traxler (Eds.), *Mobile learning: A handbook for educators and trainers* (pp. 133–138). Routledge.

Regan, M., Nash, J., & Licata, D. (2000). *Mobile learning*. Retrieved August 13, 2011from http://sll.stanford.edu/projects/mobilelearning

Roussel, S. (2011). A computer assisted method to track listening strategies in second language learning. *ReCALL, 23*(2), 98–116. doi:10.1017/S0958344011000036

Rubin, J. (1995). An overview to a guide for the teaching of second language listening. In D. J. Mendelsohn & J. Rubin (Eds.), *A Guide for the Teaching of Second Language Listening* (pp. 7–11). Dominie Press, Inc.

Sharples, M. (2000). The design of personal mobile technologies for lifelong learning. *Computers & Education, 34*(3-4), 177–193. doi:10.1016/S0360-1315(99)00044-5

Sze, P. M. (2007). Developing students' listening and speaking skills through ELT podcasts. *Education Journal, 34*, 115–134.

Tai, Y. (2012). Contextualizing a MALL: Practice design and evaluation. *Journal of Educational Technology & Society, 12*, 220–230.

Vandergrift, L., & Goh, C. C. M. (2012). *Teaching and learning second language listening: Metacognition in Action*. Taylor & Francis. doi:10.4324/9780203843376

Wolvin, A. D., & Coakley, C. G. (1985). *Listening*. Willian C. Brown Company.

Wu, W. H., Wu, Y. C., Chen, C. Y., Kao, H. Y., Lin, C. H., & Huang, S. H. (2012). Review of trends from mobile learning studies: A meta-analysis. *Computers & Education*, *59*(2), 817–827. doi:10.1016/j.compedu.2012.03.016

Zhao, Y. (1997). The effects of listeners' control of speech rate on second language comprehension. *Applied Linguistics*, *18*(1), 49–68. doi:10.1093/applin/18.1.49

KEY TERMS AND DEFINITIONS

Listening Comprehension: Second language learners' receptive oral skill, or to what extent a listening passage has been understood.

Mobile-Assisted Language Learning (MALL): Any form of linguistic teaching or study that involves a portable device (e.g., mobile phones, tablet computers, and MP3 players).

Podcast: It is a portmanteau which combines the "pod" from iPod, an MP3 player produced by Apple Inc., with "broadcast." It can be uploaded and downloaded by subscribers through *iTunes* in the format of audio or video files.

Productive Podcasts: Teacher- or student-created video or audio programs. For example, teachers upload their course recordings to make them accessible to their students; students can upload their own video or audio recordings to practice their speaking skills.

Receptive Podcasts: Video or audio programs that can be subscribed or downloaded by the students to their personal electronic devices to improve their listening skills.

Chapter 5
The Effect of Audio and Video Modality on Perception of Reduced Forms:
The Role of Web-Based Instruction – Reduced Forms in an L2 Listening Context

Kaine Gulozer
Bartin University, Turkey

Zeynep Kocoglu
Yeditepe University, Turkey

ABSTRACT

Reduced forms (RFs) spoken by native English speakers have been challenging on the part of the second language (L2) learners. This chapter aims to address suprasegmental features to Turkish preparatory language school students in relation to L2 listening comprehension. Considering the limited research on RFs in learning English as a L2 context, this pre-test post-test control group design study aimed to explore whether the instruction of five RFs in sentential level results in any difference in listening comprehension test performance. The five forms entail contraction, assimilation, flap, elision, and linking. A total of 343 were recruited, and RFs instruction was delivered through the web page designated for the study for five weeks, and the performance of the eight groups was measured twice throughout the study. The findings indicated that sentence level of RFs instruction through web-based learning facilitates the listening comprehension of RFs.

DOI: 10.4018/978-1-7998-7876-6.ch005

INTRODUCTION

Today's intelligent design of multimedia enables people to interact and communicate with other people across the globe through various apps and social media platforms. The use of multimedia in the field of English Language Teaching and its inevitable integration into the English as a Second Language (ESL) / English as a Foreign Language (EFL) classes have entailed the adoption of a variety of instructional practices into the skill-based teaching processes. Given the four components of language skills as reading, writing, and speaking, listening is high of great importance at the outset of second language acquisition as well as the perception of natural speech. Furthermore, it is the listening skill that allows listeners the right conditions not only for language acquisition but also for the development of other language skills (Hasan, 2000). The power of listening is illustrated by Feyten (1991), stating that listening provides the highest proportion of our total communication as 45%, speaking 30%, reading 16%, and writing 9%, respectively. However, this is "the least understood and the least researched skill" compared to the other language skills. (O'Bryan & Hegelheimer, 2009; Vandergrift, 2004, 2007).

In relation to language learners' development of listening skills, the notion of suprasegmental features plays a crucial role. The features are the integral parts of the language components such as contraction, assimilation, flap, elision, and linking accompanying consonants and vowels. Because these features include syllables, words, or phrases, and they are not restricted to single sounds, the perception of the natural speech along with a single, or several enunciated features is highly challenging on the part of the learners to perceive and segment the words within a sentence. Because the spoken discourse is quite fast, the listener is supposed to perceive and process it online. Therefore, instruction designed particularly on RFs would be a plausible option to improve L2 learners listening development.

In relation to the incorporation of RFs in instructional materials, the negligence of both studies and materials on RFs is clearly highlighted by several scholars such as Brouwer (2010), Ur (1984), Rosa (2002). To our knowledge, so far, no comprehensive research study has been conducted incorporating several RFs through idea units in sentences. Rosa (2002) stresses that "very little material is available on the systematic use of RFs" (p. 57). An examination of the materials teaching listening and pronunciation reveals a dearth RFs (Celce-Murcia; 2011; Lodge, 2009, Roach, 2009; Shin, Sung, 2019). If they are covered at all, most provide a few of them, not a comprehensive of all information regarding RFs. Furthermore, almost in all of them "flap", one of the RFs is missing. The coverage of the RFs in most ESL materials is composed of the amalgamation of RFs, yet various practice samples of flap, a comprehensive one including all in one instructional material, and the linguistic and/or pragmatic constraints of usage are not specifically addressed. Furthermore, flap is identified as the most difficult one to recognize by ESL learners (Matsuzawa, 2006). This is an area of research that is definitely limited with respect to covering five RFs, namely, contraction, assimilation, flap, elision and linking. The RFs illustrated above have been indicated by the research (Rosa, 2002; Brown, 2006; Brown et al., 2006; Matsuzawa, 2006; Aquil, 2012; Underwood, 2012) that L2 learners are constrained to identify words in a speech that causes obstacle in listening comprehension as well.

The integration of web-based tools into L2 listening instruction has brought about numerous advantages for learners' listening development. A web-based program on RFs instruction may be considered to be helpful for learners whose first language is Turkish for several reasons. First, the nature of Turkish is agglutinative with its inflectional, derivational morphology, vowel harmony as well as free word order of main constituents (Saygin & Wilson, 2010). Turkish bearing this topological difference from English and many other commonly spoken European languages, it is of great importance to conduct a study on the perception of RFs for language learners whose first language (L1) is Turkish. Secondly, what is notable in Turkish is that the verb stands out as an acoustic and a salient unit while one is listening even if she or he does not know Turkish (Kuntay & Slobin, 2002). Yet this is not the case in English due to the distinguishing length conducted to the articulation of content and function words in English. It is known that Turkish is a 'syllable timed' and orthographically transparent language in contrast to the stress timing characteristic of English phonology. Hence it is the syllable, not the stress, which is important for the perception of segmenting a word in a sentence in Turkish. Because each syllable in a word occupies the same amount of time, it will simply take equal time to articulate as well. Even though stress elements occur in certain cases in Turkish, it does not happen to occur as content or function word variation, but it is for signaling question, exclamation, or stressing one particular issue within a sentence. Even though stress timing nature in English is questioned in some studies (Barrera-Pardo, 2008), the general consensus in literature is that English is stress-timed language (Marks, 1999).

Another significance of the study is that poor listening ability may result from several factors, such as listeners' lack of vocabulary, lack of employing necessary listening strategies, lack of teaching methodologies, lack of emphasis on listening, but the necessarily important one is the notion of RFs that poses a substantial obstacle in parsing the elements of an uttered sentence. Yet, it might be a possible solution to the problem of implementing e-learning along with the face to face instruction in a classroom. This research, "the effect of RFsI on L2 listening comprehension", will contribute to teachers, material developers, policy makers, program designers to devise ample techniques for incorporating RFs instruction (RFsI) into the classroom, leading toward the comprehensibility of natural native English speech. Furthermore, it will enforce the importance of placing appropriate teacher training and development opportunities in the field of phonology instruction because this study will enlighten the crucial role of teacher development in their future language instruction as well. What is more, not all language teaching materials include authentic ones, which might hinder awareness of phonological modifications on the part of the learners. For example, watching TV shows or listening to an inauguration speech given by a president at a university, or listening to the lecturer in a conference given by a native speaker include a range of reduced forms that would cause disruption of listening comprehension on the part of the listener.

It is also of great importance to consider the context of diverse communicative practices and recent social relations in globalization when we look through English as a Lingua Franca (ELF). Canagarajah (2014) remarks that English is highly elucidated as setting up situational norms in specific contexts of interaction. The specific context of interaction in discrete varieties of English requires update and explicit pedagogical practice as it is more significant in accounting for one's listening comprehension development. In this regard, this study will contribute to the phonological variety in relation to its pedagogical developments in teaching ELF as well.

Last but not least, the additional benefit of RFsI is not limited to the classroom context but also beyond the classroom context (Underwood, & Wallace, 2012). Language learners will raise awareness in regard to RFs and produce comprehensible spoken English. This current research will shed light on learners' listening difficulties, pinpointing the places where comprehension is impeded on the part of L2 learners and enable teacher trainers, pre-service and in-service language teachers, testers, and material developers to understand the phenomenon of RFs clearly in language instruction, thereby endorse them to teach RFs in language courses (Ito, 2006).

In brief, further research is needed to determine whether RFsI has a positive or no effect on the perception performance of the mentioned features. Additionally, which RFs are critical to perceive and which modality of listening (audio without text or video without text) is also more critical to facilitate in recognition of RFs on the part of the learners. Assuming segmental features are perceptible, teachable and learnable to listeners, it can be hypothesized that promotion of RFs to L2 learners can further aid language learners to identify the stream of speech and in turn, develop their listening comprehension. Yet, will thc instruction of those characteristics of speech through web-based learning help L2 learners improve their perception of those features and improve L2 learners' self-regulation skills in listening comprehension. This is the fundamental question that stimulated the present research: what feature/s of RFs will be improved most by instruction?

This study aims to investigate the impact of RFs instruction on Turkish EFL learners' comprehension of RFs, namely, contraction, assimilation, flapping, elision, and linking displayed at a sentence level. Additionally, the researcher aims to examine the contribution of a web-based environment, developed by the researcher as a platform to let participants perceive RFs in two modalities: audio versus video through the designated web page entitled as "CONSPERsent", the acronym which stands for "CONnected Speech PERception through SENTtences. The ultimate objective is to contribute to the development of listening comprehension on the part of L2 English language learners.

The following research questions are formulated for the aim of the study:

1. Is there any significant difference between the experimental (+RFsI) and control (-RFsI) groups in terms of the comprehension of RFs, namely contraction, assimilation, elision, flap and linking after the treatment?
2. Is there any significant difference between AHP+RFsI group and AHP-RFsI group in terms of the comprehension of RFs?
3. Is there any significant difference between VHP+RFsI and VHP -RFsI groups in terms of the comprehension of RFs?
4. Is there any significant difference between ALP+RFsI and ALP-RFsI group in terms of the comprehension of RFs?
5. Is there any significant difference between VLP+RFsI and VLP-RFsI groups in terms of the comprehension of RFs?

Background

RFs are prevalent in natural speech, making the utterance enunciated altogether. Celce-Murcia, Brinton & Goodwin (2011) defines RFs as an idea unit or thought group that is a semantically and grammatically cohesive unit of discourse. In this vein, an utterance can be separated into an idea unit when a speaker can normally pause in the real-time of speech. To illustrate the phenomena in a sentence: "I hope you won't open the window when it's so cold and snowing" two idea units occur: "you won't open the window" and "it's so cold and snowing". However, in the sentence: "The cat, with which I used to travel, is gone", the only idea unit can be considered as "I used to travel" because the speaker is supposed to give a logical pause before the comma. It is reported that all languages have a range of RFs employed (Rosa, 2002), the reason of which is attributed to the fact that not every word of utterance in natural speech is fully articulated (Aquil, 2012; Avery & Ehrlich, 1992; Celce-Murcia, Brinton & Goodwin, 2011; Roach, 2009). Rather, the speech organs articulate sounds together to save time and energy instead of forcing speech organs to take a new position for each sound. Hence, a variety of phonological modifications occur in naturally spoken English, which causes different pronunciation of words from their phonetically pronounced forms (Buck, 2001; Mitterer, Kim & Cho, 2013; Zielinski, 2007).

The following sentence will demonstrate the phenomena of the perception in a stream of speech, where some units of the sentence undergo a number of RFs that would cause a constraint to perceive on the part of L2 learners:

"Did you tell him what you saw" (Avery, Ehrlich, 1992). (pp. 87-88)

/dɪdʒə tɛl əm watʃə sa/

1. The final /d/ of 'did' and the initial /y/ of 'you' are pronounced as the consonant /dʒ/ (assimilation).
2. The vowel of the word 'you' is reduced to schwa in the two instances. The two words 'did' and 'you' are pronounced as a single syllable: /dʒə/ (elision)
3. The word 'him' is reduced to /əm/, the initial consonant being dropped and the vowel being reduced to schwa (elision).
4. The final /t/ of 'what' and the initial /y/ of the 'you' are pronounced as the consonant sound /tʃ/ (assimilation).

Each of the above units may vary in a number of sentences as the RFs it preserve. The stream of speech will be hard to perceive on the part of L2 learners as the /h/ sound is elided, and the combination of /d/, and /y/ sounds are assimilated within one sentence. Additionally, to my knowledge, because these RFs at the sentential level are not currently available in most language instruction curriculum (Settler & Jenkins, 2005), the need to solve L2 learners' confrontation and the need to train language listeners in this field are obviously called for. Turkish being a syllable-timed language, L2 learners with L1 Turkish would be considerably influenced by the unstressed part of the stream of speech as RFs. Taking these factors into account, learners need sound and explicit instruction to raise their perception of RFs, which might also incidentally contribute to their target-like production of these features.

Cognitive Theory of Multimedia Learning (CTML)

Drawn from several cognitive theories such as Baddeley's model of working memory, Paivio's dual coding theory, and Sweller's theory of cognitive code, the CTML was developed by Richard E. Mayor who focused on promoting student understanding by multimedia instructional practices and the implementation of the cognitive strategies to enable students to learn the subject matter efficiently. Multimedia is defined as the integration of words and pictures through which mental representation is built, and active learning is actualized (Mayer, 2005). In relation to reaching optimum learning in CTML, the words can be presented in spoken or written modality and pictures can be presented through any modality of the graphical image such as illustrations, photos, animations, or videos (Sorden, 2005).

As seen in Figure 1, the framework of the theory is displayed in two sets of rows and five columns that are linked to the following sections via arrows within itself. The two rows represent the two information-processing systems as auditory/verbal system on the top and the visual/pictorial channel on the bottom. The five columns illustrated in the figure are labelled as the modes of knowledge representation that is explained in the following:

1. Physical representation refers to words or pictures that are presented to the learner.
2. Sensory representation refers to the ears and eyes of the learner

3. Shallow working memory representations are like sounds or images attained by the learner
4. Deep working memory representations are verbal and pictorial models constructed by the learner
5. Long term memory representation is the learners' relevant prior knowledge (Mayer& Moreno, 2003)

In relation to the limited capacity assumption, except for the working memory system, long-term memory and the capacity for physical word or picture representation are unlimited. Cognitive processing is represented by the arrows. Learners select words or images, constructs a coherent verbal representation through organizing words or images then integrates the verbal or pictorial model that is activated or connected through the learners' prior knowledge. In brief, active learning through the model is actualized in five cognitive processes as selecting words, selecting images, organizing words, organizing images, and integrating.

Figure 1. Cognitive Theory of Multimedia Learning (Mayer& Moreno, 2003)

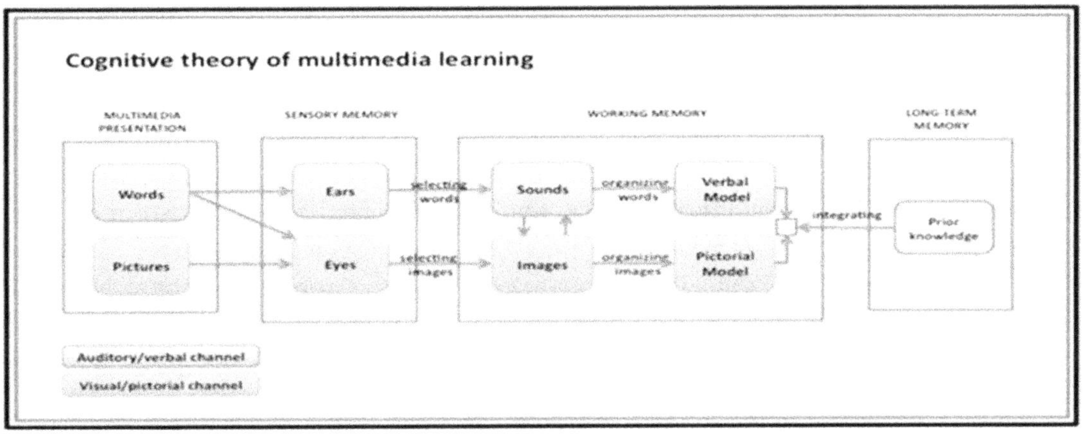

Mayer's (2001) CTML is based on cognitive science that embodies three main assumptions in relation to how the information processing system works. The three assumptions are the dual-channel, the limited capacity and the active processing (Mayer & Moreno, 2003). The first assumption refers to the idea that working memory has subsystems as auditory and visual channels. The second assumption suggests that each subsystem of working memory has a limited capacity, and the third one, the active processing assumption states that learners construct knowledge effectively when they pay attention to the subject matter, rectify and store it into a coherent mental structure, as well as integrate it with their prior knowledge (Mayer, 2009).

Review of Literature on Reduced Forms in L2 Context

The existing body of research on RFs has explored the four main factors along with its emphasis on L2 learners' listening comprehension development: perception, processing, production, and the instruction of various RFs such as contraction, assimilation, flap, elision, and linking.

In relation to the perception of RFs, Henrichsen (1984) aimed to explore how the presence and absence of reduced forms influence the comprehensibility of English input. The results showed that there was a difference between the RFs test scores of native speakers and those of high-level language learners in the presence of RFs, yet the difference was not significant in the condition without the RFs. In another perception study of RFs by Wong, Chung, Leung, Bishop, and Chow (2015), the role of perception of RFs -contraction, juncture, elision, vowel weakening, assimilation, intrusion, flapping, glottalization, and palatalization- on the part of Chinese L1 speakers of English was investigated, employing a battery of tasks such as general listening comprehension, RFs dictation, minimal pair discrimination, speech gating, phonemic awareness, receptive vocabulary, and non-word repetition. The results indicated that the ability to perceive phonologically RFs was a significant predictor of comprehension of English speech by the L2 learners. The researchers also found that receptive vocabulary and part-word recognition were crucial for successful comprehension of English RFs. Based on the phonological representations hypothesis, they further classified the phonological components that were predictive to the RFs perception in regard to their relative importance and its direct or indirect effect. As implied by the study, vocabulary and part word recognition may work independently in underpinning the comprehension of RFs perception.

Even though there has been scarce in studies investigating the effects of RFsI on listening comprehension recently, listeners' performance seems to have improved when they were delivered instruction on these peculiar features (Alameen, 2014; Ahmadian & Matour 2014; Brown, 2005; Brown & Hilferty, 2006; Ito, 2006; Matsuzawa, 2006; Yang, Lin and Chung 2009; Underwood & Wallace, 2012, Shin & Sung, 2019). Brown and Hilferty (1986) investigated the degree to which RFsI was effective for 32 intermediate level of Chinese graduate students. The results showed that RFsI had a significant effect on students' performance of the dictation task. Yet, they found no difference between the two groups in terms of listening comprehension that was measured by the UCLA listening comprehension test. One explanation given for this was that the RFs in this test represented a very small part of this subtest. Nonetheless, the results indicate some evidence that RFsI may not be helpful in overall listening comprehension. In another study dealing with the instruction of RFs, Brown, and Hilferty (2006) investigated the effectiveness of teaching RFs for improving listening comprehension among 32 EFL students in an experimental study in China. The researchers found the beneficial effect on listening comprehension in favor of the experimental group in three measures they utilized an Integrative Grammar Test, UCLA English as a Second Language Placement Examination, listening comprehension subtest, and RFs dictations. In the same vein, introducing the contrastive analysis in relation RFs, Toda (2006) dealt with the instruction of contraction, elision, contraction, and assimilation in Japanese, focusing on the instruction of contrastive analyses in English and found the similarities in both languages in relation to the reduced forms. He suggested that it would be much more beneficial for L2 learners to present contrastive samples of RFs in their L1 so that they can perceive and internalize the RFs. Matsuzawa (2006) also proposed the instruction to improve the comprehension of RFs, selecting ten RFs that seemed challenging for Japanese colleague graduates. The selected categories were assimilation, contractions, deletion, flapping, glottalization, linking, -nt reduction, palatalization, and weak forms of function words. Three versions of dictation tasks with 30 sentences in each one and a retention test were administered to 20 people with no control group in order to identify the performance scores of the reduced forms. The results of the study showed that the students' performance in all chosen categories highly improved.

In an attempt to adopt the audiovisual feedback and audio-only feedback in teaching linking and developing online materials, Alameen (2014) investigated the effectiveness of linking instruction on nonnative speech perception and production. The results indicated that both audio and audio-visual type of training were effective in developing learners' perception and production of linking. However, the method of audiovisual instruction with waveforms yielded better improvement both in perception and production of linking. The audio-visual training also received more positive learner feedback than that of audio-only training. The results also indicated that learners' improvement of linking high-frequency words was higher than that of low-frequency words. Similarly, Yang, Lin, and Chung (2009) moved the way of instruction of RFs from audio to video medium and reported the positive effect of the development of a learning program on comprehension of contraction, assimilation, elision, linking, and hybrid through movies or television programs. They recruited 32 students at the tertiary level, developing an instructional system in five several modules. Students chose an assigned film, the 10th season of an American television series, "Friends", and a subtitle source from the database of the RFs. The subtitles were imported in the Characteristic Judging Module that followed to examine the following: the characteristics of the sentence, word positioning within the sentence, the syllable position in each word, and articulation of new syllables by the nature of a reduced form. Another step to follow was to import the information to a Characteristic Annotation Module to produce an XML file in order to create a sentence pattern database of RFs in conversations. The sentence pattern became a reference resource along with a keyword function for the searching module as well as the Listening Close Module. In brief, the program helped the students observe RFs that they had learned.

In another study, Underwood and Wallace (2012) highlighted the importance of RFsI and investigated the comprehension and productive development of RFs with 52 low-proficiency level undergraduate Japanese students. The results of the study proved a significant increase in comprehension of RFs. In relation to the productive development of RFs, the researchers found out that there was an increase in the production of RFs measured by pre- and post- instruction speaking tests. Employing two experimental and a control group in a pre-test and post-test design study, Ahmadian and Matour (2014) investigated the effect of explicit instruction of RFs on listening comprehension with forty Persian speaking English learners. On the basis of the pre-test and post-test scores assessed through a listening comprehension test and a gap-filling task on RFs, the results indicated that the experimental group outperformed the control group, thereby explicit instruction of RFs was helpful for developing learners' listening comprehension skill. Erbanova (2014) examined Czech students' production of RFs and identified the aspect of the RFs as stress, intonation, and rhythm in English speech in particular. The students' RFs production skill was assessed through two recordings –reading aloud and free speech. The researcher found out that almost all participants doubled their performance in both speaking tasks of RFs and that the instruction was highly effective on the part of the learners.

In literature, there are also some contradicting findings, as shown in Gaskell and Snoeren (2008) study. They carried out two experiments to search whether there is a context effect in assimilation. In their first experiment, 20 undergraduate students were presented a set of spoken sentences, semantically neutral sentential context, with the two response choices as "run vs. "rum" to identify the target word (p.8.). A delayed response by the participants showed that assimilation created lexical ambiguity, which was also led to higher miscommunication rates in the identification task. In the second one, a different group composed of 49 participants were asked to respond to 40 item-sentences that are identical to those utilized in the first experiment with the only difference that each sentence preceded by a semantically biased sentence. The result indicated that there is no context effect in assimilation. This finding contradicts Gow and Im's study (2004), in which there is a context effect in assimilation. The contradicting results are likely due to the perceptual accounts of language-specific knowledge as they conducted a cross-linguistic study dealing with the perception of Hungarian and Korean voicing assimilation.

Flap is another phenomenon of RFs that commonly occurs in a stream of speech. Most of the work regarding flapping has investigated either its perception (Idsardi & Sung, 2003; Tucker, 2007), its production (Eddington & Elzinga 2008), or its nature in the representation of the lexicon (Connine, 2004). Idsardi and Sung (2003) focused on the comparison of acoustics and perception of American English and Korean flaps through a phoneme identification task in their cross-linguistic study. Korean flap is manifested in /r/ sound while American English flap is characterized in /t/ and /d/ sounds. Some words are presented either embedded in either a flap or its variations. Students were asked to identify the initial segment (b or p) of word or non-word variations as in "pretty- betty" (p.187). The results indicated that American English and Korean flaps are identical and have no statistically significant differences in duration and instances of burst. The researchers further suggested that listeners perceive flaps based on their pre-existing representation in their lexical memory, in other words, based on their first language background. In another study dealing with flapping, Tucker (2007) investigated the effect of flapping on processing. The results indicated that the phonetic variation in speech styles, including reduction causes variations in processing. Additionally, it was found that the processing of flapping was inhibited by weakened acoustic information and that participants used information about speech style to process the widely varying acoustic reflection of a segment in RFs. Connine (2004) also conducted a study on the American English flap and that nature of its representation in the lexicon. Similar to Idsardi and Sung (2003), Connine contrasted flapped words with real and pseudo words. The contrasted items were /t/ and /r/. The results indicated that listener responses as a highly frequent word such as pretty was contained in the lexical representation of words rather than that of the pseudo word "bretty". On the other hand, in cross-language acoustic research on both perception and production of flapping, Sung (2003b) investigated the relationship between the phonological phenomenon and the effects of the L1 Korean phonological system in perception with 10 American and 10 Korean undergraduate students. Flaps in both American English and Korean were stated to be acoustically similar though they were categorized as obstruent in American English and sonorants in Korean. A word list of the test items both in English and Korean orthography was given, and students' production was recorded. The comparison of the acoustic properties was measured by closure duration, the existence of voicing, and occurrence of bursts. The result of the production experiment yielded that there were no significant differences between American English and Korean flaps in relation to the mean values of closure duration and mean percentage of voicing instances. However, the results also showed that there was a significant difference in the mean values of burst occurrence with more burst in Korean flaps. On the basis of the findings, the researcher suggested that phonetic properties of sounds were not able to account for phonological functions of the sounds. In another study describing two corpus analyses, (Eddington & Elzinga; 2008) investigated the influence of stress and following vowel quality on word-internal flapping. More precisely, they examined the phonetic context of word-internal flapping by taking into account some entities such as stress placement, following phone, and syllabification. In the two experiments conducted, they tested whether the specific allophones of /t/, and /d/ influence judgments of syllabification. In comparison of stress and vowel quality, the researchers utilized "CMU Pronouncing Dictionary (n.d.)" (p. 247) and chose 3114 words as the representative of word-internal flapping. A total of 39 test items –flapped words with initial stress and final stress- along with nonce words were read to 26 native speakers of American and asked which one sounded like a natural American English pronunciation. Results indicated that stress and vowel quality are main indicators of whether /t/ or /d/ appear word-medially in American English.

In an investigation of linking, Melenca (2001) developed an instructional program and investigated the effectiveness of linking instruction on learners' speech production. Though the small sample size employed in the study, results indicated that learners' awareness of the use of linking while speaking seemed to have increased. In another study focusing on linking, Alameen (2014) investigated the effectiveness of using two methods of teaching linking to learners of English over a two-week period. The two methods included the audio-visual (AV) feedback and the audio-only (AO) feedback. A pre-test, post-test, and a delayed test design with two treatment groups and a control group were implemented among a group of 45 learners of English. Results of the study indicated that both of the AV and AO training were facilitative in increasing learners' perception and production of linking and that the facilitative effect was also retained one month after the training. In relation to the learners' improvement of linking high and low-frequency words, the results yielded that low-frequency words were more difficult to improve in linking than that of the high-frequency words.

As can be seen from the above mentioned studies, RFs studies are being dealt in relation to perceptive, productive skills as well as its processing. The present study aims to contribute to the field of listening comprehension research with its comprehensive and empirical focus on the features, namely, contraction, assimilation, flap, elision, and linking.

THE STUDY

This study in this chapter adopted an experimental research design -pre-test, post-test, and control group- to investigate the Turkish EFL learners' perceptions of RFs. It is an experimental study conducted in the spring semester of the academic year 2014 and 2015 at the School of Foreign Languages of a state university in Istanbul, Turkey. It lasted seven weeks. In an experimental study, conditions are under the direct control of the investigator, facilitating to determination of causal-like relationships between one or more independent variables and one or more dependent variables (Mackey & Gass, 2005). Experimental studies allow the researcher to identify the strongest evidence about the existence of a cause-effect relationship. In this vein, the experimental study approach was employed in this study on the grounds that the study aims to investigate the effect of RFsI on students' development of listening comprehension.

The study was conducted at a language school of a state university where participants were randomly selected for the research, including four treatment groups from audio high proficiency group with RFsI (AHP +RFsI) and video high proficiency group with RFsI (VHP+ RFsI) and audio, low proficiency groups with RFs instruction (ALP+RFsI) and video low proficiency groups with RFsI (VLP+RFsI) who received two different modalities of instructional packages through the web page at www.kainegulozer.com (CONSPERsentA for audio and CONSPERsentB for the video modality) and four control groups (AHP-RFsI, VHP-RFsI, ALP-RFsI, VLP-RFsI) who did not receive explicit instruction about RFs; instead, followed the syllabus in their course book's CD and DVD in which written and spoken samples regarding the target RFs were included implicitly. The study was conducted as a counterbalanced pre-test post-test only design (Brown, 2005).

The Theoretical Framework in the Study

The study adopted Mayor's (2001) cognitive theory of multimedia learning to the development of listening skills in a Turkish EFL context. CTML can provide powerful instructional design pertaining to listening comprehension performance on the part of L2 learners, RFsI in particular.

Meskill (1996) presented a comprehensive account of how multi-modal processing via multimedia would be available for listening comprehension development. Highlighting the importance of the use of multimedia for instructional purposes, Meskill remarks that learners can have the chance to process audio input, text, and video simultaneously via multimedia. Furthermore, she emphasizes that multimedia is specifically convenient to guide learners in their understanding of RFs of English due to the fact that the delivery of the written version of the fast spoken input on a computer screen allows learners to access to both written and spoken forms. She also gives the examples of "wadjagonnado?" instead of "what are you going to do?" or "whaddjasay" instead of "What did you say?" (pp.190) to compare the aural and the written text so that learners can decipher the RFs. Meskill argued that reduced forms could be delivered through the contrast forms in a written text, and perception or pronunciation of the RFs could be explicitly taught.

In the study, participants were exposed to the audio and video modality of the RFsI so as to enable learners to perceive the utterances in RFs and segment the enunciated words delivered in sentences. The modality also boosted participants' awareness on how to articulate the natural stream of speech using the RFs.

Following one of the main aims of multimedia instruction, the researcher attempts to encourage the learners to establish a coherent mental representation from the RFs presented through the sequential stages in the designated web page as well as the summary presented at the end of the instruction pertaining to the preceding instruction of the RFs delivered. What the researcher expects the novice learners of the RFs is to conceptualize the RFs as active learners, ultimately constructing the meaningful and intelligible comprehension of the RFs in L2 listening.

The Study Context

The study was carried out in a listening course offered to preparatory school students as a Foreign Language in the spring semester of the academic year 2014 and 2015. The course provided both a traditional and blended learning environment for the participants, integrating face-to-face instruction, CD players, videos as well as feature movies incorporated in the course books. Participants were following a main course book entitled "New Language Leader" by Lebeu and Rees (2004) along with its supplementary materials. Students were engaged in listening, and speaking activities throughout the course delivered two hours per week. The total amount of English classes each week was 20 hours for each proficiency level of English.

Participants

At the outset of the semester, an English language proficiency test was administered by the language school of the state university to diagnose the participants' proficiency levels. Of all the proficiency test-takers, (n=2424, 100%) the majority of the participants (n= 1200, 49, 5%) were set as an elementary level, while 768 (34%) pre-intermediate and 456 as intermediate level (18,8%). As suggested by Ito, (2007), it was both elementary and pre-intermediate learners that were the target population for this study as those in intermediate levels already had some knowledge about the RFs to be tested. Upon getting the participants consent and explanation that they can withdraw from the study at any stage they like, one thousand and two hundred elementary students were randomly distributed into 40 classes, of which ten classes were then randomly selected for the study additionally, 768 pre intermediate students were randomly distributed into thirty-two classes of which ten classes were then randomly selected for the study and randomly assigned into the following groups as:

a. Four instructional classes (AHP+RFsI, VHP+RFsI, ALP+RFsI, VLP+RFsI)
b. Four control group class (AHP+RFsI, VHP+RFsI, ALP+RFsI, VLP+RFsI)

The initial sample consisted of 357 learners, four of whom did not consent to be involved in the study, and 10 did not answer all the questions given during the pre or post-test. Thus, a total of 343 participants, 181 of whom were male (74%), 162 of whom female with an average of 19.41 years old (SD=1.36), who have normal hearing and vision, were recruited for the investigation. All participants possessed Turkish as a first language and had a mean self-reported duration of English language study of 9.18 years (SD= 2.51). None of the participants' major is English. English courses at preparatory school are four skill classes as well as grammar. The participants were preparatory language school students from eighteen different classes with a range of eight different majors. All participants were enrolled in consecutive five courses and took the pre, post-test at the computer lab of the university. The distribution of participants both in the experimental and control groups is presented in Table 1

Table 1. Distribution of participants both in the treatment and control groups

Modality	**Level**	**Treatment**		**Control**		**Total**	
		N	%	N	%	N	%
Audio	High	51	24.05	34	26.56	85	25
Audio	Low	53	25	32	25	85	25
Video	High	54	25.47	33	25.78	87	25.58
Video	Low	54	25.47	29	22.65	83	24.41
Total		212	100	128	100	340	100

Setting

The researcher carried out the study in a computer lab of a university where there were 60 PCs along with a headset and one projector-camera. The preparatory school education takes place for two semesters (one year), aiming to teach English for general academic purposes. The learners at the language school follow the coursebook "New English File" series and its supplementary materials by Pearson. Each of the learners has to take the language proficiency exam and Placement exams at the beginning of the semester and is placed into the three levels as beginner, pre-intermediate, and intermediate. Learners at upper-intermediate and advanced levels do not pursue their education in preparatory school; instead, they follow their own freshmen year program in their particular departments or faculties. There were over 130 English language teachers at the school during the specified semester. The number of learners in each class was around 28. Each class had one CD player and a loudspeaker, one overhead projector-camera. Learners at the prep school were having 25 hours of English class hour each week, generally taught by three teachers at the time of the study.

Figure 2. The Interface of the audio and video groups' separate sign into the page

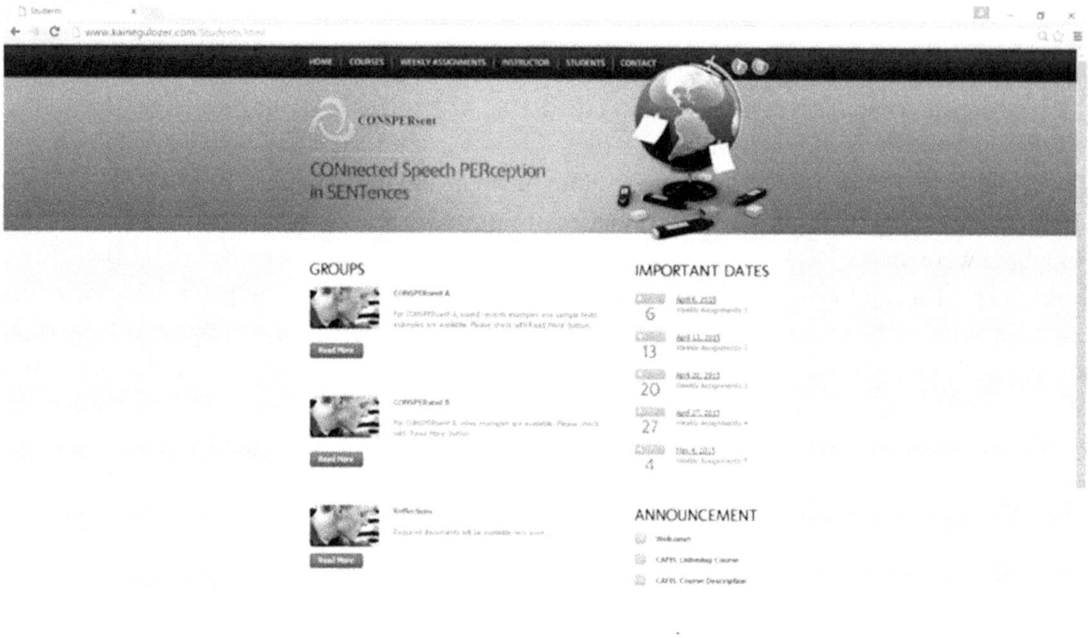

Figure 3. The Interface of audio modality

Instructional Materials for the Reduced Forms

The instructional package in audio and video form was prepared and presented on a website dedicated to deliver RFsI. The heading of CONSPERsentA served for the audio group whereas CONSPERsentB for the video modality. The audio and video materials in four instructional groups started with explicit articulation information by a native speaker of English along with the sample sentences of the target RFs. For each week, one feature is explained in detail, assigning an assignment at the end of each instruction. Learners were reminded to pay attention to the audio or visual input delivered to them in a computer lab. While learners in audio groups were allowed to listen, the ones in video groups were allowed to watch the instructional material, and both groups were delivered weekly assignments concerning the feature to have been taught. Some examples of the interface were presented below as in Figure 2, 3, 4 and 5.

Figure 4. The interface of the video modality

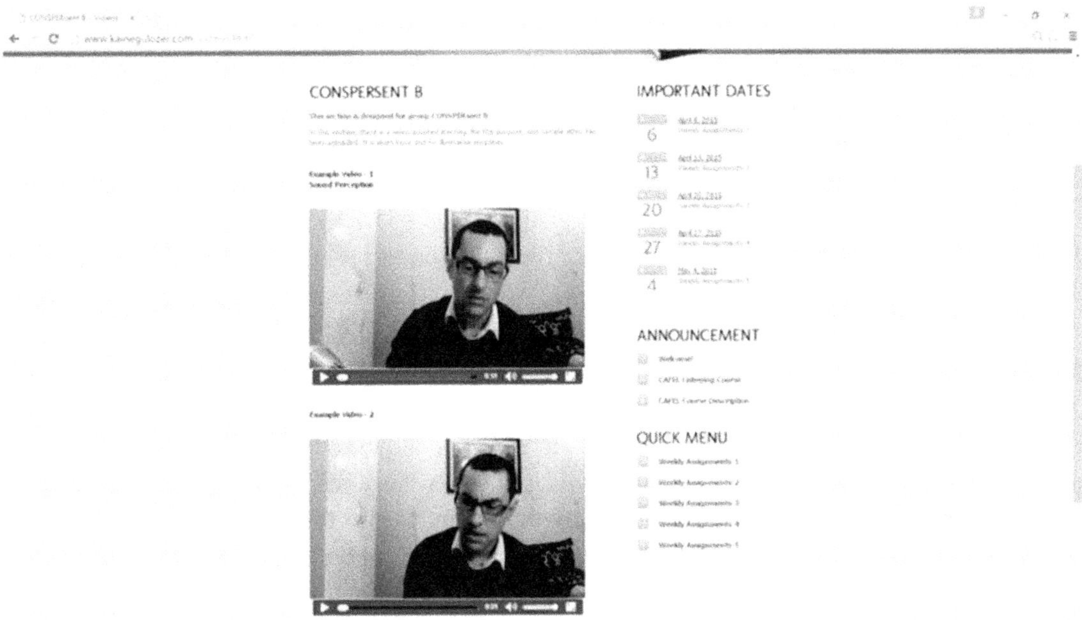

Both instructional packets included the same activities: pre, while, and post-listening sections along with a short description and anticipated problem with the target feature. Instructional packets were devised according to the procedure used by Kissling (2012), in which she developed online video modules to teach pronunciation in Spanish to English speakers. That is, the researcher adapted the modules with the purpose of teaching reduced forms and developed both audio and video versions of the same material.

A syllabus of the instructional package was prepared following Hagen (2000) and Matsuzawa (2006). Arbaugh and Benbunan- Fich's (2006) Constructivist Teaching Approach Framework was employed as it relies on the assumption that "students construct their own knowledge independently by actively interacting with the subject matter and combining information from different sources" (p.438). Sentences incorporating the target features were compiled from several resources such as Kelly (2000), as well as the course book of the learners, Lebeau, Rees (2004) (New English File by Pearson).

Testing Materials of RFs

This study consisted of three testing materials as dictation task, discrimination task and listening comprehension test, the summary of which can be seen as in the following:

Table 2. Distribution of RFs for each testing materials=5 9 30

Tests	Type of Reduced Form					Total
	Contraction	Assimilation	Flap	Elision	Linking	
Dictation	3	3	3	3	3	15
Discrimination	3	3	3	3	3	15
Listening Comp.	9	4	3	5	9	30

The distribution of the number of the RFs items in the pre-test and post-test to be tested is: a total of 15 items comprise the "Contraction", total of 10 items comprise the "Assimilation", total of 9 items comprise the "Flap", total of 11 items comprise the "Elison", and the total of 15 items comprise the "Linking" for the comprehension of RFs in English. The highest possible score is 60. The current test underwent piloting, and validity by the two experts in the field of L2 listening comprehension. In relation to the reliability of each tests, the reliability of the dictation test is computed as .76, discrimination test .88, and listening comprehension test .89. Item difficulty was also calculated using KR 20 formula to determine which sentences to keep or remove in tests. The difficulty level of the sentences was in the range of .08 and .65, meaning that the sentences were in the medium difficulty level (Larson- Hall, 2009).

The first testing instrument is the dictation task involving 15 sentences. In this type of test, participants were presented with an audio input along with a sheet of paper and asked to identify the type of RFs and fill in the blanks for each sentence. The blanks given in each sentence were representing a target reduced form. Each target feature was evenly distributed across the sentences as three different sentences for contraction, assimilation, flap, elision, and linking. Table 3 lists the sample sentences of target RFs tested.

Table 3. Sample sentences in the dictation task

RFs	Sentence
Contraction	I wish I knew her problem before. I (------) my best to solve it. *(would've done)
Assimilation	When you were really hopeless, I (--------)! *(got you)
Flap	He couldn't succeed so he (-------). *(needed a tutor)
Elision	He says today seems (----------) *(bright and beautiful)
Linking	Could you (-------) this sofa? I don't need it any more. *(take away)

The second testing instrument is a discrimination task that is similar to those used in other studies to evaluate the comprehension of relevant, targeted contrasts (e.g., Ito, 2006; Kissling, 2012; Cahill, 2006). The discrimination task was used to establish whether a difference in meaning exists between the two audio stimuli containing both reduced and unreduced forms of each targeted feature in 15 sentences as a whole. A response that is the same or different for a reduced form than an unreduced form indicates an effect of comprehension of RFs. For example, a listener might understand clearly uttered items in the same way as the reduced items. Conversely, a listener might understand unreduced items in a different way than that of RFs, meaning that the presence of RFs has an impact on the comprehension of the sentences. The AX discrimination task paired target RFs with the presence and absence of the same target in sentences with the five RFs (Contraction, Assimilation, Flap, Elision, Linking). The testing stimuli consisted of sound recordings by a female native speaker from the States who has been working at a language school at another public university. The recordings were made using the audio editing software Soundtrack Pro, at a 16-bit sampling rate of 48 KHz in a sound-proof booth. Table 4 illustrates the sample of a RFs target in a discrimination task.

Table 4. Reduced form targets tested in discrimination task

RFs	Absence of RFs	Presence of RFs
Contraction	This experience is going to be hard	This experience**'s gonna** be hard
Assimilation	This is the kind you are interested	This is the kin**d y**ou are interested
Flap	Things will settle down	Things will se**tt**le down
Elision	I want him to study	I wan**t h**im to study
Linking	How's your mother coming along?	How's your mother coming‿ along?

The discrimination test included a total of 15 pairs which were scrambled into different intra pair RFs order, and they were used for the pre-test and post-test. While recording the stimuli, both the researcher and the native speaker of English took great care that the only audible differences in the AX stimuli pairings were the target segments. The other parameters, such as volume, pitch, duration, and length of sentences, were kept as identical as possible between the presence and absence of the feature in the AX pairings.

As for the third testing instrument, a listening comprehension test consisting of thirty short dialogues was delivered. There was 10 seconds of interval after each question so that students could mark their answers on the answer sheet. Each dialogue was followed by one question consisting of comprehension questions. Each listening dialogue was approximately 10 to 13 seconds long. Five native English instructors and two native Turkish instructors who teach English took the test before it was piloted with ten students of preparatory students who were later on excluded from the study. In addition, the seven instructors were asked to identify the dialogues incorporating the target RFs. All of them identified RFs correctly. The pilot results from the preparatory class students indicated that the test was functioning appropriately. Table 5 illustrates an example of the feature "contraction" incorporated in the listening dialogue.

Table 5. An Example of contraction incorporated in listening comprehension test

Reduced Form Short Dialogue and question sample
Contraction
(woman):Somebody's been leaving this door unlocked. a) He's not the one to blame.
(man): Don't look at me! b) Somebody just left
(narrator):What does the man mean? c) He has been looking for the key
d) Somebody's knocking at the door

Table 6. Means and standard deviations for pretest and posttest results of the AHP+RFsI and AHP-RFsI Groups

Test type	Groups	N	M	S
Pre total	AHP+RFsI	51	12.52	3.73
	AHP-RFsI	34	14.20	3.07
	Total	85	13.20	3.56
Post total	AHP+RFsI	51	22.78	5.61
	AHP-RFsI	34	15.09	7.69
	Total	85	19.71	7.50

Participants listened to each short dialogue that took between 10 to 13 seconds and answered multiple-choice questions per dialogue. The listening items were allocated to measure overall listening. In order to grasp the meaning, students had to distinguish between targeted RFs incorporated in dialogues as well.

Data Analysis and Results

In this experimental study, quantitative statistics analyses were utilized, employing descriptive and inferential statistics. Descriptive analyses were conducted for the purpose of determining the frequencies in demographic and linguistic information. Statistical Package for Social Sciences program (Version 15.0) was run to compute the statistical portion of the study. Kolmogorov Smirnov was conducted to ensure that there were no pre-existing differences before the instructions among the treatment groups and the control groups. The analysis revealed no significant difference among the eight groups on listening comprehension of RFs, as the value was found to be greater than .05 ($p>0.05$). In other words, normal distribution across the eight groups of test-takers was confirmed. Therefore, it can be stated that the eight groups in the study started at the same level of knowledge of the target RFs before they received instructions, thereby leading to assume that any improvements in their performance were due to instructional treatment. Upon confirming the normal distribution across the eight groups, the data were ready to compute descriptive and inferential statistics to compare the effect of RFsI across the treatment groups. The analyses were conducted for all treatment and control groups on RFs listening comprehension test.

For research question 1, results regarding the effect of RFsI were provided, comparing the difference between the scores of pre-test and post-test across the treatment and the control groups. Whether there were any differences in audio in each group (high and low proficiency) and whether the two groups differed in their use of audio and video modality were analyzed by a two–way split-plot design ANOVA with two grouping variables (AHP+RFsI and AHP-RFsI) and modality (audio) and one within-subjects variable (RFs comprehension). Hence language proficiency (high proficiency and low proficiency) and modality (audio and video) were the independent variables with two levels, whereas the RFs Comprehension test was the dependent variable with five levels measured on a dichotomous scale. In order to support the findings through inferential statistics, reflection reports were asked to the treatment group to find out the perceptions of the participants towards web-based RFsI.

Results for the Research Question 1

The first research question was whether RFs instruction helped students improve their listening comprehension with regard to previously specified RFs. The RFs listening comprehension post-test scores

Table 7. Pretest and posttest results of the AHP+RFsI and AHP-RFsI Groups

Source	SS	Df	MS	F	p
Test type	369.60	1	369.60	13.53	.000
Error	2267.02	83	27.31		
Pretest-Posttest	1265.19	1	1265.19	46.24	.000
Pretest posttest x AHP+/-RFsI	896.01	1	896.01	32.75	.000
Error	2270.60	83	27.35		

*p<0.05

of treatment (N=51) and control groups (N=34) were compared to determine whether RFsI influenced student performance in each proficiency level of treatment and control groups on RFs listening comprehension tests (dictation, discrimination, and listening comprehension). The highest score in the dictation task is 15, discrimination task 15, and multiple choice 30, which makes 60 as the total score. A 2X2 repeated measures ANOVA was conducted to test whether the mean differences between the pre-test and post-test results as well as the differences within each group with respect to the pre-test and post-test on each proficiency level (AHP +RFsI, AHP-RFsI; VHP+RFsI,VHP-RFsI; ALP+RFsI, ALP-RFsI; VLP+RFsI, VLP-RFsI) were different.

Table 8. Means and standard deviations for pre-test and post-test results of the VHP+RFsI and VHP-RFsI Groups

Test type		N	M	S
Pre total	VHP+RFsI	54	13.07	3.44
	VHP-RFsI	33	15.63	3.27
	Total	87	14.04	3.58
Post total	VHP+RFsI	54	21.28	7.22
	VHP-RFsI	33	14.82	9.30
	Total	87	18.83	8.62

Table 9. Pretest and Posttest results of the VHP+RFsI and VHP-RFsI groups

Source	SS	Df	MS	F	p
Test type	155.55	1	155.55	4.28	.042
Error	3089.24	85	36.34		
Pretest-posttest	558.62	1	558.62	13.87	.000
VHP+/-RFsI x RFs Test	833.59	1	833.59	20.70	.000
Error	342.83	85	40.25		

Table 6 shows the means and the standard deviations for the pre-test and post-test of the AHP+RFsI and AHP-RFsI groups in RFs comprehension test. Note that the means indicate the ratio of the total number of participants in the specified groups to the total number of correct answers that participants could accomplish.

As the table indicates, the AHP+RFsI group (M=22.78, SD=5.61) had a higher mean for RFs comprehension in post-test than that of the control groups (AHP-RFsI), (M=15.09, SD=7.69), thus the common-sense notion that whether RFsI helps students improve their listening performance of RFs appears to be supported.

Prior to performing the two-way ANOVA analyses of variance tests, the sample data were checked for the assumptions of ANOVA. Descriptive statistics indicated that variables were normally distributed. In order to examine the group means regarding the pre-test RFs comprehension test, a 2x2 repeated measure ANOVA was conducted. As Table 7 indicates, based on the ANOVA results, there was a significant instruction effect for AHP+RFsI group and pre-test post-test (2X2) RFs Comprehension Test F (1-83) =32.75, p= 0.00. The significant instruction effect indicates that the treatment groups improved the comprehension RFs when the pre-test and post-test mean scores were compared in both AHP+RFsI and AHP-RFsI groups.

As for the VHP+RFsI and VHP-RFsI groups, Table 8 shows the means and the standard deviations for the pre-test and post-test of the VHP+RFsI and VHP-RFsI groups in RFs comprehension test. It should be noted that the means indicate the ratio of the total number of participants in the specified groups to the total number of correct answers that participants could accomplish.

Table 10. Shows means and standard deviations for pre-test and post-test results of the ALP+RFsI and ALP-RFsI Groups

Test type N M S
Pretest total ALP+RFsI 53 13.09 4.16
ALP-RFsI 32 16.15 3.29
Total 85 14.24 4.11
Posttest total ALP+RFs 53 14.72 5.93
ALP-RFs 32 7.75 6.15
Total 85 12.09 6.87

Table 11. Pretest and posttest results of the ALP+RFsI and ALP-RFsI groups

Source SS Df MS F p
Test type 152.13 1 152.13 5.43 .022
Error 2324.41 83 28.00
Pretest-Posttest 459.09 1 459.09 19.87 .000
ALP+/-RFsI x Test 1003.42 1 1003.43 43.44 .000
Error 1917.08 83 23.09
**p<0.05*

Table 12. Means and standard deviations for pre-test and post-test results of the VLP+RFsI and VLP-RFsI Groups

Test type N M S
Pretest total VLP+RFsI 53 13.11 3.67
VLP-RFsI 29 15.44 3.31
Total 82 13.93 3.70
Posttest total VLP+RFsI 53 14.30 6.81
VLP-RFsI 29 7.38 4.41
Total 82 11.85 6.90

As seen in Table 8, the treatment group (VHP+RFsI), (M=21.28, SD=7.22) had a higher mean for RFs comprehension in post-test than that of the control groups (VHP-RFsI), (M=14.82, SD=9.30), which means that RFsI helps students improve their listening performance of RFs appears to be supported.

The ANOVA results revealed a significant instruction effect for VHP+RFsI group and pre-test post-test RFs Comprehension Test F (1-85) =20.707, p= 0.00. The significant instruction effect indicates that the treatment groups who were delivered through web-based RFsI improved the comprehension RFs when the pre-test and post-test gain scores were compared in both VHP+RFsI and VHP-RFsI groups.

As for the ALP+RFsI and ALP-RFsI groups, the treatment group (ALP+RFsI), (M=14.72, SD=5.93) had a higher mean for RFs comprehension in post-test than that of the control groups (ALP-RFsI), (M=7.75, SD=6.15), which means that web-based RFsI helps students improve their listening performance of RFs.

The ANOVA results revealed a significant instruction effect for ALP+RFsI group, pre-test post-test RFs Comprehension Test F(1-83)=43.44, p= 0.00. The significant instruction effect indicates that the treatment groups who were delivered through web-based RFsI improved the comprehension RFs when the pre-test and post-test mean scores were compared in both ALP+RFsI and ALP-RFsI groups.

As for the VLP+RFsI and VLP-RFsI groups, the treatment group, VLP+RFsI, (M=14.30, SD=6.81) had a higher mean for RFs comprehension in post-test than that of the control group, VLP-RFsI, (M=7.38, SD=4.41), thus common-sense notion that whether RFsI helps students improve their listening performance of RFs appears to be supported (see Table 12).

Table 13. Pretest and posttest results of the VLP+RFsI and VLP-RFsI groups

Source	SS	Df	MS	F	p
Test type	197.23	1	197.23	7.51	.008
Error	2098.50	80	26.23		
Pretest-Posttest	443.65	1	443.65	18.96	.000
VLP+/-RFsI x Test	803.21	1	803.21	34.32	.000
Error	1871.98	80	23.40		

As seen in Table 13, the ANOVA results revealed, a significant instruction effect for VLP+RFsI group, pre-test, post-test RFs Comprehension Test F(1-80)=34.32, p= 0.00. The significant instruction effect indicates that the treatment groups improved the comprehension RFs when the pre-test and post-test mean scores were compared in both VLP+RFsI and VLP-RFsI groups.

As for the results of the sub-research questions, independent samples t-test was carried out to determine whether students' scores of the comprehension of RFs changed in treatment (AHP, ALP, VHP, VLP+RFsI) and control groups (AHP, ALP, VHP, VLP-RFsI), setting the alpha level at .05. In this regard, each proficiency level of treatment and control groups was compared to identify the likely difference between the groups.

1. Is there any significant difference between AHP+RFsI group and AHP-RFsI group in terms of the comprehension of RFs, namely contraction, assimilation, flap, elision, and linking?

The results revealed that the AHP+RFsI group had statistically higher performance (M= 6.74), (M=5.23), (M=3.72), (M=4.82), (6.33) at the end of the RFs instruction web-based program compared to the AHP-RFsI groups (M=5.38), (M=3,64), (M=2.97), (M=3.75), (M=4.64) in five forms of the RFs (t (83)=2.61, P =.01), (t (83)=4.91, P =.00), (t (83)=2.47, P =.01), (t (83)=1.86, P =.04), (t (83)=3.18, P =.02). In other words, there was a significant difference between the AHP+RFsI group and the AHP-RFsI group in terms of the comprehension of each RFs, namely contraction, assimilation, flap, elision, and linking.

2. Is there any significant difference between VHP+RFsI and VHP -RFsI groups in terms of the comprehension of RFs?

The results revealed that the VHP+RFsI group had statistically higher performance (M= 6.22), (M=4.74), (M=3.66), (M=4.25), (5.96) at the end of the RFs instruction web-based program compared to the VHP-RFsI groups (M=5.04), (M=3,57), (M=3.42), (M=4.75) in the four forms of the RFs (t (83)=1.14, P =.03), (t (83)=3.13, P =.00), (t (83)=2.05, P =.04), (t (83)=2.16, P =03). In other words, there was a significant difference between the AHP+RFsI group and the AHP-RFsI group in terms of the comprehension of the RFs as contraction, assimilation, elision, and linking. However, there was no significant difference between the treatment and control group in relation to the feature of "flap", (t (83) = .66, P =.05).

3. Is there any significant difference between ALP+RFsI and ALP-RFsI groups in terms of the comprehension of RFs?

The results revealed that the ALP+RFsI group had statistically higher performance (M= 6.64), (M=3.94), (M=3.73), (4.90) at the end of the web-based RFs instruction program (CONSPERsent) compared to that of AHP-RFsI groups (M=4.71), (M=2,93), (M=3.73), (M=4.90) in four of the RFs (t (83)=1.74, P =.02), (t (83)=2.47, P =.01), (t (83)=1.49, P =.04), (t (83)=2.36, P =.02). In other words, apart from the feature of "flap" there was a significant difference between the AHP+RFsI group and the AHP-RFsI group in terms of the comprehension of each RFs, namely contraction, assimilation, elision and linking. Yet, there was no significant difference between the treatment and control groups in relation to the feature of "flap", (t (83) =.57, P =.56).

4. Is there any significant difference between VLP+RFsI and VLP-RFsI groups in terms of the comprehension of RFs?

The results revealed that the VLP+RFsI group had statistically higher performance (M= 4.96), (M=3.85), (M=3.54), (4.48) at the end of the RFsI through the web compared to the VLP-RFsI groups (M=4.07), (M=2,30), (M=2.33), (M=3.16) in the four of the RFs (t (83)=.53, P =.03), (t (83)=.07, P =.02), (t (83)=.85, P =.02), (t (83)=.51, P =.01). In other words, apart from the feature of "flap" there was a significant difference between the AHP+RFsI group and the AHP-RFsI group in terms of the comprehension of each RFs, namely contraction, assimilation, elision, and linking. Yet, there was no significant difference between the treatment and control groups in relation to the feature of "flap", (t (83) =.67, P =.50).

SOLUTIONS AND RECOMMENDATIONS

The findings of this study have pedagogical and methodological and technical implications for the provision of web-based instruction for L2 listening and the design of reduced forms presentation.

Pedagogically, this study offers many ideas for language teachers. The first suggestion for language teachers may be that they could incorporate RFsI into their L2 listening classes for low proficiency and high proficiency learners through web-based audio or video modality as both equally help learners alter their default segmenting problems and gain more perceptive processing strategies pertaining to the RFs. Secondly, material developers, bearing in mind that RFsI has a beneficial effect on listening comprehension development and that there is "not enough material available on RFs" (Rogerson, 2006, p.94), could prepare materials representing extensive examples of RFs both in sentential and textual level. In this vein, Rogerson (2006) concerns about not only the need to develop more materials focusing on each feature of RFs but also the need to raise teacher trainers' awareness on the instruction of RFs. However, among the five features as contraction, assimilation, flap, elision, and linking, it can be suggested that teachers could employ less exposure on flapping than the other features as it seems flapping forms relatively less constraint on listening comprehension for Turkish EFL learners' context in particular. This was an unexpected discovery, given the challenging nature of flap (Riehl, 2003). This feeling about flapping was likely a result of learners' initial exposure to the feature through movies or the salient phonological rule of this feature whereby the phone /t/ or /d/ becomes a flap before an unstressed vowel, as in the word "water".

Methodologically, this study utilized web-based RFsI instruction for the four instructional groups- AHP+RFsI, VHP+RFsI, ALP+RFsI and VLP+RFsI in students' after class hours, and employed a pen and paper assessment test. There were also additional four groups for control groups. Employing eight different groups, though the sample size of English learners in our language school is quite high, might bear a disadvantage as the learners in these eight different groups might have shared the information with their peers in other groups; instead, sample size could be smaller or a computer-based application could be designed and employed to collect data. Additionally, RFs in the sentential level were the targets to attain listening comprehension; however, assessing learners' listening comprehension of RFs through spoken texts in a context would have given different results.

As for the design of RFs presentation, a learner corpus of authentic speech could be a potential precursor to develop as a resource, and rich examples of speech modifications can be presented through authentic speech. In the present study, all instructional and assessment materials are recorded in a controlled way such as recording in a sound proof room, controlling the rate of speech; however, the presentation of RFs both in instructional materials and assessment materials could have been elicited from an authentic spoken corpus and designed in a web-based platform for instructional as well as assessment purposes of the target RFs. In fact, when considered all spoken corpora, another issue can arise for researchers or L2 teachers as to how to determine the RFs to be included. The target RFs should be chosen based on a thorough search of RFs in comprehensive spoken corpora and their importance and usefulness in learners' listening comprehension of conversation rather than personal judgment. However, it should also be noted that some phrases in daily conversation might be frequently used whereas others are used in a specific accent; therefore, the accent of the phrase should also be checked. Apart from checking the accent in spoken corpora, pre-existing knowledge of RFs crucial to learn on the part of the learners can also be checked by having some representative learners randomly select the unknown spoken phrases in a spoken corpus.

In relation to technical concerns, captions in the video modality should also be mentioned as regards the effectiveness of RFs comprehension. Yang & Chang (2013) suggests the use of annotated keyword captions in that it assists with the learning of RFs. Employing the video modality in the present study might have yielded different results in video modality.

FUTURE RESEARCH DIRECTIONS

This study focused on comprehension of RFs in listening comprehension context, which can be used as a basis for future research in the areas of computer-assisted language instruction of RFs, perception of RFs and production of RFs. The ideas outlined below demonstrate possible directions for future research.

First of all, this research is among the first to employ RFs instruction in the L1 Turkish context. Five types of RFs –contraction, assimilation, flap, elision, and linking- were instructed to four different groups explicitly, yet the same features were instructed to the corresponding level of four control groups implicitly. Cross comparisons between the experimental and control groups were made, and the comparison within the group was administered. Random selection of the participants and their random assignment into groups within a quasi-experimental design, lack of teacher variability, allocation of the equal amount of time for instruction in each group- all these factors make the research design of the study sound enough to make reliable and to some extent generalizable results as to whether and which RFs foster learners' listening comprehension development. In the future, more replication studies could be conducted to reach more generalizable results on the effectiveness of RFsI or to confirm the findings by using the same research design. What is more, this study can be expanded, including a wide range of population as language learners from different universities in Turkey, language learners from other countries with distinct L1 context, language instructors, and the experts in material development so as to provide a more thorough depth of data and to get greater insight into the issue and to widen the scope of generalizability.

Secondly, one factor that has influenced the results is the scoring method in the dictation task. Subjects were given either full or non-credit for their responses. Partial credit was not given even if their response did reflect a very vague understanding of the original RF. For example, in flap, there was only one word to assess as a target; however, this was not the case when the target feature linking was evaluated because, by the very nature of each feature, linking was composed of several words. Future research concerning RFs should find a way to give subjects credit for various subtypes of partially correct answers depending on the nature of the feature.

Thirdly, future studies could explore the effects of different instruments of instruction on listening comprehension of RFs utilizing this experimental design. Such methodology of instruction would include the incorporation of web-based applications, mobile-assisted language learning or captions with annotated words. Captioning may help segment word boundaries. In other words, it may facilitate learners to identify what might otherwise be an incomprehensible stream of speech. In relation to data collection instruments, future research could measure the learners' performance through think-aloud protocol in qualitative design or through "E-prime" in an experimental design as it allows millisecond precision timing to ensure accurate data.

Fourthly, the role of some other cognitive and affective variables such as individual differences, working memory, listening to a paragraph level passage with RFs, anxiety, or motivation on listening comprehension development could be examined more closely. These variables may increase our understanding of how they mediate the effects of RFs in relation to listening comprehension development.

Finally, RFs not only in relation to listening comprehension but also in relation to language production could be aimed in future studies to investigate in an EFL context.

CONCLUSION

The current study presents an empirical study of the effectiveness of RFsI on second language sentential level of listening comprehension. The target RFs covered in the study were contraction, assimilation, flap, elision, and linking. After developing an instructional package of RFs addressing each feature and setting up a web-based environment in which to deliver L2 aural and visual reduced forms instruction, the participants were randomly assigned to eight reduced form groups: four experimental groups (audio instruction to high proficiency, audio instruction to low proficiency, video instruction to high proficiency and video instruction to low proficiency) and four control groups who followed their regular listening courses with no explicit instruction of RFs: (audio instruction to high proficiency, audio instruction to low proficiency, video instruction to high proficiency and video instruction to low proficiency). Based on the assessment tests –dictation task, discrimination task, and listening comprehension test-, reflection report and interview, it was found that RFsI is effective in L2 listening comprehension development. One important finding is that the features contraction, assimilation, elision, and linking facilitate listening comprehension, which is in line with the findings of previous studies; however, flap has no role in this case. The results also indicate that when audio and video modality with no captions are offered, and the beneficial effect of the modality is compared students showed almost equal performance across the assessment tests as dictation task, discrimination task, and listening comprehension test. As for the proficiency levels, the four instructional groups in both high and low proficiency levels outperformed the control group in regard to the comprehension of reduced forms and the development of listening comprehension.

REFERENCES

Ahmadian, M., & Matour, M. (2014). The effects of explicit instruction of connected speech on Iranian EFL learners' listening comprehension skill. *International Journal of Applied Linguistics and English Literature*, *3*(2), 227–236. doi:10.7575/aiac.ijalel.v.3n.2p.227

Alameen, G. (2014). *The effectiveness of linking instruction on NNS speech perception and production*. Graduate Theses and Dissertations. Paper 14004. https://lib.dr.iastate.edu/etd/14004

Altenberg, E. P. (2005). The perception of word boundaries in a second language. *Second Language Research*, *21*(4), 325–358. doi:10.1191/0267658305sr250oa

Aquil, R. (2012). Listening English connected speech: A problem and solutions. *Arab World English Journal*, *3*(2), 329-364.

Arbaugh, J., & Berbunan-Fich, R. (2006). An investigation of epistemological and social dimensions of teaching in online learning environments. *Academy of Management Learning & Education*, *5*(4), 435–447. doi:10.5465/amle.2006.23473204

Barrera-Pardo, D. (2008). The reality of stress timing. *ELT Journal*, *62*(1), 11–17. doi:10.1093/elt/ccm073

Brouwer, S. (2010). *Processing strongly reduced forms in casual speech* (PhD Dissertation). Radbaud University. Retrieved from https://repository.ubn.ru.nl/handle/2066/90675

Brown & Kondo-Brown. (Eds.). (2006). Perspectives on teaching connected speech to second language speakers. University of Hawaii Press.

Brown, J. D. (2005). *Testing in Language Programs: A comprehensive guide to English language assessment*. McGraw-Hill.

Brown, J. D. (2006). Authentic communication: Whyzit importan' ta teach reduced forms? *Authentic Communication: Proceedings of the 5th Annual JALT Pan-SIG Conference*, 13-24. Retrieved June 6, 2007 from http://jalt.org/pansig/2006/HTML/Brown.htm

Brown, J. D., & Hilferty, A. (1986a). Listening for reduced forms. *TESOL Quarterly*, *20*(4), 759–763. doi:10.2307/3586525

Brown, J. D., & Hilferty, A. (1986b). The effectiveness of teaching reduced forms for listening comprehension. *RELC Journal*, *17*(2), 59–70. doi:10.1177/003368828601700204

Brown, J. D., & Hilferty, A. (2006). The effectiveness of teaching reduced forms for listening comprehension. In J. D. Brown & K. Kondo-Brown (Eds.), *Perspectives on Teaching Connected Speech to Second Language Speakers* (pp. 51–58). University of Hawai'i, National Foreign Language Resource Center.

Buck, G. (2001). *Assessing Listening*. Cambridge University Press. doi:10.1017/CBO9780511732959

Cahill, R. (2006). Teaching reduced interrogative forms to low-level EFL students in Japan. In J.D. Brown & K. Brown, (Eds.), Perspectives on teaching connected speech to second language speakers (pp. 99-125). Honolulu, HI: University of Hawai'i, National Foreign Language Resource Center.

Canagarajah, S. (2014). In search of a new paradigm for teaching as an international language. *TESOL Quarterly*, *4*(4), 767–785. doi:10.1002/tesj.166

Celce-Murcia, M., Brinton, D. M., & Goodwin, J. M. (2011). *Teaching pronunciation: A reference guide* (2nd ed.). Cambridge University Press.

Chen, C.-H., Hung, H.-T., & Yeh, H.-C. (2021). Virtual reality in problem-based learning contexts: Effects on the problem-solving performance, vocabulary acquisition and motivation of English language learners. *Journal of Computer Assisted Learning*, *37*(3), 851–860. doi:10.1111/jcal.12528

Creswell, J. W. (2012). *Educational Research: Planning, conducting and evaluating quantitative and qualitative research*. Pearson Education, Inc.

Eddington, D., & Elzinga, D. (2008). The phonetic, Context of American English Flapping: Quantitative Evidence. *Language and Speech*, *52*(3), 245–266. doi:10.1177/0023830908098542 PMID:19626926

Erbanova, P. (2014). *Aspects of connected speech in English: Assessing students' progress after pronunciation training* (Master's Thesis). Masaryk University. Retrieved from https://is.muni.cz/th/c4mfj/Master_s_thesis.pdf

Feyten, C. M. (1991). The power of listening ability: An overlooked dimension in language acquisition. *Modern Language Journal, 75*(2), 173–180. doi:10.1111/j.1540-4781.1991.tb05348.x

Gaskell, M. G., & Snoeren, N. D. (2008). The impact of strong assimilation on the perception of connected speech. *Journal of Experimental Psychology. Human Perception and Performance, 34*(6), 1–29. doi:10.1037/a0011977 PMID:19045997

Gow, D. W. Jr, & Im, A. M. (2004). A cross-linguistic examination of assimilation context effects. *Journal of Memory and Language, 51*(2), 279–296. doi:10.1016/j.jml.2004.05.004

Hasan, A. S. (2000). Learners' perceptions of listening comprehension problems. *Culture and Curriculum, 13*(2), 137-153. . doi:10.1080/07908310008666595

Henrichen, L. E. (1984). Sandhi-variation: A filter of input for learners of ESL. *Language Learning, 34*(3), 103–126. doi:10.1111/j.1467-1770.1984.tb00343.x

Idsardi, W. J., & Sung, E.-K. (2003). Flaps in American English and Korean: An acoustic and perceptual study. Generative Grammar in a Broader Perspective. *Proceedings of the 4th GLOW in Asia*, 187-208.

Ito, Y. (2001). Effect of reduced forms on ESL learners' input-intake process. *Second Language Studies 20*(1), 99-124. Retrieved June 6, 2007 from https://www.hawaii.edu/sls/uhwpesl/20(1)/Ito.pdf

Ito, Y. (2006). The comprehension of English reduced forms by second language learners and its effect on input-intake process. In J. D. Brown & K. Kondo-Brown (Eds.), *Perspectives on teaching connected speech to second language speakers* (pp. 67–81). University of Hawai'i National Foreign Language Resource Center.

Kissling, E. M. (2012). *The effect of phonetic instruction on adult learners' perception and production* (Doctoral dissertation). Georgetown University. Retrieved from https://repository.library.georgetown.edu/handle/10822/557711

Larson-Hall, J. (2009). *A guide to doing statistics in second language research using SPSS*. Routledge. doi:10.4324/9780203875964

Lodge, K. (2009). *A Critical Introduction to Phonetics*. Continuum.

Mackey, A., & Gass, S. (2005). *Second language research: Methodology and design*. Lawrence Erlbaum.

Marks, J. (1999). Is stress timing real? *ELT Journal, 52*(3), 191–199. doi:10.1093/elt/53.3.191

Matsuzawa, T. (2006). Comprehension of English reduced forms by Japanese business people and the effectiveness of instruction. In J. D. Brown & K. Kondo Brown (Eds.), *Perspectives on teaching connected speech to second language speakers* (pp. 59–66). University of Hawaii Press. National Foreign Language Resource Center.

Mayer, R. E. (2001). *Multimedia Learning*. Cambridge University Press. doi:10.1017/CBO9781139164603

Mayer, R. E. (2005). Cognitive theory of multimedia learning. In R. E. Mayer (Ed.), *The Cambridge Handbook of Multimedia Learning*. Cambridge University Press. doi:10.1017/CBO9780511816819.004

Mayer, R. E. (2009). Multimedia learning (2nd ed.). Cambridge University Press. doi:10.1017/CBO9780511811678

Mayer, R. E., Heiser, J., & Lonn, S. (2001). Cognitive constraints on multimedia learning: When presenting more material results in less understanding. *Journal of Educational Psychology*, *93*(1), 187–198. doi:10.1037/0022-0663.93.1.187

Mayer, R. E., & Moreno, R. (2003). Nine ways to reduce cognitive load in multimedia learning. *Educational Psychologist*, *38*(1), 43–52. doi:10.1207/S15326985EP3801_6

Melenca, M. (2001). *Teaching connected speech rules to Japanese speakers of English so as to avoid a staccato speech rhythm* (Master's thesis). Concordia University. Retrieved from https://spectrum.library.concordia.ca/1389/

Meskill, C. (1996). Listening skill development through multimedia. *Journal of Educational Multimedia and Hypermedia*, *5*(2), 179–201.

Monteiro, K., & Kim, Y. (2020). The effect of input characteristics and individual differences on L2 comprehension of authentic and modified listening tasks. *System*, *94*, 102–336. doi:10.1016/j.system.2020.102336

O'Bryan, A., & Hegelheimer, V. (2009). Using a mixed methods approach to explore strategies, metacognitive awareness and the effects of task design on listening development. *Canadian Journal of Applied Linguistics*, *12*(1), 9–37.

Riehl, A. K. (2003). American English Flapping: Perceptual and acoustic evidence against paradigm uniformity with phonetic features. *Working Papers of the Cornell Phonetics Laboratory*, *15*, 271-337.

Roach, P. (2009). English Phonetics and Phonology (4th ed.). Academic Press.

Rogerson, M. (2006). Don'cha Know? A survey of ESL teachers' perspectives on reduced form instruction.Second Language Studies. In J. D. Brown & K. Kondo Brown, (Eds.), Perspectives on teaching connected speech to second language speakers (pp. 85-97). Honolulu, HI: University of Hawaii Press. National Foreign Language Resource Center.

Rosa, M. (2002). Don'cha know? A survey of ESL teachers' perspectives on reduced forms instruction. *Second Language Studies*, *21*(1), 49–78.

Saygin, A. P., & Wilson, S. M. (2002). *Paradigm reanalysis and the representation of morphologically complex words in Turkish.* Paper presented at the 38th Annual Meeting of the Chicago Linguistic.

Settler, J., & Jenkins, J. (2005). Pronunciation. State-of-the-Art Review Article. *Language Teaching*, *38*(1), 1–17. doi:10.1017/S026144480500251X

Shin, M.-H., & Sung, J. (2019). The Effect of Teaching Suprasegmental Features on Korean University Students' Listening Comprehension. *International Journal of Advanced Science and Technology, 28*(3), 1 - 7. Retrieved from http://sersc.org/journals/index.php/IJAST/article/view/256

Sorden, S. D. (2005). A cognitive approach to instructional design for multimedia learning. *Information Science Journal, 8*, 263–279. doi:10.28945/498

Sung, E.-K. (2003). Acoustics and perception of flaps in American English and Korean. *Language Research, 39*(4), 933–958.

Taylor, G. (2005). Perceived strategies of students watching captioned video processing. *Foreign Language Annals, 38*(3), 422–427. doi:10.1111/j.1944-9720.2005.tb02228.x

Toda, T. (2006). Focus on form in teaching connected speech. In J. D. Brown & K. Kondo-Brown (Eds.), *Perspectives of Teaching Connected Speech to Second Language Speakers* (pp. 189–2015). University of Hawaii, National Foreign Language Resource Center.

Tucker, B. V. (2007). *Spoken word recognition of reduced American flap* (Doctoral Dissertation). The University of Arizona. Retrieved from https://sites.ualberta.ca/~bvtucker/Papers/Tucker_2007_Dissertation.pdf

Underwood, P., & Wallace, M. (2012). *The effects of instruction of English speech on low proficiency EFL students*. Retrieved from https://www.academia.edu/854671/The_Effects_of_Instruction_in_Reduced_Forms_of_English_Speech_on_Low-proficiency_EFL_University_Students

Underwood, P. R., Bridges, S., & Hattori, T. (2010). *Speak Easy: Spoken English for Japanese Learners*. Suita Mediaisland.

Ur, P. (1984). *Teaching listening comprehension*. Cambridge University.

Vandergrift, L. (2004). Listening to learn or learning to listen? *Annual Review of Applied Linguistics, 24*, 3–25. doi:10.1017/S0267190504000017

Vandergrift, L. (2007). Recent developments in second and foreign language listening comprehension research. *Language Teaching, 40*(3), 191–210. doi:10.1017/S0261444807004338

Weinstein, N. (2001). *Whaddaya Say?* (2nd ed.). Pearson Education.

Wong, S. W. L., Mok, P. P. K., Chung, K. K.-H., Leung, V. W. H., Bishop, D. V. M., & Chow, B. W.-Y. (2015). Perception of Native English Reduced Forms in Chinese Learners: Its Role in Listening Comprehension and Its Phonological Correlates. *TESOL Quarterly*. . doi:10.1002/tesq.273

Yang, J., & Chang, C. (2013). Captions and reduced form instruction: The impact on EFL students' listening comprehension. *ReCALL, 26*(1), 44–61. doi:10.1017/S0958344013000219

Yang, J.-C., Lin, Y.-L., & Chung, C.-I. (2009). Developing a film-based learning system with English verbal reduced forms for supporting English listening comprehension. In M. Chang, R. Kuo, Kinshuk, G.-D. Chen, & M. Hirose (Eds.), Learning by playing: Gamebased education system design and development (Vol. 5670, pp. 74–84). Berlin: Springer Berlin Heidelberg. doi:10.1007/978-3-642-03364-3_9

Zielinski, B. W. (2007). The listener: No longer the silent partner in reduced intelligibility. *System, 36*(1), 69–84. doi:10.1016/j.system.2007.11.004

KEY TERMS AND DEFINITIONS

Assimilation: A phenomenon in speech when sounds modify each other when they are close by either across word boundaries or within words.

Contraction: The process that we combine two words and articulate them as one word or syllable. Some illustrations of this form are "I've", "wanna", and "would've".

Elision: Omission of a sound while enunciation.

Flapping: The instances of pronunciation of /t/ sounds as /d/ such as the words 'letter', 'literature', 'butter', 'scatter', 'pattern'.

Linking: The phenomenon that groups of words within a sentence or phrase are connected and articulated as if one word.

R educed Forms: A phenomenon in L1 natural speech when a speaker articulates a phrase or a sentence with a specific sound, combining, deleting, adding, or even turning it into a totally new form.

Chapter 6
Boosting L2 Listening Comprehension Through Web-Based Listening Activities on Reduced Forms

Kaine Gulozer
Bartin University, Turkey

ABSTRACT

In the field of second language (L2) perception, there is a common adherence to quantitative methods to examine reduced forms (RFs). This chapter extends the field by reporting on an investigation that analyzed L2 listeners' perceptions of RFs in English from a qualitative perspective. RFs instruction through web-based activities was delivered to a total of 80 learners of English of varying proficiency for five weeks. Twenty participants reflected on their performance on RFs listening tasks and provided justifications for their perceptions of the target RFs. Qualitative analysis revealed that the RFs that influenced L2 learners' perceptions of RFs were linking, pause phenomena, and assimilation. The results of using such qualitative methodology highlights the important role that RFs plays in perception judgements in syllable-timed languages such as Turkish, a factor which has not always been given much prominence in previous L2 fluency quantitative research.

INTRODUCTION

Nowadays, communication with other people across the globe has almost reached its peak level through the integration of intelligent design of multimedia. The integration of technology is inevitable almost in all fields of education as well as the skills in second language (L2) instruction. The use of variety of websites, mobile applications or specifically tailored tutorials may not always meet the needs unless the problem is clearly identified and solution based activities are delivered.

It is obvious that there are some elements that cause learners difficulty in speech perception. Apart from the physical circumstances such as loudness and the setting, rate of speech topic familiarity, size

DOI: 10.4018/978-1-7998-7876-6.ch006

of vocabulary, length of sentences including the duration of speech are all important elements in speech perception. There are some features that often cause misunderstanding on the part of the L2 listeners. Those features as called reduced forms, connected speech. This phenomenon is also termed in several ways such as 'sandhi-variation' (Henrichen, 1984), modifications in speech or reduced forms (Matsuzawa, 2006). Reduced forms (henceforth abbreviated to RFs) are prevalent in real life speech or in an idea unit of a sentence in particular, making the utterance pronounced altogether. An idea unit, in other words, thought group is defined by Celce-Murcia, Brinton & Goodwin (2011) as a distinct chunk of speech that makes a semantically and grammatically cohesive unit of discourse. They further state that an utterance can be separated into an idea unit when a speaker can normally pause in the real time of speech. For example, in a sentence: "I hope you won't open the window when it's so cold and snowing" two idea units occur: "you won't open the window" and "it's so cold and snowing". However, in the sentence: "The cat, with which I used to travel, is gone", the only idea unit can be considered as "I used to travel" because the speaker is supposed to give a logical pause before comma. It is reported that all languages have a range of RFs employed (Rosa, 2002), the reason of which is attributed to the fact that not every word of utterance in natural speech is fully articulated (Aquil, 2012; Avery & Ehrlich, 1992; Roach, 2009). Rather, the speech organs articulate sounds together to save time and energy instead of forcing speech organs to take a new position for each sound. Hence, a variety of phonological modifications occur in naturally spoken English, which causes different pronunciation of words from their phonetically pronounced forms (Buck, 1995; Mitterer, Kim & Cho, 2013; Zielinski, 2007).

RFs are difficult for second language (L2) learners to perceive and identify the boundaries between the words, which in turn cause disruption in listening comprehension (Ur 1984; Underwood 2012; Brown, 2006). Listening to the sentences articulated in isolation is markedly different from those articulated in RFs (Aquil, 2012; Toro, Sebastian- Galles & Mattys, 2009). When ESL learners are asked to judge the intelligibility of a native speaker's (NS) speech through audio, they tend to focus more on global comprehension rather than that of the modifications in real life speech. Because ESL learners can easily perceive content words, they may not be able to determine what RFs impede their perception in the audio. With this in mind, the RFs such as contraction, assimilation, flap, elision and linking could be promoted to raise awareness of perception on the part of ESL learner through explicit instruction. Moreover, the limited training or the lack of the incorporation of technology in listening instruction may result in experiencing misperception and not catching up with the instructor in the class on the part of listener (Seferoğlu & Uzakgören, 2004). Thanks to the recent accessibility of online or offline tools, videos, podcasts or a range innovative mobile applications are increasingly incorporated in second language classes in second language (L2) listening instruction. What is more, course delivery through technology allows learners independent learning, in which they can regulate their learning independently apart from the classroom context (Bayyurt & Karatas 2011; Chang, 2005; Hsu et al., 2013). By doing so, learners can be exposed to hear and see a stream of speech on a video and imitate the native speaker as a part of natural language learning situation (Kuhl et al, 2008).

This may also provide learners follow their own pacing in that they can check themselves over and over again at their own availability. Hence, learners become autonomous and enhance ownership of learning the reduced forms. Learners' monitoring their own learning through visual entities is specifically important when the input is slightly beyond their linguistic competence (Danan, 2004; Erçetin, 2011). With regard to the present study, this is preserved as the video modality, (yet with no captions) in that it may serve to increase language perception, specifically the RFs, by connecting auditory to visual input. As mentioned before, audiovisual materials through technology are potential tools that are believed to improve second language listening skills (O'Bryan & Hegelheimer, 2007; Winke, Gass & Sydorenko; 2010).

The present study has derived from the researcher's classroom observations during the fall term of 2013-2014 academic years in the field of listening instruction. The following quotes from the students who took class listening and speaking course in their preparatory year of English language learning illustrated the real time impediments during the class: "The speech is too fast to keep up with because there is no isolation between the words", "I cannot determine the word boundaries within a statement, it sounds that all words run together", "I can hear only the subject of the sentence, the rest is impossible to pick up". These are the statements characterizing the phenomenon on the part of non-native English language learners whose first language (L1) is Turkish in Turkey. Apart from the L2 context of Turkish speakers, It has also been reported in a range of studies that L2 learners from many other countries such as Japan, France and China have difficulty in understanding naturally spoken English such as Japan, France (Hasan, 2000; Field 2008; Ito, 2006; Yeh, 2013). This is likely to be the case due to the fact that listeners' segmentation of words relies on procedures determined by the phonological characteristics of their first language (L1). Turkish has been said to hold the characteristic of "sound-phoneme correspondence and syllable timing" in contrast to the "stress- timing" and "no sound-phoneme correspondence" characteristic of English (Otake, Hatano, Cutler & Mehler, 1993). Since the rhythmic properties of Turkish are substantially different from English in which rhythm is based on stress units, Turkish learners of English may be influenced by some features of L2 speech such as linking, elision, and assimilation. For example, the change of the sound in assimilation may cause learners not to be able to identify the boundary between the words. Taking such factors into account in instruction will help learners identify and perceive accurately on the part of the language learner.

The current study presents a qualitative study of the effectiveness of reduced form instruction on second language sentential level listening comprehension. The target reduced forms covered in the study are contraction, assimilation, flap, elision and linking. After developing an instructional package of RFs on a web-based environment, where L2 aural and visual input is provided, the participants were randomly assigned to eight reduced form groups: four experimental groups (audio instruction to high proficiency (A+hp), audio instruction to low proficiency A+lp), video instruction to high proficiency (V+hp) and video instruction to low proficiency (V+lp).

This study aims to investigate the impact of RFs instruction on Turkish EFL learners' comprehension of RFs, namely, contraction, assimilation, flapping, elision, and linking displayed at a sentence level. The following research question is formulated for the aim of the study:

What are the students' perspectives on the instruction of reduced forms through web-based learning?

LITERATURE REVIEW

Since the seminal study conducted by Henrichsen (1984) on RFs, the body research dispersed on a variety of angels, especially, in relation to the RFs' effect on listening comprehension. Henrichsen (1984) aimed to determine how presence and absence of reduced forms influence the comprehensibility of English input. Recruiting 65 subjects with three distinct English proficiency levels (high, low, native speakers), and employing two treatment conditions as presence or absence of RFs, subjects were given three tests as RFs exercises in two versions as one with the presence and another one without the presence of RFs as well as an Integrative Grammar Test. RFs exercise test included the RFs of "contraction" and "assimilation". The result of the study showed that there was difference between the RFs test scores of native speakers and those of high- level language learners in the presence of RFs, yet the difference was not significant in the condition without the RFs. When the low level and high level language learners were compared in relation to the comprehension of RFs, high- level learners performed better than the low-level learners in the presence of RFs.

Even though there has been scarce in studies investigating the effects of reduced forms instruction (RFsI) on listening comprehension recently, listeners' performance seems to have improved when they were delivered instruction on these peculiar features (Brown, 2006; Brown & Hilferty, 1986; Ito, 2006; Masahiro, 2012; Matsuzawa, 2006; Underwood & Wallace, 2012).

Brown and Hilferty (1986) investigated the degree to which reduced forms instruction (RFsI) was effective for 32 intermediate level of Chinese graduate students. Treatment group received a four weeks of daily RFsI in ten minutes on each day, and control group received the same amount of instruction only on discrimination of minimal pairs. The researchers administered three tests: Integrative Grammar Test, a version of UCLA English as a Second Language Placement Examination listening comprehension subtest, and a RFs dictation task. The results showed that RFsI had a significant effect on students' performance of the dictation task. Yet, they found no difference between the two groups in terms of listening comprehension that is measured by UCLA listening comprehension test. One explanation given for this was that the RFs in this test represented a very small part of this subtest. Nonetheless, the results indicate some evidence that RFsI may not be helpful in overall listening comprehension.

In another study, Underwood and Wallace (2012) highlighted the importance of RFsI and investigated the comprehension and productive development of reduced form instruction with 52 low-proficiency level undergraduate Japanese students. An eleven-week instruction of reduced form on "elision, assimilation, and elision/assimilation" was provided in two parts, the first of which consisted of 10- 20 minutes weekly instruction to develop conversation along with the RFs. The second part of instruction covered through a textbook entitled "Speak Easy: Spoken English for Japanese Learners by Underwood, Bridges & Hattori (2010) to expose students in reduced forms through the dialogues in the book. The results of the study proved significant increase in comprehension of RFs. In relation to the productive development of RFs, the researchers found out that there was increase in the production of RFs measured by pre- and post- instruction speaking test.

Erbanova (2014) examined Czech students' production of RFs and identified the aspect of the RFs as stress, intonation and rhythm in English speech in particular. The researcher delivered instruction on sentence stress, rhythm, linking and elision to 11 students at intermediate language level through the methods of communicative learning and task based instruction. The students' RFs production skill was assessed through two recordings –reading aloud, free speech. The researcher found out that almost all participants doubled their performance in both speaking tasks of RFs and that the instruction was highly effective on the part of the learners.

Matsuzawa (2006) also proposed the instruction to improve the comprehension reduced forms, selecting ten RFs that seemed challenging for Japanese colleague graduates. The selected categories were assimilation, contractions, deletion, flapping, glottalization, linking, -nt reduction, palatalization and weak forms of function words. Three version of dictation tasks with 30 sentences in each one and a retention test were administered to 20 people with no control group in order to identify the performance scores of the reduced forms. The results of the study showed that the students' performance in all chosen categories highly improved.

THE USE OF TECHNOLOGY IN L2 LISTENING CONTEXT

The popularity of digital technology has enabled the additional modalities to be integrated both in the instruction and testing of listening comprehension. A significant body of research comparing modalities in the context of enhancing listening comprehension has emerged since the 1970s. The types of modality studied in earlier investigations ranged from audio (such as traditional audio CDs that came with the course textbook, radios and podcasts) to visual (such as video and multimedia or the amalgamation of both or even multiple modalities).

Aural input is considered to be a main stimulus in L2 language learning as well as listening comprehension. Rosell-Aguilar (2007) presented a comprehensive overview of podcast materials both in audio and video form and asserted that effective practice in podcasting was worthwhile for language learners. Many researchers utilized audio modality mainly for providing target language input or supporting learners' independent learning (Chang, 2009; Cross, 2013; Yeh, 2013).

Employing two modalities of auditory input, Chang (2009) compared the impact of "reading while listening (R/L)" and "listening only (L/O)" modality on listeners' listening comprehension performance. A total of 84 college students were assigned to read two story books in one of the modalities that followed the listening comprehension tests with 95 items in the form of sequencing and gap filling. A questionnaire was also administered so as to get students' perceptions regarding the two stories. Results indicated that students gained only 10% more with R/L modality and that they opted for the R/L modality rather than that of L/O modality in their L2 listening.

In another study, in an attempt to present pedagogical insights into independent listening of podcasts implemented as a case study of a Japanese learner of English as a foreign language, Cross (2013) found that the autonomous use of the BBC's online podcasts could improve students' L2 listening ability. The researcher explored the effect of delivering guidance and feedback on meta-textual skills and aspects of metacognitive instruction on an L2 learner's autonomous use of podcasts over a period of 9 weeks. Podcasts were presented by a multimedia file delivered through the internet using syndication feeds, for playback on mobile devices and/or computers or on portable players after downloading the one in interest via the internet. Meta-textual skills is regarded as a phenomenon that enables students' improvement from lower-order text comprehension processes such as word recognition and syntactic parsing to higher level ones such as drawing conclusion about the purpose and content of the text (Bull & Anstey, 2010). Data collection tools were listening journals in which the student recorded aspects concerning the listening podcasts that she had chosen to listen from the BBC and an interview. The results of the weekly journal entries and transcribed interview responses indicated that the learner could improve her metacognitive capacity and enhance her understanding of what listening comprehension in an L2 entails, showing improvements in her listening performance.

The importance of podcasts in developing language learners' listening skills and investigated the extent to which CALL could be integrated into an academic English as a Second Language course along with the instruction of listening strategies was also highlighted by O'Bryan and Hegelheimer (2007). Both graduate and undergraduate students from a range of majors such as Economics and Engineering participated in the study over fifteen weeks with two hours per session of instruction. A total of fourteen podcasts were delivered in two modalities designed for the listening course as two video podcasts and twelve audio podcasts. Students received explicit strategy instruction on a range of academic and general listening strategies in which students were listening to the lectures, taking notes and coping with the difficulties they encountered in their audio and video listening materials. On the basis of the reflective journals collected from instructors and structured interviews and surveys with the students, results of the study showed that podcasts were potentially facilitative for developing listening comprehension skills and that the delivery of the video podcast – the notetaking in this case - was reported to be more appealing and informative than the audio counterparts on the part of the students.

A number of studies combined more than one modality to compare the effectiveness of the audio and video modalities in an identical listening comprehension test (Başal, Gülözer & Demir, 2015; Batty, 2015; Hayati & Mohmedi, 2011; Pardo-Ballester, 2016; Wagner, 2010). Hayati & Mohmedi (2011) focused on the video modality in relation to listening comprehension, examining the efficacy of subtitles under the three treatment modalities as English subtitles, Persian subtitles and no subtitles with 90 undergraduate students majoring in ELT department. The students watched the six episodes of a documentary film under one of the randomly assigned modality, which ensued the six sets of multiple choice tests to assess listening comprehension of the learners. Results indicated that the English subtitles group outperformed the Persian subtitles group. Of all the groups, the group with no subtitles achieved the least performance in listening comprehension test. Batty (2015) also compared the audio and video modalities, investigating interactions between format -multiple choice listening test forms- and the three text types – monologue, conversation, and lecture. A three minute listening tests comprised from a placement test along with a four option multiple choice questions in the two modalities were delivered to a 164 Japanese students majoring in English in four different proficiency level. The delivery of audio modality included a still picture of a speaker along with the name of the test. Many-facet Rasch modeling was utilized to compare the difficulties of the audio and video format in which two main concepts were focused: interaction between audio and video modality and text type. In this study, this modeling referred to the multiple choice test data, a one or a zero as being the product of the difficulties of the facets. Results of the study indicated that modality had no effect on students' responses, that is, both audio and video modality were equally effective on the part of the students to answer the listening comprehension test questions. Similarly, Başal, Gülözer and Demir (2015) also compared audio and video modality in a post-test only control group design study in which they examined how the use of audio and video modality would affect L2 listening test-takers' performance. A total of 57 EFL learners majoring in ELT program participated in the study. The instruments used consisted of four lectures chosen from a TOEFL textbook and five items of multiple choice comprehension questions for each lectures. The questions inquired about the main idea, vocabulary as well as inferred meaning based on a clear reference from the text. Results revealed that students answering the questions in each modality performed in idiosyncratic ways, so it was difficult to draw conclusions that would apply to all students in one modality. Some students in video modality could perform better in two of the lectures whereas some students in audio modality could perform better in audio modality depending on the topic familiarity. As such it could be argued that the use of video might also be potentially distracting on the part of the language learners, which could lead to inhibit comprehension. On the other hand, the findings of Batty (2015) and Başal et al.'s study (2015) contradict with what Wagner (2010) has found in which he asserted that the video group outperformed the audio group on the overall posttest. Designing a quasi-experimental pretest and posttest study, he investigated the effect of non-verbal information in a video listening text on learners' listening test performance, comparing the performance of two groups on a listening test through audio only or and video only modality. The set of short dialogue texts and a three to four minute lecture type of listening along with multiple-choice and open-ended comprehension questions were administered to the total of 112 students assigned in experimental and control groups.

In the same vein, learners had better listening comprehension scores in video modality than in audio modality in another study. Pardo-Ballester (2016) compared the effects of listening support in audio and video modality adding the redundancy elements for the both conditions for the total of 246 intermediate level of Spanish learners who enrolled in two distinct format of the courses as online – hybrid and face2face- blended. A cross-over design was adopted and a battery of eight web-based listening tests with monologues and multiple choice questions for each one in audio and video modality were administered. While the video modality included word fillers such as umm, eh, the audio modality was in audible form with non-word filler.

Nowadays, many recent researchers have shifted from audio to multimedia in which texts or captions along with audio input are integrated into a source of listening material. Yang and Chang (2013) conducted an experimental study with a pretest and post-test design and they tested whether different modes of captions could increase listening comprehension. Three modes of captions are full, keyword-only, and annotated keyword captions. They also investigated the possible contribution of the three captions to the learning of RFs and overall listening comprehension. A total of 44 Taiwanese students were recruited and randomly assigned to one of the three modes of caption groups. The source of captions were animation, cartons, movies and TV series with a total of 51 video clips to be utilized in the test. Target RFs were assimilation, liaison, contraction and elision. Pictorial symbols for each type of RFs were added to facilitate learning in annotated keyword captions mode. For example, while a blue dot indicated an assimilated sound, a grey letter marked a contracted sound. Testing instruments included a total of 41 questions of listening comprehension test covering tasks of dictation cloze, short dialogue comprehension, reduced form recognition as well as RFs marking. Results of the study indicated that the annotated keyword caption group got significantly higher performance in reduced form recognition when compared to that of the two other groups. Furthermore, incorporating the full caption was the least beneficial factor on the part of the listeners. The researchers suggested that the addition of the pictorial symbols provided the maximum effect on learners' listening comprehension.

THE STUDY

Participants

The initial sample consisted of 357 learners, four of whom did not consent to be involved into the study, and 10 did not answer all the questions given during the pre or posttest. Thus, a total of 343 participants, 181 of whom were male (74%), 162 of whom female with an average of 19.41 years old (SD=1.36), who have normal hearing and vision were recruited for the investigation. All participants possessed Turkish as a first language and had a mean self-reported duration of English language study of 9.18 years (SD= 2.51). At neither faculty of the participants is English a possible major. English courses at preparatory school is four skill classes as well as grammar. The participants were preparatory language school students from eighteen different classes at the Language School of Yıldız Technical University with a range of eight different majors.

Setting

The study was conducted at a language school of a state university where participants were randomly selected for the research. The participants received two different modalities of instructional packages through the web page at www.kainegulozer.com (CONSPERsentA for audio and CONSPERsentB for the video modality), all of whom comprised of four groups (A+hp, V+hp, A+lp, V+lp). They all received explicit instruction on RFs in a computer lab of the university. Learners were following a main course book entitled "New Language Leader" by Lebeau, Rees, (2004) along with its supplementary materials. They were taking 20 English class hours and two hours of extra speaking courses a week at the time of the pilot study.

Figure 1.

Instruments

The distribution of the number of the RFs items are as follows: 15 "Contraction", 10 "Assimilation, 9 "Flap", 11 "Elision", and a total of 15 items of "Linking" for the comprehension of RFs in English. The highest possible score the participants can achieve is 60. The data collection was provided through the reflection reports written right after the web-based courses delivered, additionally an interview was conducted to get the depth of understanding in relation to their reflections.

Procedure

After getting permission for the research from the director of the language school, the piloting research process started. First, the content of the school syllabus was checked and confirmed that target features RFs were not explicitly explained beforehand. After that, learners were randomly selected and then assigned into the four instructional groups. Before the experiment started, learners responded to a background questionnaire, filled a consent form.

The test was made up 3 sections: a) dictation task, b) discrimination task, c) listening comprehension test. Instructions for each part was written in English yet verbal Turkish explanation for the instructions was also delivered so that all participants could follow the instructions properly. The experiment took place in a computer lab of the university over four days with four different sessions for each treatment and control groups. The researcher herself conducted the tests that were administered to 85 students per day. The researcher used an overhead camera along with computer demonstration monitors to give instructions before administering the test. Then each participant got his/her headset and filled in the spaces given to them for each sentence on their spreadsheet of paper. Due to the poor speed of typing on a keyboard, and the ease of typing on a sheet of paper reportedly by the students in piloting, the participants typed their answers on a separate spreadsheet given to them along with the sentences. The students were allowed to listen to each input only once then they typed their answers to the given spaces on their spreadsheet. Students had 60 seconds for each questions in the three types of assessment tasks to type their responses after listening to each input. All the tests took 45 minutes to conduct in one session.

After the pretest, the forty students received five- class hours of instruction from the same teacher in the form of audio and video. All the experiments took place in learners' regular classes apart from the class hour. When the researcher was in the class to give instruction, another instructor, a native speaker of American English, also stayed in and used an observation sheet to control whether any bias to one of the instructions was raised. The teacher also checked the time for the each web-based instructional activity. The teacher also checked the time for each instructional activity. Upon completion of the instructions, the groups were asked to write a reflection report. Then the following week after the instruction students were interviewed.

In order to support the researcher's understanding of the students' perceptions regarding the RFs instruction, the study group of 20 students was asked an open-ended question: What is your opinion about the RF instruction that you were delivered through the webpage during the five weeks?

Data Analysis

In order to support the results obtained from the quantitative analyses, the data obtained from the students' reflection reports and the interview were analyzed by means of content analysis following Patton's (2002) evaluation methods. Written consents were obtained in advance of the self-reflection reports and face-to-face interviews with the four instructional group of learners. The interview data were transcribed verbatim in either English or Turkish by the researcher, and double-checked for accuracy by another expert in the field. As for the self –reflection reports, the data were categorized into organizational themes based on the questions provided in weekly reflection reports. The excerpts under each organizational theme were coded through content analysis. Each participants were coded by number.

Results

What are the students' perspectives on the instruction of RFs through web-based learning?

A qualitative investigation through the reflection reports and interviews provided in depth information pertaining the participants' perspectives on the instruction of RFs through web based learning and their experience of listening comprehension development through raising awareness on RFs.

All the participants in the four instructional groups – A+hp, V+hp, A+lp, V+lp- who were interviewed indicated that they enjoyed the experience of listening comprehension concerning the RFs: it was different from listening activities they did in class. More importantly, they reported the web-based RFs listening comprehension experience as "comprehensible", "easier", and "facilitating". Additionally, findings from reflection reports also indicated parallel results of the students' views in the interview.

Firstly, the results from interview and reflection report data from the A+hp group are described and participant 1 this group described her experience as follows:

"You learn what part in a sentence you miss while listening. I can grasp the words much more easily than the times when I used to miss after getting the information about the feature. You have chance to listen to the parts you miss in a sentence over and over, which helps you understand the little parts you miss. In a regular class, we do not have this kind of chance. I have not had any difficulty while practicing the web- based listening activities on RFs. RFs were quite new entities for me to improve my listening comprehension".

Participant 2 provided a similar response:

"In a class setting, we listen to a text only once and most of the times I am not able to realize the whole sentence at all. There's always a part that I miss. Yet, there is always somebody who is better in listening and answering the questions so I just keep being silent and try to understand from my friends' answers. But here, I have the chance to keep the icon back and forth when I could not pick up the expression. I like following my own pacing while listening to a series of sentences. Once I clearly learn about what I've missed, I feel secure, which is the case here, in web based application."

Thus, the emphasis in these comments was the usefulness of the RFs instruction through the web-based activities as well as effectiveness of comprehension of the RFs on the part of the learners.

Participants were asked to write reflection reports on each type RFs they had learned. The majority of the A+hp group reported that having listening practice of the RFs –except for flap - were useful to improve their listening comprehension. This group stated that because they had already known about the feature "flap", it was much easier for them to pick up a flapped word while listening. In an attempt to answer the question whether they had any challenges during the listening practice through the web, they said that the instructions were clear enough and the instructor along with another teacher in the computer lab were helpful in each step of the course. Thus the reports suggest that A+hp learners perceived the RFs instruction to be useful for the listening comprehension.

Secondly, the results from interview and reflection report data from the V+hp are described and participant 3 in this group described her experience as follows:

"You can watch the video and follow the sentences while listening. It is much easier to catch up with the way how the person utters certain parts of a sentence. In fact, I have learned a lot about the features as contraction, assimilation, elision and linking. For example, I did not know that native speakers

were omitting the "t" sound while they were pronouncing the auxiliary verb "wouldn't", "couldn't" in natural speech. I also remember giving mistaken answers to simple questions in the quiz just because I did not know the difference of pronunciation in contracted form of a natural speech. I have not had any problem practicing the exercises through the website. "would've" kind of contraction was that I had not initially known".

Participant 4 in this group found the RFsI beneficial:

You may already know the meaning of a sentence when you read the sentence. However, it is almost impossible to catch up with the speed and the meaning while listening. Getting the tips for each feature through the web based instruction was so practical and even kind of life saving for our listening comprehension. I wish you had taught us these features at the very beginning of the semester, which would have helped us in improving our speaking fluency as well. Thus, it seems that there was consensus on usefulness and the benefit of RFs Instruction for both V+hp and A+hp groups.

Participants in the V+hp group were also asked to write reflection reports on each type RFs they had learned. The majority of the group reported that having listening practice of the RFs were useful to improve their listening comprehension. They added that having comprehensive information about the features would advance their real time comprehension of utterances and sentences. Because they had the notion that some sounds may change, some others may be omitted or linked to each other in this web-based RFs instruction program, the blur of the utterances dispersed and the chance of comprehension of a phrase, utterance or expression would be high on the part of the listener. Thus, the reflection reports of the V+hp group suggest that learners perceived the RFs instruction to be useful and illuminating for the listening comprehension.

Thirdly, the A+lp group's interview and reflection report data revealed the following results. Participant 5 in this group reported his experience as follows:

"Initially, we had had a lot of listening classes, yet they were different in the several ways: First, we have rarely had sentential forms in a listening course, instead we were listening a whole text or dialogue then answering the comprehension questions in which I used to miss a lot of words and easily get blocked in the first few missing or the very beginning of the listening comprehension activity. But now, thanks to the web-based RFsI program, I know the reasons why I did miss the words. Secondly, I could realize that there have been several features employed in a sentence by a native speaker of English in real time speech. Some examples of features like contracted "would've" or flapped "water", assimilated in "did you" and linked "may I" even in a simple statement or utterance. Unless the statement is too advanced for me, I can clearly identify the features, which enables me further understand the subject matter. Thirdly, it facilitated me with autonomous and a sort of entertaining learning as I could follow my own pacing, giving pauses when I miss specific parts in audio".

Participant 6 in this group found the RFsI beneficial:

"I used to think that speakers were just swallowing the words while speaking. I was dealing with vocabulary, the speed as well as the real time of fluent speech while listening to native speaker. Actually, what I could not understand was what unit of a sentence they were swallowing during the speech. I usually watch movies with subtitles and I had just bare information about the natural speech but thanks to the program I could clearly identify some types of utterances are representing certain type of pronunciation,

which further helped me identifying the borderlines of words in a sentence. I feel more secure when I start listening on a sentential basis and realize what causes my misunderstanding or even missing the whole bunch of a phrase or a sentence. I got a real relief when I started catching up with the sentences in listening, which was a real problem beforehand. I did not suffer from any problems yet I just wanted to skip practicing flap as I already heard that sound change in most American movies".

Two other participants reported that they knew some samples from flap such as "letter", "literature" or "navigator" yet they did not know the other features especially contraction, assimilation, elision and linking.

Participant 7 stated that the easiest feature included the examples of "letter", "literature", navigator" as he had heard those examples on movies before. Nevertheless, the rest were all new for me.

Participant 8 reported that she could not believe what she saw when the sentence uttered was exactly different from what she could guess as it was contracted, elided or assimilated. She also added that of course, they should also know the sentence structure or a meaning of the word, yet learning the way how it was uttered in a natural speech helps us save time in understanding the subject matter.

The overall message from the A+lp group concerning their perspectives on RFs instruction was that they heavily favored the instruction as facilitating, and comprehensive. The feature of "flap" was the only one that had prior information in their personal listening practice.

As for the reflection report data in this group, participant 9 highlighted the importance of learning the features, stating that:

"This experience was like getting vitamin pills for me in the concept of listening comprehension cause I could consciously pick up the words while listening, at least I could draw a clear conclusion on the basis of what I have learned and what I should take care when I may have missed in certain units of a sentence. I should have more practice with other sample sentences though. I tried to do weekly assignments, yet they were composed of just seven sentences. I wish we had extra assignments as a free time activity. I believe that the more practice I have on the listening of these features, the more competent I will be in both listening and speaking. What is more, I really enjoy going back and forth with the mouse to catch up with the problematic unit of the chunk in a sentence thanks to the delivery of the information about the features".

Thus, the reflection reports of the A+lp group suggest that learners perceived the RFs instruction to be beneficial and practical for listening comprehension

Finally, the V+lp group's interview and reflection report data revealed the following results. Participant 10 in this group reported her experience as follows:

"I was thinking that I was poor in listening comprehension due to the lack of structural knowledge in a sentence, yet I realized that I could understand the sentence when I read the uttered sentence so my problem was hindered in some features of natural speech. Because I had a chance to learn them through the video examples along with the PDF form of its parallel sentences, I could grasp the sentences more easily. This web-based listening experience was like opening the curtains in your ears and being able to see the beyond much more clearly. It helped me improve in my listening comprehension whatever I listen either on youtube or in the class. I think this should be one of the first steps to be taught in listening comprehension as I had a real difficulty of comprehension while listening to the text in the former

semester. It could have been even more fun than the torture if we had had this type of listening exercise beforehand".

Participant 11 in this group found the RFs instruction beneficial in different ways:

"You may watch a lot of movies or listen to popular TED talks, yet you may get only a few messages of the whole listening text without the knowledge of the features delivered in web based RFs instructional program. Watching American movies, I can now realize the light shed on us through the program. The instruction given to us on the basis of sentences fulfilled our gaps in particular units of a sentence or an utterance. Otherwise, I just used to ignore many parts in listening, which caused me to misunderstand and simply get discouraged in many cases. In brief, rigorous practice for each feature fueled my listening comprehension".

As was the case for all other groups, participants in V+lp group were also asked to write reflection reports on each type RFs they had learned. Participant 12 had a clear view of RFs instruction as a beneficial resource, stating that neither listening exercise in any other resources such as listening books or internet provided this kind of sentential practice on RFs. He also stated that he was getting lost right after listening to a few sentences and he did not know that those features would have been one of the major entities that he should have learned to improve his listening comprehension. The forms "contraction" and "flap" might be an easy feature to pick up in listening yet the feature that a sound is deleted or words are linked together is almost impossible for me to get the real message in listening. Yet, I think I will be overcome this problem as I have a chance to watch the video in a sentential form and learn the notion of RFs.

Participant 13 in this group underlined the importance of learning these features however he did not view of the benefit because he thought that they were not asked the listening questions through videos so it could be fun watching them but also waste of time to watch while we are expected to listen and answer the question in a real exam. Thus, the reflection reports of the V+lp group suggest that learners perceived the RFs instruction to be useful and important for the listening comprehension. Yet, the fact that the form of the delivery of a listening exam in a real class setting led some of the participants hesitation on the effectiveness of the video modality.

Discussion

This study elicited the instructional group of learners', namely A+hp, A+lp, V+hp and V+lp, perspectives on the instruction of RFs through web-based learning. Illustration for each group will be given in the same order as seen above.

It is worth noting that a small number of participants volunteered for self-reflection reports and interviews related to the qualitative portion of participants' perception on RFs instruction. It is also possible that they may have overlooked some of their important thoughts in their reflection reports. Therefore findings reported for this research question and the ensuing discussion and conclusions apply solely to this sample and should not be generalized. Yet, what appears certain is that, based on student responses by A+hp group through reflection reports and interviews, RFs instruction contributed to their listening comprehension. Although some students found it easy to identify the feature of flapping in a speech, others reported this feature as novice and facilitating as the other RFs in listening comprehension. Apart from this feature, the students' opinion about the instruction of contraction, assimilation, flap, elision and linking was positive as they all reported that the instruction of these features was helpful in developing their listening comprehension.

As a means of study, a majority of students embraced the web-based instruction of RFs instruction as a comprehensive, important and alternative study option. Cross (2013) promoted an autonomous use of BBC's online podcasts outside the classroom to improve learners' listening ability, introducing meta-textual skills and activities for metacognitive instruction. Collecting data through journal entries and interviews, he indicates self-regulation (giving learners control over) of L2 listening materials using podcasts for listening comprehension purposes helps the learner develop his/her own independent listening outside the classroom, thus promoting comprehension. The findings of this study are in line with the studies conducted by Cross (2013) who indicated that self-regulation during listening affects the learners' listening comprehension and Lai, Yeung and Hu (2016) who dealt with student and teacher perceptions regarding the promotion of autonomous language learning through technology.

Secondly, the VHP+RFsI group stated that they were satisfied with the web-based RFs instruction, which fostered their listening comprehension development. However, some of them pointed out that this approach to instruction was different from watching films in the sense that they have listened to the sentences. It seems that learners expect watching movies or at least short clips as it is normal behavior for L2 learners. Therefore, they described watching sentential level of RFs did not fit in their usual expectations. As such, some of the students in this group had independently decided to watch the video in full to identify the RFs in the video content. However, the first three weeks of the application of this web-based RFs video listening was effective and enjoyable on the part of the learners. They reportedly, enjoyed listening yet the following four weeks, they stated that watching a video on the basis of sentences turned out to be facilitative for their listening comprehension yet not oriented for independent watching. Because they had to deal with bottom up (detailed listening) skills in listening through the sentential examples of RFs with a structured sequence of tasks, students may not have maintained interest and motivation. Obviously, learners in this group had different listening goals, level of interest and motivation towards improving listening comprehension through web-based designs. This perception may be attributed to the fact that learners are exposed to top down skills rather than that of bottom up skills in their listening comprehension courses and exams. Another factor could be the fact that learners are highly motivated by exam-oriented type of listening tasks rather than that of the detailed listening tasks that they may not be responsible in a real exam of a listening comprehension test. Considering the motivational point of view and level of interest, this study is consistent with Cross (2013) as he suggests that though learners have "different listening goals, time available, level of interest and motivation", (p. 17) online tools do provide learners with useful options to meet the learners' specific developmental needs.

Thirdly, the A+lp group's comments showed that they believed the web-based RFs instruction course was helpful for their listening comprehension ability and the perception of RFs. When the students responded to the reflection reports, they reported all the RF listening tasks through web-based instruction in a positive manner. Words positively used to describe their perspectives of the RFs instruction included "awesome, very useful, beneficial, practical, informative, comprehensive and entertaining, essential". These utterances are at the core of the learners' reactions in their reflection papers. The learners had positive opinions about the RFsI's effect on their learning behavior as they stated that they could get feedback after learning each feature and that they could monitor their own progress. This finding echoes previous research with Mathews & O'Toole (2015) that provided learners with computer-mediated aural modality in which they incorporated a computer application to enhance leaners' word recognition for speech in L2 context. The learners in this modality could listen to input over and over again as termed by the researcher "self-determined exposure" (p.364). The researchers indicated significant differences between word recognition scores in pretest and posttest.

Finally, the V+lp group made positive remarks on their listening comprehension in this study, which indicates that they enjoyed having engaged in web-based RFs learning so they could better gauge their listening comprehension of the target RFs. In fact, they also informed that flap was already the feature they had been familiar with due to the frequent exposure to the feature through movies they had initially watched with subtitles. Apart from this perception, in overall, they reported that they optimized their listening comprehension experience through the web-based RFs instruction tool which was, reportedly, a valid study option based on the evidence from student interviews. Furthermore, the students reported that RFs, particularly, elision, assimilation and linking were the ones that met their needs not only in listening comprehension but also in speaking comprehension. As they stated, being able to see the speaker's articulation of contraction and see the exact units in the articulation of the RFs a number of times enabled them to foster their listening comprehension. The findings of the present study are consistent with Yang and Chang (2014) as their study found that annotated key word captions were beneficial for learners in the sense that it guided them to perceive RFs in listening comprehension texts.

In brief, concurring with current research findings on learners' perceptions of web- based RFs instruction, the participants in this study perceived the instruction as being a beneficial learning experience for developing their listening comprehension.

Limitations

The present study has several limitations, perhaps the most consequential of which was its limited control of proficiency levels. The proficiency level of the learners was identified on the basis of the exam developed and run by the language school at the very beginning of the semester. However, this is clearly not a perfect measure of L2 proficiency for overall listening ability. Additionally, the study did not include advanced learners, as almost all students reported having no problem in listening comprehension.

The second limitation was the controlled experiment of RFs at a sentence-level of listening comprehension. Listening comprehension of RFs at extended sentence level or a paragraph-level with a consistent context was not employed in the study. It was also challenging to ensure that the students keep listening and answering all the questions in the given tasks throughout the pretest and posttest.

Lastly, the number of students who volunteered to write reflection reports in the qualitative analysis limits the generalizability of the results. This study was carried out in an EFL setting with Turkish participants who were at the preparatory school of a state University in Istanbul, Turkey. The targeted RFs were contraction, assimilation, flap, elision and linking at sentence level and thus the effect of instruction of these features on listening comprehension at the sentence-level was investigated. For this reason, its findings may not be generalizable to other L2 learners, other RFs and a paragraph level of listening comprehension in a context. Web based RFsI was designed to help students master challenging sound modifications by offering them a scaffolded and instructional platform. Due to the qualitative nature of reflection reports and interviews, and the insufficient number of participants in reflection reports, learners' reports regarding the perceptions on the web based RFs instruction was limited, but such a data might be collected if the research were replicated and expanded to include a larger student sample. In a similar vein, because the data for this study came from the preparatory language year of students from various fields in a technical university, it cannot be guaranteed that they all had the same degree of listening comprehension ability in RFs, and its results may not be generalizable to young EFL learners.

FUTURE RESEARCH DIRECTIONS

This study focused on comprehension of RFs in L2 listening comprehension context, which can be used as a basis for future research in the areas of web-based language instruction of RFs, perception of RFs and production of RFs. Future research can be implemented through the use of subtitles that can be extracted from movies. Hence both context and the RFs can be imbedded and both bottom-up and top down comprehension skills can be crosschecked on the part of the L2 learners.

CONCLUSION

The current study presents a qualitative study of the effectiveness of reduced form instruction on second language sentential level listening comprehension. The target reduced forms covered in the study were contraction, assimilation, flap, elision and linking. After developing an instructional package of RFs on a web-based environment, where L2 aural and visual imput is provided, the participants were randomly assigned to eight reduced form groups: four experimental groups (audio instruction to high proficiency (A+hp), audio instruction to low proficiency A+lp), video instruction to high proficiency (V+hp) and video instruction to low proficiency (V+lp). Based on the assessment tests (dictation task, discrimination task, listening comprehension test, reflection report and interview), it was found that the instruction of RFs is effective in L2 listening comprehension development. One important finding is that the features of contraction, assimilation, elision and linking facilitate listening comprehension, which is in line with the findings of previous studies; however, flap has no role in this case. The results also indicate that when audio and video modality with no captions are offered and the beneficial effect of the modality is compared, students showed almost equal performance across the assessment tests (dictation task, discrimination task and listening comprehension test). As for the proficiency levels, the four instructional groups in both high and low proficiency levels outperformed the control groups in regard to the comprehension of reduced forms and the development of listening comprehension.

REFERENCES

Aquil, R. (2012). Listening English connected speech: A problem and solutions. *Arab World English Journal, 3*(2), 329–364.

Avery, P., & Ehrlich, S. (1992). *Teaching American English Pronunciation*. Oxford University Press.

Başal, A., Gülözer, K., & Demir, İ. (2015). Use of video and audio in EFL listening test. *Journal of Education and Training Studies, 3*(6), 83-89. . doi:10.11114/jets.vi6.1001

Batty, A. O. (2015). A comparison of video- and audio-mediated listening tests with many-facet Rasch modeling and differential distractor functioning. *Language Testing, 32*(1), 3–20. doi:10.1177/0265532214531254

Bayyurt, Y., & Karataş, N. B. (2011). Needs Analysis: The first step in content development for mobile learning. *Boğaziçi Journal of Education*, *28*(1), 13–22.

Brown, J. D. (2006). Authentic communication: Why is it important' ta teach reduced forms? *Authentic Communication: Proceedings of the 5th Annual JALT Pan-SIG Conference*, 13-24. Retrieved June 6, 2007 from http://jalt.org/pansig/2006/HTML/Brown.htm

Brown, J. D., & Hilferty, A. (1986). The effectiveness of teaching reduced forms for listening comprehension. *RELC Journal*, *17*(2), 59–70. doi:10.1177/003368828601700204

Buck, G. (2001). *Assessing Listening*. Cambridge University Press. doi:10.1017/CBO9780511732959

Celce-Murcia, M., Brinton, D. M., & Goodwin, J. M. (2011). *Teaching pronunciation: A reference guide* (2nd ed.). Cambridge University Press.

Chang, C.-S. (2009). Gains to L2 listeners from reading while listening versus listening only in comprehending short stories. *System*, *37*(4), 652–663. doi:10.1016/j.system.2009.09.009

Creswell, J. W. (2012). *Educational Research: Planning, conducting and evaluating quantitative and qualitative research.* Pearson Education, Inc.

Cross, J. (2013). Promoting autonomous listening to podcasts: A case study. *Language Teaching Research*, *18*(1), 8–32. doi:10.1177/1362168813505394

Danan, M. (2004). Captioning and subtitling: Undervalued language learning strategies. *Meta*, *49*(1), 67–77. doi:10.7202/009021ar

Erbanova, P. (2014). *Aspects of connected speech in English: Assessing students' progress after pronunciation training* (Master's Thesis). Masaryk University. Retrieved from https://is.muni.cz/th/c4mfj/Master_s_thesis.pdf

Erçetin, G. (2011). Pedagogical issues in developing mobile assisted language learning materials. *Boğaziçi Journal of Education*, *28*(1), 22–31.

Field, J. (2008). Revising segmentation hypotheses in first and second language listening. *System*, *36*(1), 35–51. doi:10.1016/j.system.2007.10.003

Hasan, A. S. (2000). Learners' perceptions of listening comprehension problems. *Culture and Curriculum, 13*(2), 137-153. . doi:10.1080/07908310008666595

Hayati, A., & Mohmedi, F. (2011). The effect of films with and without subtitles on listening comprehension of ESL learners. *British Journal of Educational Technology*, *42*(1), 181–192. doi:10.1111/j.1467-8535.2009.01004.x

Henrichen, L. E. (1984). Sandhi-variation: A filter of input for learners of ESL. *Language Learning*, *34*(3), 103–126. doi:10.1111/j.1467-1770.1984.tb00343.x

Hsu, C.-K., Hwank, G.-J., Chang, Y.-T., & Chang, C.-K. (2013). Effects of video caption modes on English listening comprehension and vocabulary acquisition using handheld devices. *Journal of Educational Technology & Society*, *16*(1), 403–414.

Ito, Y. (2006). The comprehension of English reduced forms by second language learners and its effect on input-intake process. In J. D. Brown & K. Kondo-Brown (Eds.), *Perspectives on teaching connected speech to second language speakers* (pp. 67–81). University of Hawai'i National Foreign Language Resource Center.

Kuhl, P. K., Conboy, B. T., Coffey-Corina, S., Padden, D., Rivera-Gaxiola, M., & Nelson, T. (2008). Phonetic Learning as a pathway to language: New data and native language magnet theory expanded (NLM-e). *Philosophical Transactions of the Royal Society of London. Series B, Biological Sciences, 363*(1493), 979–1000. doi:10.1098/rstb.2007.2154 PMID:17846016

Levis, G. M., & Levis, J. (2010). Authentic speech and teaching sentence focus In J. Levis, & K. LeVelle (Eds.), *Proceedings of 1st Pronunciation in Second Language Learning and Teaching Conference* (pp. 135-144). Ames, IA: Iowa State University.

Matsuzawa, T. (2006). Comprehension of English reduced forms by Japanese business people and the effectiveness of instruction. In J. D. Brown & K. Kondo Brown (Eds.), *Perspectives on teaching connected speech to second language speakers* (pp. 59–66). University of Hawaii Press. National Foreign Language Resource Center.

Mitterer, H., Kim, S., & Cho, T. (2013). Compensation for complete assimilation in speech perception: The case of Korean labial- to – velar assimilation. *Journal of Memory and Language, 69*(1), 59–83. doi:10.1016/j.jml.2013.02.001

O'Bryan, A., & Hegelheimer, V. (2007). *Integrating CALL into the classroom: the role of podcasting in an ESL listening strategies course*. European Association for Computer Assisted Language Learning. doi:10.1017/S0958344007000523

Otake, T., Hatano, G., Cutler, A., & Mehler, J. (1993). Mora or Syllable? Speech segmentation in Japanese. *Journal of Memory and Language, 32*(2), 258–278. doi:10.1006/jmla.1993.1014

Pardo-Ballester, C. (2016). Using video in web-based listening tests. *New Approaches in Educational Research, 5*(2), 91–98. doi:10.7821/naer.2016.7.170

Rosa, M. (2002). Don'cha know? A survey of ESL teachers' perspectives on reduced forms instruction. *Second Language Studies, 21*(1), 49–78.

Rosell-Aguilar, F. (2007). Top of the Pods: In search of podcasting 'pedagogy' for language learning. *Computer Assisted Language Learning, 20*(5), 471–492. doi:10.1080/09588220701746047

Seferoğlu, G., & Uzakgören, S. (2004). Equipping learners with listening strategies in English language classes. *Hacettepe Üniversitesi Eğitim Fakültesi Dergisi, 27*, 223–231.

Toro, J. M., Sebastian-Galles, N., & Mattys, S. L. (2009). The rule of perceptual salience during the segmentation of connected speech. *The European Journal of Cognitive Psychology, 21*(5), 786–800. doi:10.1080/09541440802405584

Underwood, P., & Wallace, M. (2012). *The effects of instruction of English speech on low proficiency EFL students*. Retrieved from https://www.academia.edu/854671/The_Effects_of_Instruction_in_Reduced_Forms_of_English_Speech_on_Low-proficiency_EFL_University_Students

Ur, P. (1984). *Teaching listening comprehension*. Cambridge University.

Wagner, E. (2010). The effect of the use of video texts on ESL listening test-taker performance. *Language Testing*, *27*(4), 493–513. doi:10.1177/0265532209355668

Weinstein, N. (2001). *Whaddaya Say?* (2nd ed.). Pearson Education.

Yang, J., & Chang, C. (2013). Captions and reduced form instruction: The impact on EFL students' listening comprehension. *ReCALL*, *26*(1), 44–61. doi:10.1017/S0958344013000219

Yeh, C.-C. (2013). An investigation of a podcast learning project for extensive listening. *Language Education in Asia*, *4*(2), 135–149. doi:10.5746/LEiA/13/V4/I2/A04/Yeh

Zielinski, B. W. (2007). The listener: No longer the silent partner in reduced intelligibility. *System*, *36*(1), 69–84. doi:10.1016/j.system.2007.11.004

KEY TERMS AND DEFINITIONS

Assimilation: A phenomenon in speech when sounds modify each other when they are close by either across word boundaries or within words.

Contraction: The process that we combine two words and articulate them as one word or syllable. Some illustrations of this form are "I've", "wanna", and "would've".

Elision: Omission of a sound while enunciation.

Flapping: The instances of pronunciation of /t/ sounds as /d/ such as the words 'letter', 'literature', 'butter', 'scatter', 'pattern'.

Linking: The phenomenon that groups of words within a sentence or phrase are connected and articulated as if one word.

Reduced Forms: A phenomenon in L1 natural speech when a speaker articulates a phrase or a sentence with a specific sound, combining, deleting, adding, or even turning it into a totally new form.

Chapter 7
The Impact of Online and Face-to-Face Peer Feedback on Pre-Service Teachers in EFL Academic Writing

Işıl Günseli Kaçar
Middle East Technical University, Turkey

ABSTRACT

This mixed-method 14-week case study investigated the impact of peer feedback on Turkish English as a foreign language (EFL) prospective teachers' attitudes to and performance in a blended academic writing course in the tertiary setting. The peer feedback provision process involved online and face-to-face written peer feedback on weblogs, enhanced with a six-hour training phase for peer feedback. Weekly student journals, group feedback conferences, the end-of-the-semester survey, and the pre- and post-writing tests were used in the data collection. The qualitative data were analyzed through content analysis and the quantitative data via descriptive statistics and an independent paired sample t-test. The study indicated that the use of face-to-face and online peer feedback together was perceived as beneficial and effective by EFL pre-service teachers and that it led to a relatively successful performance in academic writing. It is suggested that participants unfamiliar with peer revision be provided with training and continuous scaffolding/guidance to ensure the effectiveness of the process.

INTRODUCTION

Feedback has been acknowledged to play a significant role in instruction (Mory, 2004; Topping, 1998). In fact, as suggested by Hattie and Timperly (2007) and Driscoll (2000), it is stated to have a very powerful impact on learning. It informs the learners of their current state of learning or performance in their zone of proximal development, promoting the students' capacity for self-regulating their own performance to achieve the goal set at the beginning of the learning process (Narciss, 2008, 2012; Nicol & Mcfarlane-Dick, 2006; Shute, 2008). Instructional feedback leads learners to confirm their existing

DOI: 10.4018/978-1-7998-7876-6.ch007

knowledge and beliefs or to modify them. It is regarded as a formative assessment tool to ameliorate or facilitate learning (Nicol & Macfarlane-Dick, 2006). Meaningful, high quality and timely feedback is likely to foster learner engagement in the content as well as in the learning process (Higgins, Hartley & Skelton, 2002). Learners can obtain feedback from diverse sources, one of which is peer feedback.

Peer feedback has been considered a vital component of the first language (L1) and second/foreign language (L2) writing instruction (Alshuraida & Storch, 2019; Lei, 2017; Rollinson, 2005) owing to the widespread student-centered learning practices. It can be defined as a collaborative activity where learners are engaged in reading, criticizing, commenting, and providing feedback on each other's writing to foster their writing competence through the provision of mutual scaffolding (Hu, 2005; Tsui & Ng, 2000; Zhu, 2001). Providing numerous affordances for enhancing L2 writing development, peer feedback is an area of growing interest for various scholars in L2 writing (Ferris, 2003; Hu, 2005; Lam, 2010; Min, 2016). In this study the terms *peer feedback*, *peer response* and *peer revision* were used interchangeably.

Peer feedback is closely associated with the process approach to writing (Zhang, 1995; Khalil, 2018), which is considered "an important developmental tool moving learners through multiple drafts towards the capability for effective self-expression" (Hyland & Hyland, 2006, p.83). A large number of scholars and educators concur with the learning affordances of classroom/face-to-face peer feedback in the English as a Second Language (ESL) and Englsih as a Foreign Language (EFL) context (Allen & Katayama, 2015; Amores, 1997; Kuyyogsuy, 2019; Min, 2005, 2006; Shrum & Glisan, 2005). However, it is also widely recognized that a multitude of factors are likely to affect the quality and the implementation of the face-to-face peer feedback process as well as its impact on learners (Ferris, 2006; Liu & Hansen, 2002; Nelson & Carson, 2006; Villamil & De Guerrero, 2006). The most significant issues regarding face-to-face feedback include the following: Learners' unfamiliarity, the lack of learner confidence and motivation to provide feedback (Chen, 2016; Speck, 2000), learners' lack of response to feedback from their peers (Cho & Schunn, 2007), and a feeling of discomfort and anxiety (Lu & Bol, 2007; Wu, Petit, & Chen, 2015).

Considering the abovementioned issues and challenges concerning face-to-face peer feedback on writing, the integration of technology in ESL/EFL writing is conducive to learners' feedback provision on writing (Warschauer & Ware, 2006). In fact, the use of synchronous and asynchronous tools can promote learner engagement in peer feedback on writing (Bradley, 2014; Ho, 2015; Liou & Peng, 2009).

Although nowadays digital technology has been indicated to provide some affordances for peer feedback provision (Warschauer & Ware, 2006), there are also bound to be certain concerns with the use of digital technology in the peer review process as the sole medium, which are pointed out in previous literature (See Ertmer et al., 2007). Therefore, the blended approach to peer feedback with a combination of online and face-to-face feedback on weblogs, enhanced with a learner training phase, is likely to enhance the quality of the peer feedback in ESL/EFL writing classes at the tertiary level. Taking into consideration the adoption of peer feedback as an integral part of mainstream ESL/EFL writing instruction all over the world in the 21st-century classrooms, it is crucial to incorporate a triangulation of methodologies and data sources to shed light on the various issues related to peer feedback and to better inform ESL/EFL classroom practices in this regard. *Online written feedback* is defined in the study as feedback responding to the students' written work on a weblog entry asychronously in a delayed time frame while *face-to-face oral feedback* is defined as the feedback provided orally in a real-time frame.

Although the impact of face-to-face and digital (online) peer feedback on learners' attitudes and their performance in writing classes has been investigated in different ESL/EFL contexts, research findings are inconclusive as to the effectiveness of different types of peer feedback. The research on the impact of both modes of feedback on EFL pre-service teachers' attitudes and their writing performances in the same context is scarce to date. Considering the context-bound nature of peer feedback as well as the challenges and issues regarding the face-to-face and online peer feedback provision, the current study, with a blended approach to peer feedback facilitated by a training program for Turkish freshman pre-service EFL teachers in the tertiary context, is likely to be of significance for the EFL teacher educators. The study attempted to highlight the impact of a face-to-face and asynchronous peer feedback in a weblog-enhanced learning environment on the freshman EFL pre-service teachers' attitudes towards writing and on their writing performance. Hence, this study aimed to address the following research questions:

1. What is the impact of the face-to-face and online written peer feedback on Turkish EFL pre-service teachers' attitudes to writing in the Turkish tertiary context?
2. What is the impact of the face-to-face and online written peer feedback on Turkish EFL pre-service teachers' writing performance in the Turkish tertiary context?

BACKGROUND

This study adopted Vygotsky's (1978) socio-cultural theory and the process writing theory (Flower & Hayes, 1980; Hayes, 2012) as the theoretical framework. Following Vygotsky's (1978) socio-cultural theory, learning and the construction of knowledge take place via mediated interaction or within the zone of proximal development (ZDP). Vygotsky's (1978) socio-cultural theory is regarded as important in the investigation of peer feedback as mediation and peer scaffolding play a crucial role in the development of writing skills for learners (Hyland & Hyland, 2006). Mediated interaction enables learners to spot and solve problems through scaffolding in their joint tasks. Interactional feedback facilitates a mutual exchange of scaffolding among learners in the peer feedback process. That is to say, learners both receive and provide scaffolding for one another in writing (Bradley, 2014; Chang, 2012; Ho, 2015). When engaged in a reciprocal exchange of peer feedback, learners tend to better understand and refine their written texts (Saeed, Ghazali, & Aljaberi, 2018).

The process writing theory (Flower & Hayes, 1980; Hayes, 2012) regards writing as a dynamic and recursive process and peer feedback is considered to be a significant stage in this process where learners fine-tune their texts via feedback and text revisions. In line with this theory, peer feedback plays a facilitative role in the peer revision activity. It provides learners with an opportunity to express their ideas, play a more active role in the text revision, and to engage in negotiation with other learners to effectively revise their texts (Hu, 2005).

Recent research has revealed that learners address global issues (content/idea development, organization and purpose) and local issues (wording, grammar, and punctuation) issues in their written texts (Bradley, 2014; Cha & Park, 2010; Chang, 2012; DiGiovanni & Nagaswami, 2001, Liang, 2010, Liou & Peng, 2009; Saeed et al, 2018). The learners are also found to exchange global revision-oriented comments more often than local revision-oriented ones in online peer feedback (Saeed, et al., 2018).

The rate of learner engagement in the revision-oriented feedback varies in line with different modes: Face-to-face and online (synchronous versus asychronous) peer feedback. While some researchers maintained that synchronous tools provide an interactive environment for peer feedback by allowing immediate or spontaneous learner response (Chang, 2012; Liang, 2010), others claimed that asynchronous tools enable learners to reflect on their feedback, fostering their feedback exchange (Liu & Sadler,

2003). Nonetheless, a majority of previous studies (Bradley, 2014; Cha & Park, 2010; Chang, 2012; Ho, 2015; Liou & Peng, 2009) underlined the major role online peer feedback plays in the promotion of learner engagement in the revision-oriented discourse of feedback. The abovementioned researchers also classified revisions in learners' drafts of writing into two categories: global and local text revisions. The former aim to improve the text in terms of of content, organization and purpose-related aspects while the latter enhance the text in terms of vocabulary and grammar usage.

MAIN FOCUS OF THE CHAPTER

Affordances and Challenges of Peer Feedback

The previous peer feedback literature reported contradictory evidence as to the impact of peer feedback on ESL/EFL learners. While some studies emphasized the affordances (McGroarty & Zhu, 1997; Mendonca & Johnson, 1994; Nelson & Carson, 1998), others pointed out some reservations and caveats associated with it (Tsui Ng, 2000; Wang, 2014). This section is devoted to the research studies related to affordances of and challenges with peer feedback. Peer feedback is reported to have multiple affordances for learners (Chaudron, 1984). The affordances are presented in three main categories in this section: cognitive, social, and methodological.

As regards the cognitive affordances, peer feedback promotes learners to take responsibility for their own learning and fosters the negotiation process, reinforcing the sense of audience and the sense of ownership, paving the way for the learner autonomy (Cazden, 1988; Ekşi, 2012; Yang, Badger, & Yu, 2006). It tends to establish a low-anxiety or anxiety-free, motivating learning environment (Sato, 2013; Yang et al., 2006), and helps develop learners' writing skills (Diab, 2010; Lundstrom & Baker, 2009).

Peer interaction during peer feedback enables students to be engaged in "exploratory talk" as they can experiment with new ideas using the emergent language (Barnes, 1976, p. 200). Concerning the linguistic benefits of peer feedback, peer interactions are considered to represent an important component of effective L2 writing instruction (Leki, 1990; Mangelsdorf & Schumberger, 1992). As Mangelsdorf (1992) argues, peer interaction facilitates L2 students' communication and L2 development in general. Peer feedback also fosters L2 acquisition by providing learners with an opportunity to test and revise their hypotheses about L2 in meaningful contexts through audience feedback, raising their awareness towards their own weaknesses in their written work (Mangelsdorf, 1992). Peer interactions promote discovery learning, self-exploration, and self-expression, and interaction with the audience, which play an important role in second language acquisition and cognitive growth (Mangelsdorf, 1992). Being in a position to provide feedback may have the impact of reducing learners' dependence on the teacher as the sole source of knowledge, which may prompt a shift towards learner autonomy and independence to allow each learner to develop self-assessment skills (McConnell, 2002).

Classroom/face-to-face peer feedback is also demonstrated to be beneficial for learners of ESL/EFL in terms of improving their collaborative learning skills, raising their awareness as readers, developing their sense of autonomy as well as their quality of writing (De Guerrero & Villamil, 1994; Hanjani & Li, 2014; Hansen & Liu, 2005). It also promotes active learner engagement, social interaction, and feedback provision (Liu & Carless 2006). Involvement in peer feedback provision may prepare learners for their lives after higher education in that it enables learners to sharpen their assessment skills and to properly evaluate the quality and quantity of not only their own but also others' work (Van Popta, Kral, Camp, Martens, & Simons, 2017).

As far as the social affordances of peer feedback are concerned, from the socio-constructivist perspective of learning (Carson & Nelson, 1994; Koschman, 1996; Vygotsky, 1978), peer feedback highlights the value of interaction and collaboration among L2 students and their writing skills development. Peer feedback promotes learners' contributions to others' learning processes, raising their awareness towards their peers' experiences and perspectives and elevating their level of understanding and appreciation in this regard (Corgan, Hammer, Margolies, & Crossley, 2004). Learners are provided with an opportunity to reflect on their peers' as well as their work, which is gradually likely to result in enhanced learning (Ertmer et al., 2007). It was indicated that the instances of interaction through peer response foster L2 acquisition (Carson & Nelson, 1996; De Guerrerro, & Villamil, 1994; Nelson & Carson, 1998; Villamil & De Guerrero, 1996). Being engaged in the revision process with a critical perspective, learners become critical readers of others' writing as well as the revisors of their writing (Rolinson, 1998). Peer feedback also entails a real audience providing immediate feedback, encouraging the writer to formulate his or her writing in line with the characteristics and demands of his/her readers. As Mendonca and Johnson (1994) indicated, peer revision can give the learners an opportunity for effective revision based on peer readers' comments. As Hyland and Hyland (2006) argued, effective peer response enables novice writers to be familiar with the readers' perspective. Some learners mentioned finding peer feedback on content useful (McGroarty & Zhu, 1997; Mendonca & Johnson, 1994). The peer audience is considered more sympathetic and less judgemental than the teacher audience. Peer response also finds support from the interactionist theories of SLA, advocating the need to push learners for the negotiation of meaning (Long & Porter, 1985) needed for the successful communication of the intended message, reinforcing the concept of audience in writing (Hu, 2005; Hu & Lam, 2010; Jacobs, Curtis, & Braine, 1998; Tsui & Ng, 2000, Yang, et al., 2006; Zhan, 2016). It was reported to promote a sense of ownership and a decline in the levels of writing apprehension among ESL/EFL learners via the provision of a supportive learning environment (Hu & Lam, 2010; Jacobs, Curtis, Braine, & Juan, 1998; Jahin, 2012; Yang et al., 2006; Zhao, 2010, 2014).

The social dimension of feedback provision was also found to enhance students' favourable attitudes towards writing (Chaudron, 1984). As regards the student attitudes towards peer feedback, students in general exhibited quite favourable views (Mendonça & Johnson, 1994; Nelson & Murphy, 1992) Peer review facilitates students' learning from each other through interaction via a socially-mediated process called *scaffolding* (Kayi-Aydar, 2013; Ruecker, 2010; Weissberg, 2006), which is in accordance with Vygotsky's (1978) Zone of Proximal Development (ZPD). As Liu and Hansen (2002) pointed out, "cognitive development is a result of social interaction in which an individual learns to extend her or his current competence through the guidance of a more experienced individual "(p.5), thereby promoting his progress in his or her zone of proximal development (Altstaedter & Doolittle, 2014). Peer feedback also provides opportunities for students to reflect on the linguistic and rhetorical features of texts for communicative purposes (Hu, 2005; Willams, 2005), facilitating their understanding of the local (linguistic features) and the global (discursive features) aspects of the text (Altstaedter & Doolittle, 2014). It also enables learners to get engaged in a social mediation situated in an authentic learning environment where they are provided with an opportunity to become self-regulated, self-mediated, and self-aware language users and to be reflective on their understanding and experience (Doolittle & Hicks, 2003, Shrum & Glisan, 2005) Thus, as argued by Donato (2004), Lantolf (2000), Lantolf and Thorne (2006), peer collaboration is regarded as a viable approach to foster EFL learners' language development via interaction.

From the methodological perspective, peer feedback is regarded as part of the formative assessment practices (See Hyland, 2000). While exchanging feedback, student authors are granted an opportunity of a role reversal, from the author to the reader (Beggren, 2015). In addition, being an essential component of the process approach which involves a series of stages such as brainstorming, planning, drafting, and revising (Wette, 2015), peer feedback promotes the development of students' creative and critical thinking skills, particularly the higher-order thinking skills (Mangelsdorf, 1992). Peer feedback provision practices also foster learner autonomy (Hyland, 2000; Villamil & De Guerrero, 1996).

Despite numerous affordances associated with peer feedback revealed in the above-mentioned studies, an array of caveats are reported related to the peer-review process. To begin with, time constraints are considered an issue. Whether it is oral or written, the peer response process is likely to be a lengthy one. As pointed out in Rollinson (2005), completing the full feedback cycle (reading a draft multiple times, making notes, collaborating with another reader and exchanging ideas with him or her to reach a consensus and writing comments, as well as revising the draft) is regarded as somewhat time-consuming. Secondly, student characteristics might also interfere with the effectiveness of the peer feedback. The student profile (the age and the proficiency level of the students) may impact the effectiveness of the peer-review process (Rolinson, 2005).

A lack of trust in their peers' competence (Bergren, 2015; Nelson & Murphy, 1992; Paulus, 1999; Razı, 2014; Rinehart & Chen, 2012; Rollinson, 2005; Ruecker, 2010; Saito & Fujita 2004; Yang, et al., 2006; Zhao, 2014), and a preference for teacher feedback, regarding the teachers' comments as more authoritative and more quality (Hu & Lam, 2010; Jacobs et al., 1998; Zhao, 2010), were also pointed out as challenges in peer feedback. Additionally, learner frustration and demotivation were also reported due to the learners' failure to attach due importance to peer feedback (Best, Jones-Katz, Smolarek, Stolzenburg, & Williamson, 2014; Wang, 2014). Pham and Gillies (2010), in a study in the EFL context, pointed out that it is necessary to consider students' culture during the implementation of peer revision. They indicated that due to a lack of self-confidence in feedback provision, students may show reluctance to became involved in direct and face-to-face peer assessment while they seem to be willing to participate in an intergroup peer assessment (i.e., "where group members evaluated one another individually") that is not face-threatening (Pham & Gillies, 2010, p.75). Learner avoidance from making critical comments to maintain group harmony and learner reluctance to claim authority were enumerated as some other factors that constitute further issues in peer feedback literature (Liou & Peng, 2009).

In some other contexts, the teacher's role as the sole feedback source for student writing (Ezza, 2010), learners' inclination to excessively focus on error corrections in the feedback (Al-Hazmi & Schofield, 2007) as well as time constraints in writing courses (Razak & Saeed, 2014) were reported as major challenges with peer feedback . Still, other studies pointed out the concerns with the quality of peer feedback such as unspecific peer responses, a lack of elaboration or explanations, failing to facilitate the revision process of ESL/EFL learners (Tsui & Ng, 2000; Wang, 2014). The learners' perceptions regarding the value of peer feedback is a further area that received a great deal of attention in the literature. While some learners found peer feedback on content useful (McGroarty & Zhu, 1997; Mendonca & Johnson, 1994), others revealed their dissatisfaction with their peers' overemphasis on the surface errors (Nelson & Carson, 1998; Tsui & Ng, 2000). A further concern that was raised against peer feedback is that it is not so influential as teacher feedback in developing EFL students' writing (Ruegg, 2015).

Last but not least, the teachers' reluctance to abandon their traditional role as the feedback provider and to shift the responsibility of giving feedback to the students, assuming new roles such as the trainer or supervisor in peer feedback was cited as an additional challenge in terms of peer feedback (Rolinson, 2005).

Different Perspectives on Peer Feedback

In the last three decades, different comparative perspectives have evolved related to peer feedback. In this section these perspectives are delineated. The following perspectives regarding peer feedback are discussed here respectively: Peer versus teacher feedback, online versus traditional peer feedback, and written versus oral feedback.

Peer Versus Teacher Feedback

As regards peer versus teacher feedback, research studies indicated inconclusive evidence. Some studies revealed findings in favour of teacher feedback (e.g., Tsui & Ng, 2000; Zhang, 1995). In his study on learners' perceptions on teacher and peer feedback in the US ESL context, Zhang (1995) pointed out that learners held more favourable perceptions of teacher feedback. Sengupta's (1998) study also revealed findings in line with the previous studies in this regard, which pointed out that students who are reluctant to be involved in the peer-review process may feel reassured to know that their teachers are continually monitoring them.

On the other hand, feedback literature also involves studies that showed the student's preference for peer feedback. Unlike teacher feedback (Zhang, 1995), peer feedback is likely to foster a sense of audience and textual ownership and collaborative learning (Tsui & Ng, 2000), playing a complementary role in feedback practice (Jacobs et al, 1998). As compared to teacher feedback which may be unspecific, contradictory, inconsistent, inaccurate, vague, over-general, abstract, formulaic, and idiosyncratic (Zamel,1995), peer response comments were found to lead to meaningful revisions in terms of vocabulary, organization and content (Paulus, 1999). Peers can spend much more time providing feedback on an individual draft than the overworked teacher. There can be a quick turnover time between finishing a draft and receiving feedback. Apart from the studies that prioritize either teacher feedback or peer feedback, some studies such as Jacobs et al. (1998) suggested taking " a middle path" where teacher feedback assists peer feedback (p.309). Similarly, Hu (2005) also indicated that peer feedback plays a complementary role to teacher feedback.

The interactive process at work in peer feedback displays some differences from the traditional feedback provision. During the one-way interaction between the student and the teacher in the traditional feedback provision, students tend to make revisions without necessarily agreeing with or even understanding the teacher's authoritative comments. The exchange that takes place on a more informal level in the case of peer response is likely to be more encouraging and motivating for writers and they retain the right to accept or decline the comments they receive from their peers. As Ferris (2002) claimed, "it is more engaging and motivating to work on a peer's paper under construction than to always look at models, especially because the author is usually working on the same general assignment that the editor is" (p.103).

On the other hand, some comparative studies put forward "a middle path" approach which advocates a variety of feedback options to be given to the learners (See Jacobs et al, 1998; Hu & Lam, 2010; Zhao, 2014). The results of these studies revealed the learners' preference towards a combined feedback mode. They indicated various advantages and shortcomings related to different modes of feedback. To illustrate, as opposed to teacher feedback, which is perceived to be more authoritative, more accurate and more helpful than peer feedback, peer feedback was likely to promote a rise in learners' language awareness, and to provide more social and affective affordances for learners in comparison to teacher feedback (Lee, 2015). It created more common ground among learners during their exchange of ideas with their peers (Yang et al, 2006, p.193). Thus, it allowed "greater freedom" for peers during feedback discussions (Hu & Lam, 1995, p. 385).

Online Peer Feedback and Traditional Peer Feedback

In the last two decades, online feedback provision has been widely promoted (Van Popta et al, 2017). Asynchronous written peer feedback is the most preferred version of online feedback (van Popta, et al., 2017). Peer feedback plays a much more significant role in online learning environments than in traditional face-to-face ones (Lynch, 2002; Palloff & Pratt, 2013). A lack of feedback may cause students in online courses to become disoriented from the instructional materials or the learning environment more frequently than those in face-to-face courses (Ko & Rossen, 2001). It is reported to be the most common motive behind the student withdrawal from the online courses (Ertmer, et al., 2007). Studies (e.g, Ko & Rossen, 2001) indicated that feedback is a stimulator for students' online activities. Taking into consideration the critical nature of feedback in online educational settings, the characteristics of effective feedback were defined as "diagnostic and prescriptive, formative and iterative and involving both peer and group assessment" (Notar, et al., 2002, p.646). It is argued that the feedback should focus on the process rather than on the end product (Ertmer, et al., 2007). As indicated in Schwarz and White (cited in Ertmer et al., 2007), students would like the online feedback to be "prompt, timely, and thorough, formative and summative, constructive, supportive, and substantive specific, objective and individual and consistent" (p.414). However, the provision of such a high level of feedback in online courses requires the instructors to make a significant investment in time and effort, which means an increased workload. Besides, to be able to provide prompt and ongoing feedback online, the instructor needs to be online almost all the time, which is not a sustainable practice nor the autonomous, self-directed learning principles advocated in online courses (Dunlap & Grabinger, 1996). It is also hard to promote the dialogue between the teacher and the students due to the physical distance in online learning environments (van Popta et al., 2017).

In online peer feedback, while reviewing their writing synchronously and asychronously, learners exchange in revision-oriented feedback that addresses the global (e.g., idea development) or local (e.g., language) issues (Cha & Park, 2010; Chang, 2012; Liu & Sadler, 2003). They also exchange non-revision-oriented comments that aim to establish a friendly atmosphere and an amiable social context (Bradley, 2014; Liang, 2010). Such non-revision-oriented feedback in sychronous and asychronous peer discussions serve to establish cohesion among learners online (Jen & Shah, 1997).

Group cohesion is regarded as important, considering the voluntary student participation in the written peer feedback process beyond the writing courses (Razak & Saeed, 2014). The integration of technology into the peer feedback process is likely to promote EFL learners' engagement in peer feedback outside the classroom (Chen, 2016).

The introduction of peer feedback in online learning environments can be regarded as a viable instructional strategy to alleviate the problem of quality, constructive, and prompt teacher feedback provision encountered in mainstream education (Ertmer et al., 2007; Van Popta et al., 2017). Being engaged in providing feedback to one another in online environments facilitates meaningful interaction among students, which is likely to enhance community building, as suggested by Corgan, et al., (2004). The students' interaction with the content is also likely to increase their satisfaction with the course (Fulford

& Zhang 1998; Richardson & Swan, 2003). Provided that it is used effectively, the use of teacher and peer feedback is likely to facilitate the quality of online learning (Ertmer et al., 2007). Peer feedback promotes learners' contribution to others' learning process, raising their awareness towards their peers' experiences and perspectives and elevating their level of understanding and appreciation in this regard (Corgan et al., 2004). Learners are provided with an opportunity to reflect on their peers' as well as their work, which is gradually likely to result in enhanced learning (Ertmer et al., 2007).

Although the learner perspectives on the peer and teacher feedback in the ESL / EFL context is extensively researched, the learners' perceptions on the traditional and online peer feedback from a comparative perspective is underexplored. With the advent of Computer Assisted Language Learning (CALL), as a common feature of ESL/EFL writing classes with its motivational impact on students, online peer review was acknowledged to become a valuable asset to and for and sometimes an alternative to the face-to-face peer feedback (FFPR), available to both L2 and EFL teachers and students in 21st-century classrooms (DiGiovanni & Nagaswami, 2001). In comparison with face-to-face peer review, online peer review was reported to provide certain affordances for learners such as long and focused on-task behavior as well as for the teachers such as total access to every peer dyad's complete interaction (Sengupta, 1998). It also promoted negotiation of meaning and student interaction (Chang, Chen, & Hsu, 2011). However, to date, there is no consensus on the superiority of online or the traditional mode (See Elwood & Bode, 2014). Some studies indicated evidence for the favorable impact of online feedback on academic writing skills in the tertiary context (See Razı, 2014). Razı (2014, 2016) suggested that the anonymous peer review process from multiple resources in a digital environment provides valuable contributions towards improved academic writing skills. The anonymous multi-mediated writing model developed by Razı (2015) aimed to provide a remedy towards the students' reluctance to provide feedback for one another. The model, with the distribution of asymmetrical and symmetrical feedback, matches the students' papers with three peers having different writing proficiency levels, upon their draft submissions, which is followed by the anonymous task submission (Hanjani & Li, 2014). On the other hand, some other studies pointed out that several factors may limit the impact of peer review when the feedback provision is not planned carefully (Robinson, 2002) or when students lack motivation to be engaged in peer feedback (Hojeij & Baroudi, 2018) and have negative perceptions towards peer feedback (Kaufman & Schunn, 2011). Hence, the impact of online peer feedback on the quality of revisions was reported to be inconclusive (Fitze, 2006; Guardado & Shi, 2007).

Even though peer feedback is a two-way process, the majority of the studies on peer feedback in online learning environments focused on the impact of peer feedback on the performance of the receiver and not on the learning gains of the peer feedback provider (e.g., Ertmer et al., 2007; Gurdado & Shi, 2007; Smits, Boon, Sluijmans, & Van Gogh, 2008). Some studies focused on the differences in terms of their impact on the learning gains of the feedback providers and receivers (See Narciss, 2013). However, the overemphasis on the value for the receiver might underestimate the reciprocal nature of the peer feedback process and the value of the feedback for the provider. It is not clear whether the challenges the peer feedback providers and receivers encounter in the traditional learning environments will escalate or diminish in online learning environments.

Regarding the challenges of peer feedback in virtual instructional settings, Tunison and Noonan (2001) revealed the difficulty students encounter while expressing sophisticated views in such learning environments and the boundaries they face while addressing their inquiries and comprehending detailed explanations due to the absence of face-to-face interaction. Despite the more even distribution of student participation in online course discussions and the higher level of student participation than in traditional learning environments, these online interactions are not likely to be as effective as face-to-face interactions, until the participants establish a comfort zone in their community of practice and feel at ease with each other, forming and strengthening an affective bond among them (Arbaugh, 2000). Additionally, some students find the use of technology in peer review distracting (Kessler & Bukowski, 2010; Ho, 2015) and some students may fail to meet what their teachers expect of them during the online peer-review process (Kessler & Bukowski, 2010).

Other studies related to peer feedback in the virtual learning environments have investigated the benefits of providing thorough and detailed feedback. To illustrate, Van der Pol, Van der Berg, Admiral, and Simons (2008) pointed out that the learning benefits of online peer feedback provision could be felt only if learners are willing to devote their time and effort to the active construction of content-oriented responses. In addition, Lin, Liu, and Yuan (2001) argued that when engaged in providing online written peer feedback, learners were involved in a variety of cognitive processes such as reading, comparing, questioning ideas, suggesting certain modifications as well as reflecting on to what extent and in what ways their work is better than others.

Although there is a surge of interest in the integration of technology into educational settings, online peer feedback is likely to pose problems for L2 school learners than those in the tertiary settings, where the research is mainly focused (Lee, 2017). L2 school contexts are mainly characterized by an exam-oriented culture, where the students are expected to write in a product-oriented manner and pass the paper and pencil-based public examinations (Lee, 2017).

Peer Feedback Research

The final perspective related to peer review is concerned with the quantitative versus qualitative methods to investigate the impact of peer revision. Generally speaking, initially quantitative methods were used to investigate the impact of peer revision on the quality of student writing (Hedgecock & Lefkowitz, 1992, Zhang, 1995), using the students' composition scores as the sole data collection instrument. Later on, peer feedback studies started to adopt qualitative methods, particularly regarding the exploration of student talk and interaction during peer revision (Gere & Abbot, 1985; Lockhart & Ng 1995; Villamil & De Guerrero, 1996), the impact of peer feedback on the student revision (Connor & Asenavage, 1994; Herrington & Cadman, 1992; Mendonça & Johnson, 1994; Nelson & Murphy, 1993; Sommers & Lawrence, 1992), and students' reactions to peer revision (Carson & Nelson, 1996; Jacobs, 1987: Mangelsdorf, 1992). Relatively few studies have integrated different methods, data sources, measures and theoretical perspectives. Generally speaking, researchers have used quantitative methods to investigate the affective advantages of peer feedback (Zhang, 1995), peer revision's effects on the quality of student writing using scores on student compositions as the single source of data (Chaudron, 1983; Hedgcock & Lefkowitz, 1992. Only a few studies on peer revision have integrated measurements, data, and methods (Freedman, 1992; Nelson & Murphy, 1992).

The current mixed-method case study investigated the impact of peer feedback on a group of freshman EFL pre-service teachers' attitudes towards their writing skills development and on their writing skills performance in a blended learning (online and traditional) environment. The study also contributed to the enhancement of the EFL majors' metacognitive skills in writing as it incorporated a training program for the EFL majors on how to provide feedback prior to their involvement in the study.

Written Versus Oral Peer Feedback in the Tertiary Context

There is a paucity of research studies as regards the investigation of the ESL/EFL learners' perceptions on written and oral peer feedback from a comparative perspective in the university context (Van den Berg, Admiraal, & Pilot 2006). As they argued in a study in relation to the analysis of written and oral peer feedback in higher education, students tended to focus on different aspects in their oral and written feedback. In comparison between two kinds of feedback, the findings of the study pointed out "in their written feedback students commented on structure more than in their oral feedback. In their oral feedback students commented more on style" (p.145). In a similar study in the tertiary context, Krych-Appelbaum and Musial (2007) concluded that the study provided some evidence showing that students valued both types of feedback, but they indicated their preference for the oral peer feedback to precede the written one. In another study on the higher education students' perceptions of different forms of feedback, Cartney (2010) reinforced the findings of Krych-Appelbaum and Musial (2007). Although students in the latter study regarded both types of feedback as beneficial, they indicated a distinct preference for the oral feedback process complementing the written feedback, a result which might be attributed to the cultural background of the participants and their lack of experience with the e-learning practices. On the other hand, Abdalla Salih (2013), in a study on the actual focus of peer review and the types of corrective feedback provided in L2 composing process, indicated a mismatch between the peer reviewer's focus (the clarity of feedback) and the writer's focus (the grammar corrections). Taking into consideration the relative lack of research on the EFL majors' perceptions on different types of feedback, the current study explored the impact of various feedback types on students' attitudes and their writing performance in a tertiary academic writing context.

THE STUDY

Context

The study was conducted in the Department of Foreign Language Education at the Faculty of Education in a large English medium university in central Anatolia in the fall semester of the academic year 2014-2015, which lasted 14 weeks. It was integrated into the Advanced Reading and Writing course, which is offerred to freshman EFL major students. The course was designed to assist EFL majors in developing critical reading, thinking and academic writing skills. Classroom instruction in the course focused on the development of both reading and writing skills, with a considerable emphasis placed on the genre-based paragraph writing. Following the process approach in the course, the participants were required to produce multiple drafts of paragraphs and peer revision was an integral component of the course.

A blended learning approach was adopted in the course, which involved a face-to-face learning environment in the form of class sessions as well as an online learning environment in the form of blog entries. The course was based on genre-based pedagogy, which required the EFL majors to write blog entries in different genres (i.e., expository, narrative, and descriptive) in line with the process approach. The blog entries were composed of paragraphs between 300 and 350 words.

Participants

The participants were 30 Turkish freshman EFL major students enrolled in the Advanced Reading and Writing course at the department of Foreign Language Education at a large urban English-medium university in Turkey. Purposeful sampling procedures were adopted in the selection of the participants. *Purposeful sampling* was preferred as the sampling strategy in the study as intentionally selected participants for the case (i.e., EFL majors) were considered the most informative group about the research problem under investigation (Creswell, 2013).

As the aim of the study was to investigate the impact of face-to-face and online peer feedback on the attitudes and academic writing performances of EFL majors in their first year at the department, EFL majors taking the course for the first time were included as the cohort in the study. The researcher was also the instructor of the course. The participants were composed of 20 female and 10 male freshman students between 18 and 21. They all came from various teacher training high schools in Turkey. They reported having studied English for about seven years prior to their undergraduate education. They all had a B2 to C1 level of proficiency in English, receiving a minimum of 70 points in the English Proficiency Exam prepared by the institution which is the equivalent of 86 on TOEFL IBT and 7 on IELTS exam.

The course participants were also provided with some technology training on how to create and upload various online gadgets to their own blogs prior to their involvement in the study. Prior to their involvement in the study, all the participants possessed a basic level of digital literacy as a result of a compulsory basic information technnology skills course they took at the university.

Research Design

The study is a mixed method case study. The term *case study* can be defined as a "systematic inquiry into an event or a set of related events which aims to describe and explain the phenomenon of interest" (Bromley, 1990, p.302). According to Yin (2003), in a case study, the researcher simply explores individuals or organizations through complex interventions, relationships, communities, or programs and the researcher cannot manipulate the behaviors of the participants. The phenomenon is explored within its context using different data sources (Baxter & Jack, 2008). The case study was chosen as the research design as it is concerned with a systematic investigation of the impact of the incorporation of a particular methodological approach (the blended approach to peer review) on a particular group of participants' perceptions towards a certain phenomenon of interest (the freshman EFL majors' attitudes towards academic writing) and their academic writing performances. The mixed method design, with a quantitative and a qualitative perspective, was adopted in the study to provide more in-depth explanations and investigate the research questions satisfactorily.

TRAINING PROGRAM IN THE STUDY

Pre-Training Phase

Prior to the implementation of the training program, a pre-training phase was conducted. Prior to the implementation of the training program, a pre-training phase was conducted, which aimed to raise the EFL majors' awareness towards various feedback techniques and various comment types, to inform them of how to achieve productive group interaction through collaboration and how to to establish an effective reader-writer dialogue. The training program started in the third week of the course.

The pre-training program in the study was developed based on the peer feedback training programs utilized in previous research studies (See Hansen & Liu, 2005; Keh, 1990; Rollinson, 2005). It was composed of a six-hour intensive face-to-face training program in the writing class which incorporated certain themes. The first theme the participants were introduced to was the lower-order and higher-order concerns. The former were related to mechanical errors, punctuation and spelling whereas the latter were related to the development of ideas, organization, unity, coherence and cohesion of their written work and the overall focus of their written work. At this point the participants were introduced to the lower-order (local issues) and higher-order concerns (global issues). The former were related to grammar mistakes, punctuation and spelling, which are incorporated in the grammar checklist, and appropriate vocabulary choice in a particular content. However, the latter were related to the unity and coherence (e.g., the topic sentence and controlling ideas, the summary sentence, and the irrelevant sentence) as well as the cohesion of their written work (e.g., transitions). The teacher showed the points that the participants needed to be careful about while giving constructive feedback. She also focused on how to write text-specific comments as well as a summative comment related to a particular written student's work. She also pointed out how to stay away from using judgemental language during the peer feedback provision process. She distributed a hard copy of a sample student paragraph to the participants and pointed out how to review it through explicit instruction focusing on content, organization (issues such as unity, coherence, and cohesion) as well as language use (including mechanics) and vocabulary in the form of questions, using checklists. The second theme in the training program was concerned with the notion of audience. The importance of audience was emphasized during the training process. The EFL pre-service teachers were shown that writing for an audience provides them with a perspective in writing and shapes the language they use, the content they produce as well as their writing style. The third theme was the awareness of purpose. During the training program, the students were made aware of the purpose of peer reviews, which was to help them revise their paragraphs by receiving different points of views about their drafts. The various perspectives the students gained through the peer reviews enabled them to make rhetorical choices to select from when they revised their paragraphs. Finally, the participans were introduced to three different roles of an ESL/EFL writers. These roles were "a reader", "a writing teacher," and "a grammarian" (Keh, 1990, p. 301). As readers they interacted with the writer and responded to the content with comments such as "a good point" or "I agree" while as writing teachers, they were concerned with points of confusion and breaks in logic, but they still maintained their reader role (Keh, 1990, p.301) . On the other hand, as grammarians, they were engaged in writing comments concerning the appropriacy of a particular grammatical form preferred by the students, as in the case of the tense choice. Although there were three separate roles identified for the ESL/EFL writers in the training program, it was difficult to separate these roles from one another during the peer feedback provision. Therefore, the EFL pre-service teachers in the study were instructed to keep a balanced focus on the overall problems as well as the strengths of their peers' writing. They were also introduced to the concepts related to paragraph writing such as the topic sentence, controlling ideas, transitions, the summary sentence, and the irrelevant sentence. They were asked to use the peer response sheet prepared by the researcher during the peer revision activities in the study (See Appendix A).

Training Phase

Following the pre-training phase, a six-hour training program was integrated into the study to enhance the effectiveness of the peer feedback activity in English writing classes for the participants (Berg, 1999; Bitchener & Ferris, 2012; Crinon, 2012; Ferris, Liu, Sinha, & Senna, 2013; Hu, 2005; Hu & Lam, 2010; Lam, 2010; Min, 2005, 2006, 2016; Rahimi, 2013; Rollinson, 2005). A three-hour period was allocated to face-to-face peer feedback training. The program was composed of three main stages: "awareness-raising (principles and objectives of peer response), productive group interaction (collaboration, supportiveness, tact, etiquette), and productive response and revision (basic procedures, effective commenting, reader-writer dialogue, effective revision)" (Rollinson, 2005, p. 27). The awareness-raising stage involved informing the audience of the purpose of the peer review activity explicitly (i.e., to help them revise their paragraphs by receiving different points of views about their drafts), clarifying the expectations and promoting students' exchange of experiences and concerns. The participants were informed that various perspectives they would gain through the peer reviews would enable them to make rhetorical choices to select from when they revised their paragraphs. At this stage, the participants were introduced to the notion of audience in the peer feedback process through the jigsaw peer feedback technique, a group feedback practice where four students work as a group to provide feedback on their peers' writing (Ferris, 2003). As it was important for students to work in a pair or group where they feel comfortable and their need to develop trust in one another (Lee, 2015), the group members chose their group members on their own (Hansen & Liu, 2005).

Throughout the training program, the participants raised their awareness towards multiple perspectives provided by a diverse audience (Ferris, 2003). Each pre-service teacher was provided feedback to a first draft produced by his/her peer in a different group regarding a different . The jigsaw format was adopted in peer feedback provision process as adopting a collaborative stance towards peer feedback was shown to be beneficial for ESL university students' text revisions in the literature (See Yu & Lee, 2015). The participants were involved in a sample collaborative oral peer response activity in groups of four in the jigsaw format, using sample student drafts and the peer response sheet. In this collaborative peer response activity, each group member focused on a different feedback concern (i.e., content, organization, language, and vocabulary use) while evaluating the strengths and weaknesses of his/her peer's draft and decided on what should be done to revise it in line with the peer response sheet (See Appendix A). Group members were also asked to write a summative constructive comment for the paragraph, accompanied with suggestions for revision. The course instructor monitored the pre-service teachers' peer feedback process, offering advice in cases when they failed to follow the feedback procedures properly (Min, 2016). In order to guide participants in the peer revision process, she ensured that everybody assumed a different role in the process. The importance of establishing an effective reader-writer dialogue was emphasized in the peer feedback training program. The pre-service teachers were provided with instructins in appropriate response behavior (e.g., the need to give credit to the strengths and to provide constructive feedback and text-specific comments).

As the participants in the study were not familiar with the strategies to negotiate meaning among group members during the collaborative peer revision activitiy, the course instructor taught them some expressions to communicate clearly and to fine-tune their meaning, thereby extending their linguistic repertoire. In order to ensure that the participants had a clear focus on how to make relevant comments, the instructor trained them to ask questions to generate a response from the writer and to start a fruitful discussion on the contextual, rhetorical, and linguistic aspects of the grammar based on the purpose of the activity. important to pre-service teachers were trained to ask questions that were revision-oriented so that there was a meaningful discussion about the content, rhetoric or grammar of the paper, depending on the purpose of the activity. They were also introduced to clarification questions (e.g., 'What do you mean when you say…?) or elaboration questions (e.g., 'Could you explain this point in more detail?' or 'Can you give any examples?') so that they could be engaged in a meaningful discussion, negotiation of meaning, and revision. They were discouraged from making evaluative statements (e.g., 'This is not a good topic sentence'). Since one of the goals of the Advanced Reading and Writing course was to make students autonomous writers, pre-service teachers were given the authority to retain certain aspects of their texts and not to make certain changes advised by students and the teacher. Participants were also trained to ask revision-oriented questions to lead a meaningful discussion about the content, rhetoric or grammar of the paper, depending on the purpose of the activity. They practiced the clarification, elaboration and negotiation strategies during the activity and felt confident about providing feedback for others.

Having established the face-to-face peer feedback procedures, which can be seen above, the participants also received training on how to provide online written feedback. The online peer feedback training session lasted three hours, which involved online hands-on practice on giving feedback to a blog entry in the computer lab in the department. The pre-service teachers were provided with written guideliness related to the online written peer feedback procedures. They were instructed to provide feedback in terms of three main areas: content, organization, as well as language use and vocabulary. The researcher also modelled the online peer feedback provision process on the weblogs. She first showed students a sample student paragraph draft posted on the blog and discussed with the class members how to provide feedback to this draft. After a whole class disccussion, she posted a sample peer commentary on the class blog and the students checked to what extent the writer took the suggestions for improvement into consideration. Then, the pre-service teachers worked together in groups of four in the jigsaw format to review another sample student paragraph draft on the class blog, using different checklists and prompted to make specific comments and suggestions for revisions on the class blog. The pre-service teachers worked on their comments and revisions via a shared Google document. Later on, students exchanged their ideas on the aforementioned revision categories suggested by each group as a whole class. After the in-class session, they were asked to write the final version of their revision and upload it to their classmates' blog entry in the form of a comment. The participants were warned against the issue of anonymity while posting their comments on their classmates' blogs, which was suggested by Razı (2014, 2016). Therefore they removed their names while posting their comments to the blogs.

While writing comments in the comment section of the blog, pre-service teachers were told to write both text-specific comments as well as a summative comment at the end of the blog including the overall strengths and weaknesses of their peers' written work, followed by some suggestions for improvement and a new goal for the revised version. They were reminded to employ a positive, encouraging, and non-judgemental feedback language and make constructive comments. They were also recommended to act as an advisor while giving feedback. They were instructed some politeness strategies such as the use of modals and practiced how to make polite suggestions for revision through the teacher's guidance.

Some of the prompts that the instructor used during the training session on how to instruct the students to make polite suggestions are as follows:

Prompt 1: You can think of changing / I recommend you to change /your topic sentence so that it can cover all the major supports in the paragraph.

Prompt 2: It is a good idea to change your introduction to make it more interesting. For instance, you can talk about a personal experience.

During the face-to–face and online peer feeback training sessions, the instructor emphasized that making mistakes is a natural part of the learning process and she tried to create a non-face threatening learning environment by adopting a collaborative jigsaw format for both types of sessions. Taking into consideration the differences in thc proficiency level of participants, the pre-service teachers were asked to assume different roles and work with different group members during the training sessions.

Data Collection Procedures

The data were collected in the fall semester of the academic year 2014 and 2015 for a period of 14 weeks. Three different data collection tools were employed in this study: Weekly student peer feedback journals, focused group interviews, the end-of-the-semester questionnaire, the pre- and post-test.

Peer Feedback Journals

On a weekly basis, pre-service teachers were asked to keep journals where they expressed their perceptions of their progress concerning both face-to-face and online peer feedback processes and the difficulties they faced during both processes, their feelings, the insights that they gained regarding the provision and reception of peer feedback The length of each journal entry ranged between a paragraph of 350 to 500 words to a full A-4 page. The participants wrote six journals during the study.

Focus Group Interviews

After each face-to-face and online feedback provision session, one-and-a-half-hour long focus grup interviews were held with the participants in the researcher's office to gain insights into their perceptions concerning the feedback provision process, the challenges they faced and the benefits they gained from the face-to-face and online peer revision activities. The interviews were held in English and lasted for 45 minutes. They were audiotaped with the permission of the participants and transcribed later on.

The End-of-the-Semester Questionnaire

At the end of the study, the participants were administered a questionnaire probing their face-to- face and online peer feedback experiences, their perceptions regarding their experiences as feedback providers and receivers as well as their feedback-related challenges. The questionnaire was designed in the form of Likert-scale items with 5 options (1= not effective at all, 2= not effective, 3= of average effectiveness, 4= effective, 5= very effective). Apart from the 10 Likert-scale items, there were four open-ended items on the questionnaire too. In the design phase of the questionnaire, the previous questionnaires utilized in peer feedback literature were taken into consideration (See Altstaedter & Doolittle, 2014). The reliability of the questionnaire in the study was calculated to be .90.

Pre- and Post-Test

The participants' first draft paragraphs in three different genres (expository, narrative, and descriptive) were regarded as the pre-test in the study while their third drafts were considered the post-test. All the drafts were evaluated using a holistic scale adapted from Gilles and Snook (1990). in the measurement of the quality of the participants' written works due to the multi-faceted nature of L2 writing, incorporating developmental aspects of meaning-making in the specific genre in focus (Manchon, 2012), and the difficulty of observing a significant improvement in the writing skills over a short period of 14 weeks (See Dyson and Freedman, 1991). (See also Dyson & Freedman, 1991; McGroarty & Zhu, 1997). The holistic scale can be seen in Appendix B.

Eighty-six paragraphs were incorporated in the study as the first and third drafts. Four participants who failed to submit their third drafts were excluded from the study. The researcher and a departmental colleague, an experienced writing instructor with an M.A. degree in English Language Teaching, independently evaluated all the participant drafts, using the holistic scoring criteria in Gilles and Snook (1990). Then they compared the emerging codes in the data together and tried to reach an agreement. The interrelater reliability in the study was found to be .90.

The Peer-Review Activities

The participants were asked to write on their blogs their first draft paragraphs in three different genres: expository, narrative, and descriptive. In the study the expository paragraph was defined as a paragraph that explains or analyzes a topic while the narrative paragraph was referred to a paragraph that requires a chronological development or one that is concerned with the relating of an experience.

The expository genre was incorporated into the study as the participants were studying English for academic purposes. The narrative and descriptive genres were included due to the important role these genres played in self-expression and interpersonal communication in EFL academic writing contexts. Hence the inclusion of these genres in the study were considered to contribute to the development of the pre-service teachers' communicative competence. They were asked to write their first paragraph drafts in class, which was followed by a a face-to-face peer feedback session in class. They worked on the revisions in groups of three in the jigsaw format during the revision phase. Every pre-service teacher in each group was engaged in providing feedback on one specific aspect (content, organization, language use, and vocabulary) to the first drafts produced by their peers in a different group. Each member was responsible for one aspect of revision. They were asked to collaborate in groups of three on sample student essays. Through collaborative reviews, they discussed the strengths and weaknesses of the paragraphs and decided on what comments to give and how to express themselves while providing feedback. The pre-service teachers negotiated their ideas with each other in collaboration, which also enabled them to feel more confident about the feedback they provided for their peers.

As far as the written online feedback is concerned, the participants were asked to bring to the class session a hard copy of their draft paragraphs that thcy uploaded to their blogs in the form of a blog entry. They first exchanged their drafts and read them, followed by an oral peer assessment using the prompts on the peer response sheet prepared by the researcher beforehand. They were also encouraged to provide any suggestions or recommendations that they thought might help improve their peers' drafts in the face-to-face feedback session. Each group member wrote his/her suggestions for revision on a shared Google document. After the session, the participants were asked to prepare the revised versions of their draft paragraphs and upload them to their blogs, taking into consideration their peers' suggestions for improvement.

After the participants uploaded to the blog the revised versions of their draft paragraphs, the second drafts, the online peer review process started. The group members were involved in the written peer review of the second drafts utilizing the online peer review guidelines prepared by the researcher, who was also the classroom teacher. They provided their feedbacks in the form of comments using the comment facility of the weblogs. While providing online written peer feedback, the participants were asked to review their peers' drafts from three different perspectives: content, organization, the language use, and vocabulary. They were asked to provide constructive feedback for their peers, by pointing out the strengths and weaknesses of their peers' drafts. Having received the online written feedbacks from their peers, the participants were asked to write the third draft paragraphs and upload them to their weblogs. Thirty participants produced three draft paragraphs in three genres (expository, narrative, and descriptive) in the study.

Utmost care was taken in the study to provide the participants with scaffolding throughout the peer feedback provision process. They were asked to upload their blog entries in the aforementioned genres. Each participant received one face-to-face and one online written peer feedback for his/her blog entry for each genre covered in the course. After each type of feedback they were required to revise their draft blog entries on the weblog. After receiving both types of peer feedback and revising their blog entries after each feedback, they were asked to upload their final versions of their blog entries.

Data Analysis

The quantitative data from the end-of-the-semester questionnaire were analyzed via descriptive statistics. The quantitative data from the students' first drafts and third drafts were analyzed via the paired sample t-test. On the other hand, the qualitative data in the study from weekly peer feedback journals, the open-ended items on the questionnaire, and the focus group interviews were analyzed via the constant comparison method, which is "a method of analyzing qualitative data which combines inductive category coding with a simultaneous comparison of all units of meaning obtained" (Maykut & Morehouse, 1994, p.134). Following this method of analysis, each unit of meaning selected for analysis was compared to all the other units of meaning, which is then categorized and coded. Afterwards, similar units of meaning were grouped. In cases where there were no similar units of meaning, a new category was created. In this process the categories were continuously refined; initial categories changed. New categories emerged and new links were discovered between the categories, leading to the emergence of new themes (Maykut & Morehouse, 1994). The analysis process followed the following stages: coding, categorizing, and theming. To be able to address the research questions, the codes with similar ideas were grouped to create categories, leading to the emergence of the main themes in the study. The participants were sent consent forms before the study to obtain permission to utilize their data for research purposes.

Results

In relation to Research Question 1, the impact of face-to-face and online peer revision on the writing performance of freshman Turkish EFL pre-service teachers, the results of the independent paired sample t-test based on the writing scores on the first and the third drafts revealed that although there was no statistically significant difference observed between the writing scores on the Turkish EFL pre-service teachers' first drafts and those on the third drafts ($p>.05$), there is a slight improvement in the writing performance of the EFL pre-service teachers. The results can be seen in Table 1 .

Genres	Draft No	Mean (out of 4)	Standard Deviation	T value and the level of significance
Expository	Draft 1 (N=32)	3.05	.94	.32 (p>.05)
	Draft 2 (N=30)	3.12	.87	
Narrative	Draft 1 (N=32)	3.32	.52	.87 (p>.05)
	Draft 2 (N=30)	3.37	.73	
Descriptive	Draft 1 (N=32)	3.20	.48	.75 (p>.05)
	Draft 2 (N=30)	3.30	.45	.75 (p.>.05)

The findings indicated that the peer revision training was likely to have a positive impact on the pre-service teachers' cumulative writing development, which is in line with De Guerro and Villamil (1994), Villamil and De Guerrero (1996), Carson and Nelson (1994) and Nelson and Carson (1998) and Chaudron (2004). The drafts were evaluated using a holistic scale by Gilles and Snoook (1990), which can be seen in Appendix B.

With respect to the impact of the blended approach to the peer feedback on the EFL pre-service teachers' perceptions of peer revision activities in writing, the results of the quantitative analysis of the data on the end-of-the-semester questionnaire (See Table 2 in Appendix C) indicated that the majority of the participants (65.50%) developed favourable attitudes towards face-to-face and online written peer revision activities in general. These findings were consistent with the results of the previous studies by Lee (2015), Jacobs et al. (1998), Yang, et al. (1995), and Tsui and Ng (2000). The participants reported finding peer feedback training sessions beneficial on the whole (65.52%). They also underscored the benefits of face-to-face collaborative peer feedback sessions for them as prospective teachers (72%) in general (Yang et al., 2006). Although slightly over one-third of the students (35%) reported encountering certain difficulties in providing face-to-face oral feedback, almost half of the participants (49.60%) expressed their challenges with the online written feedback provision, as indicated in Nelson and Carson (1998) and Tsui and Ng (2000). In fact, slightly over half of the participants (53%) mentioned consulting their peers for the confusing written feedback, which was echoed by Tsui and Ng (2000) and Wang (2014). The majority of EFL pre-service teachers in the study revealed that they found a combination of face-to-face oral and online written feedback highly beneficial (65.50%). The participants emphasized that face-to-face oral peer feedback should precede online written peer feedback, which was also advocated by Krych-Appelbaum and Musial (2007) and Cartney (2010). The findings showed that the use of online written feedback along with oral face-to-face oral feedback was likely to enhance EFL pre-service teachers' academic writing performance although the study results were not statistically significant. The results might be attributable to the participants' cultural and educational backgrounds. The participants were not used to receiving and providing online written feedback and they were not accustomed to online learning practices in their previous schooling experiences.

Regarding the second research question in the study, the attitudes of Turkish EFL pre-service teachers towards peer revision activities in the tertiary context, the qualitative analysis of the open-ended items on the questionnaire provided valuable insights into the EFL pre-service teachers' attitudes towards peer revision activities. The findings revealed the factors influencing EFL pre-service teachers' attitudes towards peer feedback, including the level of self-confidence in their ability to write in EFL, and the existence of a strong bond between group members, favourable group dynamics, the effective use of communication skills, and negotiation of meaning strategies. The qualitative analysis results were quite in line with those of the quantitative ones. Both quantitative and qualitative data analyses in the study demonstrated that training for face-to-face and online peer feedback improved the pre-service teachers' attitudes towards peer feedback and hence towards EFL academic writing.

The findings indicated quite positive views concerning the blended approach to peer feedback activities (i.e., the combination of the face-to-face and online written feedback). As far as the pre-service teachers' perceptions of the peer response activities are concerned, they emphasized the cognitive benefits of the peer feeedback activities. They indicated that they regarded these activites as an opportunity to reflect on their writing, which was also argued by Mangelsdorf (1989). The analysis of the qualitative data suggested that the peer feedback experience helped them see their own mistakes, raising their awareness towards different aspects involved in the writing process and their use of language (particularly, the kinds of errors they and their peers committed). The findings also revealed that the peer feedback provision experience (both face-to-face and online) was conducive to the development of the pre-service teachers' metacognitive awareness, and their quality of writing in general. These findings were in accordance with Altstaedter and Doolittle (2014), Barnes (1976), Kayi-Aydar (2013), Weissberg (2006), and Ruecker (2010). Participants reported that the experience led to the improvement of their paragraph organization, the use of transitions, and the flow of their writing, (Altstaedter & Doolittle, 2014). They expressed their content with the detailed feedback they received in terms of global aspects and indicated their enhanced metacognitive awareness, which is illustrated in the following extract by P10 (See also Alstaedter & Doolittle, 2014):

Before my partner pointed out my inappropriate use of transitions and the irrelevant sentences in my paragraph, I wasn't aware of how awkward my paragraph seemed. Now in the revised version, the sentences flow better and the new version seems more meaningful and proper.

The quote above is also in line with the previous research (Tsui & Ng, 2000), which indicates the P10's acknowledgement of the impact of peer feedback experience on the provision of a real audience and the enhancement of the participants' metacognitive awareness. The provision of feedback on the local and global aspects of their paragraphs enabled them to develop different perspectives towards their writing. On the other hand, the presence of different audiences was another aspect of the peer feedback experience that participants were content with (Alstaedter & Doolittle, 2014). The participants added that they gained insights into the error patterns of native speakers of Turkish while learning English, which was a further benefit of their engagement with the peer feedback. The peer feedack involvement enabled them to enhance their critical thinking skills. They developed a critical perspective into their language use, which echoes Liu and Hansen (2002).

In addition to the abovementioned favourable aspects of the peer feedback provision experience, the pre-service teachers indicated that they realized the importance of concentration while composing a text. The engagement in peer response activities enabled them to improve their writing in the EFL context. They mentioned their involvement in peer feedback experience provided them with an opportunity to brush up their knowledge of grammar and to use their knowledge of grammar functionally (Hansen & Liu, 2005). All these affordances pointed out above also contributed to the affective affordances of the experience, in the form of an increase in their level of motivation towards EFL writing as these activities added a variety into the routine writing activities.

Peer response activities were also reported to add to the development of a teacher identity in the participants. The pre-service teachers indicated that these activities granted them a professional development opportunity and to take a professional stance. They felt that learning how to evaluate student papers prepared them for their future teacher role. The peer response activities provided the prospective teachers with an opportunity to get engaged not only in formative but also in summative assessment practices, thereby contributing to the pedagogical content knowledge and their procedural knowledge as prospective teachers.

In relation to the perceptions of the pre-service teachers on the face-to-face collaborative peer feedback sessions, they pointed out a variety of benefits in affective and cognitive terms. As for the affective gains, the participants indicated that their engagement in such interactive and collaborative feedback sessions enhanced their team spirit. They developed their interpersonal communication skills thanks to their exchange of ideas during the conferences, which facilitated the establishment of a strong bond among group members and the development of a sense of trust among group members, which was in line with Rolinson (2005), such collaborative feedback sessions reinforced the classroom dynamics as well. As the study was conducted with the freshman EFL pre-service teachers in the first semester of the academic year, they also served as an ice-breaker and created a relaxed atmosphere with a low affective filter in class, which was conducive to learning. The participants mentioned that they felt empowered to write paragraphs through their classmates' suggestions for improvement. With respect to the contribution of the face-to-face collaborative feedback sessions to the participants from a cognitive perspective, the participants reported that the conferences helped them activate their critical thinking skills. They pointed out that they acted as complementary to the online written feedback and they saw such interactive sessions as an opportunity to reflect on the online written feedback they received from their peers and clarify the misunderstandings that arose from it. They also improved their writing skills. The conferences allowed them to get instant feedback from their peers through face-to-face interaction. To illustrate, they had an opportunity to find out irrelevant information and identify major and minor supports.

Pre-service Teachers' Challenges With Peer Feedback

While reporting their perceptions on the impact of peer feedback experience on their attitude towards writing, the pre-service teachers also mentioned their difficulties with the online written feedback. To illustrate, they pointed out that the feedback providers tended to lack awareness concerning the evaluation of a paragraph from a variety of aspects. They mostly chose to focus on language accuracy and language use in their feedbacks, disregarding the content, coherence, and unity, which was not consistent with Van den Berg, et al. (2006) and Abdalla Salih (2013). At the initial stages of the peer feedback provision experience they lacked confidence in their ability to give proper written feedback, which might be attributed to a relative lack of familiarity with the feedback provision conventions in the academic setting or to their previous schooling background. The participants reported a lack of engagement in the written feedback provision activities prior to their engagement in academic studies at university. They reported feeling insecure while providing online written feedback due to a lack of personal contact. They added that they could not be sure whether they should show all the mistakes in the paragraph or not. P3's comments on his first feedback provision experience is representative in this respect.

The first time I gave online written feedback, I found it really difficult because I didn't know where to start. It was also difficult to give constructive feedback when the paragraph needs to be developed in many ways and there are too few things to say about the strong aspects of the paragraph.

The participants reported having a hard time spotting the errors and irrelevant sentences and they remarked that in some cases they offered misleading explanations. They indicated that the feedback they received mostly focused on the language use and that their peers did not tend to provide feedback on coherence and unity, which is also indicated in Paulus (1999) and Reinhart and Chen (2012), Saito and Fujita (2004), Yang, et al., (2006); Zhao (2014). They also pointed out that they found the feedback provision on content as unnecessary and unhelpful and that they regarded content feedback as a disrespectful attitude towards the author. They indicated that at the initial stages of the study due to a lack of close ties between the group members in class, they found it face-threatening to share their honest opinions regarding the content of their peers' works and they felt content-related feedback provision might disrupt the group harmony (Pham and Gillies, 2010; Liou & Peng, 2009). In the initial stages of the study (particularly in the training period) in the face-to-face group feedback conferences, some participants expressed their worries about the possible hostile reactions of their peers if they pointed out the weak aspects of their writing. They also pointed out the face-threatening nature of such feedback provision. The following quote by P14 might be revealing in this respect: "I find myself uneasy when I think of telling all the mistakes to my friend's face. What if he or she has too many mistakes!". However, as they got accustomed to the peer feedback experience, they started to feel more confident to provide their peers with content-related and organization-related feedback, as they established a strong bond with their classmates in the later stages of the study.

Apart from the face-threatening nature of oral feedback provision, the participants also reported having a hard time finding a common meeting time with their peers and communicating with some classmates during the face-to-face collaborative feedback sessionss. They also emphasized superficial and misleading comments from their peers, leading to confusion and misunderstandings on the part of those who received the feedback, which was also mentioned in Beggren (2015), Nelson and Murphy (1993), Paulus (1999), Rinehart and Chen (2012).

Even though the participants in general put forward their agreement with the benefits of the provision of both oral face-to-face feedback and online written feedback provision for EFL learners from a variety of perspectives, they also indicated that they had their preferences for the type of feedback provision. To illustrate, unlike those who did not feel at ease with the online written feedback provision, some participants indicated that they preferred written feedback as they had more planning time to prepare their feedback. Consequently, they could provide more detailed feedback from multiple perspectives. They also pointed out the non-face threatening feature of the written feedback due to a lack of interpersonal contact as the feedback receivers. The following comment by P8 is quite relevant at this point:

"I like written feedback. I don't need to rush about what to say. I have more time to think about what to write. I can evaluate the paragraph from different perspectives. I don't have to worry about how to give feedback to my classmate's face, which is quite hard for me particularly if I need to point out the errors. I have to think quickly and form my ideas correctly and logically in English. This is not always easy for me".

Although the data analysis of the semi-structured interviews revealed the participants' mixed feelings about their peer feedback preferences, they were, in general, of the opinion that both types of feedback (online and face-to-face) were useful in different ways and benefitted from each mode in varying degrees (Cartney, 2010; Krych-Appelbaum, & Musial, 2007). Therefore, they suggested that both types of peer feedback should be integrated into EFL writing instruction. The participants tended to prioritize one type of feedback over the other, depending on their learning style. However, they emphasized that their involvement in both types of peer feedback activities not only as receivers but also as providers raised their level of language awareness, which is an important component of linguistic competence, as well as facilitating their interpersonal skills, which is conducive to the development of their communicative competence. They emphasized that their engagement in face-to-face and online peer feedback enriched their pedagogical content knowledge as prospective teachers, particularly in terms of developing their feedback provision style, and varying their techniques of feedback provision in line with the learner profile.. They indicated that their involvement in providing and receiving oral and online written peer feedback facilitated their progress as EFL learners in line with their zone of proximal development (Altstaedter & Doolittle, 2014).

CONCLUSION

The study revealed that participants developed a favourable attitude towards EFL writing as a result of their involvement in the online written peer feedback and the oral feedback process. The combined use of online feedback and oral feeback was regarded as highly beneficial by the EFL pre-service teachers in the study in terms of its impact on the writing skills development. The use of online feedback along with oral feedback led to a relatively enhanced student performance in EFL writing, despite statistically insignificant results.

The overall evaluation of peer feedback from the participants' rather than the researcher's point of view added an emic perspective to the study (Davis, 1995). Triangulation through multiple data sources (both qualitative and quantitative) enabled the verification of the research findings. Pre-service teachers' engagement in online written peer feedback sessions, face-to-face group feedback sessions, and weekly journal-keeping practices impacted their views regarding EFL academic writing favourably. The analysis of the EFL academic writing performances of the pre-service teachers on the first and the third drafts indicated that they exhibited a relatively enhanced performance concerning paragraph writing in the EFL context, although the results did not prove to be statistically significant.

The findings of the study also indicated that beneficial as it might be, the online written peer review process might not prove to be a smooth one for the participants who came from teacher-centered educational backgrounds where the writing skills development was simply downgraded or disregarded in the mainstream L2 instruction. Launching the oral feedback initially and then proceeding to the online written feedback, which was indicated in the current study as well as in other studies (See Krych-Appelbaum, & Musial, 2007; Cartney, 2010), might be suggested to enhance the quality of peer feedback activities and to familarize the participants with the conventions of both types of feedback. The students might also be asked to bring a printout of the written feedback they received from their peers on their blogs to the face-to-face collaborative peer feedback sessions, where they might ask questions about the points that they were confused in class (DiGiovanni & Nagaswami, 2001).

In brief, although the results of the current study suggested that the integration of face-to-face and online peer review might produce fruitful results in terms of the development of favourable student attitudes towards EFL writing, a gradual improvement of students' writing skills in the EFL context, the findings need to be interpreted with caution due to the limited number of participants and the relatively short duration of the study. It is, in fact, difficult to demonstrate a significant improvement over a 14-week semester, owing to the developmental aspects of EFL writing (Dyson & Freedman, 1991). To make more certain remarks about the real pedagogical potential of peer feedback, the study needs to be extended longitudinally and with a larger sample, including the investigation of the impact of the online written and face-to-face oral peer review on students' revisions.

SOLUTIONS AND RECOMMENDATIONS

The provision of a training program, which is like the one illustrated in the study, is likely to facilitate the smooth flow of the process and contribute to the quality of the revisions and the negotiation of meaning between the group members during the face-to-face interactive peer feedback sessions. The implementation of a comprehensive training program prior to the participants' engagement in peer feedback might ensure the smooth flow of the process and alleviate the hostile reaction of some participants towards peer revision who comments "I am not the teacher" (Sengupta, 1998). Such students might feel assured to know that their teachers are constantly monitoring them. Therefore, it is recommendable to have an intensive face-to-face and online peer feedback provision training program to address the potential student concerns in this regard. It is also of great importance to provide continuous scafffolding to students throughout the peer feedback provision experience.

The participants might be asked to use the comment feature of the blogs more actively, and ask questions about the confusing points on the blog, which might provide an online interactive exchange with others. Additionally, to provide further practice for the students at the beginning of the study, the whole class might be asked to work on the printouts of sample student-written peer feedbacks and critically analyze the feedbacks, making further suggestions of revisions in groups in the oral group feedback conferences. The teacher can also elicit from the students how to comment precisely and concisely and train them to be more effective peer reviewers. In cases where the students have difficulty shifting from the face-to-face to online written feedback, it might be considered to initiate online feedback in class followed by the interaction of the peer group in the computer lab (DiGiovanni and Nagaswami, 2001)

Taking into consideration the participants' lack of engagement in L2 writing activities as well as lack of awareness concerning peer revision activities in K-12 education, teachers need to monitor the implementation of the peer revision activities step by step to ensure the effectiveness of the whole process. It is of utmost importance to provide scaffolding and guidance for the participants engaged in the peer revison activity for the first time or for those who are not quite familiar with it. For such participants, it should be taken into consideration that they may need a relatively long period of training time and experience to be able to review a piece of written work properly. The use of guided peer response sheets, such as the ones used in this study, might have the impact of facilitating the oral peer review process and may provide guidelines for the novice reviewers who do not know which aspects of the written work to attend to and how to express their feedback in a constructive manner. It might be a good idea to conduct a demo face-to-face peer feedback session in class with the supervision of the teacher before the students are involved in the oral peer review on their own. It is also recommendable for teachers to model how to provide online written peer feedback in class, which was the case in this study, and have a demo feedback session with the students. The modelling activity and the demo feedback session are likely to be useful for the ESL /EFL students and for the writing instructors as well. The former will be given a chance to ask questions while the latter will be able to obtain feedback from the former about the challenges they faced during the demo session and try to provide some remedies for them.

Peer feedback and self-feedback could be used together to enhance the quality of student learning (Lee, 2015). The students can first be engaged in "self-feedback" (reflecting on and evaluating their own writing, determining their strengths and weaknesses, etc) (Lee, 2015, p.97). Based on their self-feedback, they can make specific peer feedback requests, asking their peers to focus on specific areas of their writing. This turns peer feedback into a goal-oriented, focused and personalized activity. Alternatively, they can reflect on and evaluate their own writing based on the peer feedback they have received. The peer feedback and self-feedback can be used concurrently to promote student learning.

It is important to define the student roles in the peer feedback process. It is important to raise the students' awareness that they not only give but also receive feedback. As the feedback receivers, they reflect on their peers' comments on the strong and weak points of their work, assess their own writing, and work on improving their writing. As feedback providers, students learn from their peers' writing, raising their own awareness towards the strong and weak points of their peers as well as how to incorporate these points into their own work. Students engaged in the peer feedback process should be aware of the active role they play in the process and hence they should be reminded that they do not need to comply with all the comments that they receive; they can show partial agreement with the suggestions or reject them totally. The students may also play an active role in an "intra-feedback" activity, a procedure where students write comments individually and then get together in groups to discuss all the comments on the same written work (Lee, 2015 p.3). During the intra-feedback, the participants critically analyze the differences, inconsistencies between peer comments, negotiate any unclear points, by revisiting their original peer comments and revise them accordingly.

Most of the problems mentioned earlier regarding both pedagogical and practical issues can be alleviated by properly setting up the group and establishing effective procedures and adequate training, coaching students in terms of the principles and practices of effective peer group interaction and response. Without such training, it is more likely that student responses will be inappropriate, demotivating, and tactless (or conversely overgenerous or uncritical). In the absence of training, students may also tend to focus more on local aspects (surface matters such as punctuation and spelling) than on global aspects (discursive features including meaning and content). Student comments may be prescriptive and authoritarian rather than collaborative and supportive (Mangelsdorf & Schlumberger, 1992; Stanley, 1992).

FUTURE RESEARCH DIRECTIONS

There are many aspects of peer revision that await empirical investigation. More research is necessary to explore the factors having an impact on the students' attitudes towards various types of peer feedback and the impact of various types of peer feedback on the writing performance of different learner profiles. Also, the adoption of a comparative perspective to the feedback research, the exploration of the impact of peer feedback training on ESL/EFL learners' writing performance and their attitudes towards writing, as well as the degree of ESL/EFL learners' consideration of peer feedback in their subsequent revisions could be enumerated among some other areas of further research.

REFERENCES

Abdalla Salih, A. R. (2013). Peer response to L2 student writing: Patterns and expectations. *English Language Teaching*, *6*(3), 20–42. doi:10.5539/elt.v6n3p42

Al-Hazmi, S., & Schofield, P. (2007). Enforced revision with checklist and peer feedback in EFL writing: The example of Saudi university students. *Scientific Journal of King Faisal University (Humanities and Management Sciences*, *8*(2), 237-267.

Allen, D., & Katamaya, A. (2016). Relative second language proficiency and the giving and receiving of written peer feedback. *System*, *56*, 96–106. doi:10.1016/j.system.2015.12.002

Alshuraidah, A., & Storch, N. (2019). Investigating a collaborative approach to peer feedback. *ELT Journal*, *73*(2), 166–174. doi:10.1093/elt/ccy057

Altstaedter, L. L., & Doolittle, P. (2014). Students' perceptions of peer feedback. *Argentinian Journal of Applied Linguistics*, *2*(2), 60–76.

Amores, M. J. (1997). A new perspective on peer-editing. *Foreign Language Annals*, *30*(4), 513–522. doi:10.1111/j.1944-9720.1997.tb00858.x

Arbaugh, J. (2000b). Virtual classroom versus physical classroom: An exploratory study of class discussions patterns and student learning in an asynchronous Internet-based MBA course. *Journal of Management Education*, *24*(2), 207–227. doi:10.1177/105256290002400206

Barnes, D. (1976). *From communication to curriculum*. Harmondsworth, England: Penguin.

Baxter, P., & Jack, S. (2008). Qualitative case study methodology: Study design and implementation for novice researchers. *Qualitative Report*, *13*(4), 544–559.

Beggren, J. (2015). Learning from giving feedback: A study of secondary level students. *ELT Journal*, *69*(1), 58–70. doi:10.1093/elt/ccu036

Best, K., Jones-Katz, L., Smolarek, B., Stolzenburg, M., & Williamson, D. (2014). Listening to our students: An exploratory practice study of ESL writing students' views of feedback. *TESOL Journal*, *6*(2), 332–357. doi:10.1002/tesj.152

Bitchener, J., & Ferris, D. R. (2012). *Written corrective feedback in second language acquisition and writing*. New York: Routledge. doi:10.4324/9780203832400

Bradley, L. (2014). Peer-reviewing in an intercultural wiki environment: Student interaction and reflections. *Computers and Composition*, *34*, 80–95. doi:10.1016/j.compcom.2014.09.008

Bromley, D. B. (1990). Academic contributions to psychological counselling: A philosophy of science for the study of individual cases. *Counselling Psychology Quarterly*, *3*(3), 299–307. doi:10.1080/09515079008254261

Carson, J. G., & Nelson, G. (1994). Writing groups: Cross-cultural issues. *Journal of Second Language Writing*, *3*(1), 17–30. doi:10.1016/1060-3743(94)90003-5

Carson, J. G., & Nelson, G. L. (1996). Chinese students' perceptions of ESL peer feedback group interaction. *Journal of Second Language Writing*, *5*(1), 1–19. doi:10.1016/S1060-3743(96)90012-0

Cartney, P. (2010). Exploring the use of peer assessment as a vehicle for closing the gap between feedback given and feedback used. *Assessment & Evaluation in Higher Education*, *35*(5), 551–564. doi:10.1080/02602931003632381

Cazden, C. (1988). *Classroom discourse: The language of teaching and learning*. Porltsmouth, NH: Heinemann.

Cha, Y. J., & Park, L. E. (2010). An analysis of synchronous interaction and its influence on EFL writers' revisions. *Multimedia-Assisted Language Learning*, *13*(2), 9–36. doi:10.15702/mall.2010.13.2.9

Chang, C. F. (2012). Peer review via three modes in an EFL writing course. *Computers and Composition*, *29*(1), 63–78. doi:10.1016/j.compcom.2012.01.001

Chang, C. K., Chen, G. D., & Hsu, C. K. (2011). Providing adequate interactions in online discussion forums using few teaching assistants. *The Turkish Online Journal of Educational Technology*, *10*(3), 193–202.

Chaudron, C. (1984). The effects of feedback on students' composition revisions. *RELC Journal*, *15*(2), 1–14. doi:10.1177/003368828401500201

Chen, T. (2016). Technology-supported peer feedback in ESL/EFL writing classes: A research synthesis. *Computer Assisted Language Learning*, *29*(2), 365–397. doi:10.1080/09588221.2014.960942

Cho, K., & Schunn, C. D. (2007). Scaffolded writing and rewriting the discipline: A web-based reciprocal peer review system. *Computers & Education*, *48*(3), 409–426. doi:10.1016/j.compedu.2005.02.004

Connor, U., & Asenavage, K. (1994). Peer response groups in ESL writing classes: How much impact on revision? *Journal of Second Language Writing*, *3*(3), 257–276. doi:10.1016/1060-3743(94)90019-1

Corgan, R., Hammer, V., Margolies, M., & Crossley, C. (2004). Making your online course successful. *Business Education Forum*, *58*(3), 51–53.

Creswell, J. W. (2013). *Qualitative inquiry and research design: Choosing among five approaches*. Thousand Oaks, California: Sage Publications, Inc.

Crinon, J. (2012). The dynamics of writing and peer review at primary school. *Journal of Writing Research*, *4*(2), 121–154. doi:10.17239/jowr-2012.04.02.2

Davis, K. (1995). Qualitative theory and methods in applied linguistics research. *TESOL Quarterly*, *29*(3), 427–453. doi:10.2307/3588070

De Guerrero, M., & Villamil, O. (1994). Social-cognitive dimensions of interaction in L2 peer revision. *Modern Language Journal*, *78*(4), 484–496. doi:10.1111/j.1540-4781.1994.tb02065.x

Diab, N. M. (2010). Effects of peer- versus self-editing on students' revision of language errors in revised drafts. *System*, *38*(1), 85–95. doi:10.1016/j.system.2009.12.008

DiGiovanni, E., & Nagaswani, G. (2001). Online peer review: An alternative to face-to-face? *ELT Journal*, *55*(3), 263–272. doi:10.1093/elt/55.3.263

Donato, R. (2004). Aspects of collaboration in pedagogical discourse. *Annual Review of Applied Linguistics*, *24*, 284–302. doi:10.1017/S026719050400011X

Doolittle, P. E., & Hicks, D. (2003). Constructivism as a theoretical foundation for the use of technology in social studies. *Theory and Research in Social Education*, *31*(1), 72–104. doi:10.1080/00933104.2003.10473216

Driscoll, M. P. (2000). *Psychology of learning for instruction* (2nd ed.). Needham Heights, MA: Allyn & Bacon.

Dunlap, J. C. (2005). Workload reduction in online courses: Getting some shuteye. *Performance Improvement*, *44*(5), 18–25. doi:10.1002/pfi.4140440507

Dunlop, J. C., & Grabinger, R. S. (1996). Rich environment for the active learning in the higher education. In B. G. Wilson (Ed.), *Constructing learning environments: Case studies in instructional design* (pp. 65–82). Englewood Cliiffs, NJ: Educational Technology Publications.

Dyson, A. H., & Freedman, S. W. (1991). Writing. In *Handbook of research on teaching the English language arts*. Macmillan.

Ekşi, G. Y. (2012). Peer review versus teacher feedback in process writing: How effective? *IJAES*, *13*(1), 33–48.

Elwood, J. A., & Bode, J. (2014). Student preferences vis-à-vis teacher feedback in university EFL writing classes in Japan. *System*, *42*, 333–343. doi:10.1016/j.system.2013.12.023

Ertmer, P. A., Richardson, J. C., Belland, B., Camin, D., Connolly, P., Coulthard, G., Lei, K., & Mong, C. (2007). Using peer feedback to enhance the quality of student online postings: An exploratory study. *Journal of Computer-Mediated Communication*, *12*(2), 412–433. doi:10.1111/j.1083-6101.2007.00331.x

Ezza, E. S. (2010). Arab EFL learners' writing dilemma at tertiary level. *English Language Teaching*, *3*(4), 33–39. doi:10.5539/elt.v3n4p33

Ferris, D. (2002). *Treatment of error in second language student writing*. The University of Michigan Press.

Ferris, D. (2006). Does error feedback help student writers? New evidence on the short- and long- term effect of written error correction. In K. Hyland & F. Hyland (Eds.), *Peer feedback in second language writing: Context and issues* (pp. 81–104). Cambridge University Press. doi:10.1017/CBO9781139524742.007

Ferris, D. R. (2003). *Response to student writing: Implications for second language students*. Lawrence Erlbaum Associates. doi:10.4324/9781410607201

Ferris, D. R., Liu, H., Sinha, A., & Senna, M. (2013). Written corrective feedback for individual L2 writers. *Journal of Second Language Writing*, *22*(3), 307–329. doi:10.1016/j.jslw.2012.09.009

Fitze, M. (2006). Discourse and participation in ESL face-to-face and written electronic conferences. *Language Learning & Technology*, *10*(1), 67–86.

Flower, L., & Hayes, J. R. (1980). The cognition of discovery: Defining a rhetorical problem. *College Composition and Communication*, *31*(1), 21–32. doi:10.2307/356630

Freedman, S. W. (1992). Outside-in and inside-out: Peer response groups intwo ninth-grade classes. *Research in the Teaching of English*, *26*(1), 71–107.

Fulford, A., & Zhang, S. (1998). Assessing distance learners' satisfaction with instruction: A quantitative and qualitative measure. *American Journal of Distance Education*, *12*(3), 4–27. doi:10.1080/08923649809527002

Gere, A. R., & Abbott, R. D. (1985). Talking about writing: The language of writing groups. *Research in the Teaching of English*, *10*(4), 362–381.

Gilles, R., & Snook, L. (Eds.). (1990). *A student guide to freshman composition* (11th ed.). University of Arizona Bellwether Press.

Guardado, M., & Shi, L. (2007). ESL students' experiences of online peer feedback. *Computers and Composition*, *24*(4), 443–461. doi:10.1016/j.compcom.2007.03.002

Hanjani, A. M., & Li, L. (2014). Exploring L2 writers' collaborative revision interactions and their writing performance. *System*, *44*, 101–114. doi:10.1016/j.system.2014.03.004

Hansen, J. G., & Liu, J. (2005). Guiding principles for effective peer response. *ELT Journal*, *59*(1), 31–38. doi:10.1093/elt/cci004

Hattie, J., & Timperley, H. (2007). The power of feedback. *Review of Educational Research*, *77*(1), 81–112. doi:10.3102/003465430298487

Hayes, J. R. (2012). Modeling and remodeling writing. *Written Communication*, *29*(3), 369–388. doi:10.1177/0741088312451260

Hedgecock, J., & Lefkowitz, N. (1992). Collaborative oral/aural revision in foreign language writing instruction. *Journal of Second Language Writing*, *1*(2), 255–276. doi:10.1016/1060-3743(92)90006-B

Herrington, A., & Cadman, D. (1991). Peer review and revising in an anthropology course: Lessons for learning. *College Composition and Communication*, *42*(2), 184–199. doi:10.2307/358198

Higgins, R., Hartley, P., & Skelton, A. (2002). The conscientious consumer: Reconsidering the role of assessment feedback in student learning. *Studies in Higher Education*, *27*(1), 53–64. doi:10.1080/03075070120099368

Ho, M., & Savignon, S. (2007). Face-to-face and computer-mediated peer review in EFL writing. *CALICO Journal*, *24*(2), 269–290. doi:10.1558/cj.v24i2.269-290

Ho, M. C. (2015). The effects of face-to-face and computer-mediated peer review on EFL writers' comments and revisions. *Australasian Journal of Educational Technology*, *31*(1), 1–15. doi:10.14742/ajet.495

Hojeij, Z., & Baroudi, S. (2018). Student perceptions on peer feedback training using a blended method: A UAE case. *Issues in Educational Research*, *28*(3), 655–678. http://www.iier.org.au/iier28/hojeij.pdf

Hu, G. W. E., & Lam, S. T. E. (2010). Issues of cultural appropriateness and pedagogical efficacy: Exploring peer review in a second language writing class. *Instructional Science*, *38*(4), 371–384. doi:10.100711251-008-9086-1

Hyland, F. (2000). ESL Writers and Feedback: Giving more autonomy to students. *Language Teaching Research*, *4*(1), 33–54. doi:10.1177/136216880000400103

Hyland, K., & Hyland, F. (2006). Feedback on second language students' writing. *Language Teaching*, *39*(2), 83–101. doi:10.1017/S0261444806003399

Jacobs, G. M., Curtis, A., Braine, G., & Juan, S. Y. (1998). Feedback on student writing: Taking the middle path. *Journal of Second Language Writing*, *7*(3), 307–317. doi:10.1016/S1060-3743(98)90019-4

Jahin, J. H. (2012). The effect of peer reviewing on writing apprehension and essay writing ability of prospective EFL teachers. *The Australian Journal of Teacher Education*, *37*(1), 60–84. doi:10.14221/ajte.2012v37n11.3

Jehn, K. A., & Shah, P. P. (1997). Interpersonal relationships and task performance: An examination of mediation processes in friendship and acquaintance groups. *Journal of Personality and Social Psychology*, *72*(4), 775–790. doi:10.1037/0022-3514.72.4.775

Kaufman, J. H., & Schunn, C. D. (2011). Students' perceptions about peer assessment for writing: Their origin and impact on revision work. *Instructional Science*, *39*(3), 387–406. doi:10.100711251-010-9133-6

Kayi-Aydar, H. (2013). Scaffolding language learning in an academic ESL classroom. *ELT Journal*, *67*(3), 324–335. doi:10.1093/elt/cct016

Keh, C. L. (1990). Feedback in the writing process: A model and methods for implementation. *ELT Journal*, *44*(4), 294–394. doi:10.1093/elt/44.4.294

Kessler, G., & Bukowski, D. (2010). Developing collaborative autonomous learning abilities in computer mediated language learning: Attention to meaning among students in wiki space. *Computer Assisted Language Learning*, *23*(1), 41–58. doi:10.1080/09588220903467335

Khalil, E. (2018). The efficacy of peer feedback in Turkish EFL students' writing performance. *Journal of Literature and Art Studies*, *8*(6), 920–931.

Ko, S., & Rossen, S. (2001). *Teaching online: A practical guide*. Houghton Mifflin.

Koschmann, T. (1996). Paradigm shifts and instructional technology: An introduction. In T. Koschmann (Ed.), *CSCL: Theory and practice of an emerging paradigm* (pp. 1–23). Lawrence Erlbaum Associates.

Krych-Appelbaum, M., & Musial, J. (2007). Students' perception of value of interactive oral communication as part of writing course papers. *Journal of Instructional Psychology*, *34*(3), 131–136.

Kuyyogsuy, S. (2019). Students' attitudes toward peer feedback: Paving a way for students' English writing improvement. *English Language Teaching*, *12*(7), 107–119. doi:10.5539/elt.v12n7p107

Lam, R. (2010). A peer review training workshop: Coaching students to give and evaluate peer feedback. *TESL Canada Journal*, *27*(2), 114–127. doi:10.18806/tesl.v27i2.1052

Lantolf, J. (Ed.). (2000). *Sociocultural theory and second language learning*. Oxford University Press.

Lantolf, J., & Thorne, S. (2006). *Sociocultural theory and the genesis of second language development*. Oxford University Press.

Lee, I. (2017). *Classroom writing assessment and feedback in L2 school contexts*. Springer. doi:10.1007/978-981-10-3924-9

Lee, M. (2015). Peer feedback in second language writing: Investigating junior secondary students' perspectives on inter-feedback and intra-feedback. *System*, *55*, 1–10. doi:10.1016/j.system.2015.08.003

Lei, Z. (2017). Salience of student written feedback by peer revision in EFL writing class. *English Language Teaching*, *10*(12), 151–157. doi:10.5539/elt.v10n12p151

Leki, I. (1990). Potential problems with peer responding in ESL writing classes. *The CATESOL Journal*, *3*, 5–19.

Liang, M. (2010). Using synchronous online peer response groups in EFL writing: Revision-related discourse. *Language Learning & Technology*, *14*(1), 45–64.

Lin, S., Liu, E., & Yuan, S. (2001). Web-based peer assessment: Feedback for students with various thinking styles. *Journal of Computer Assisted Learning*, *17*(4), 420–432. doi:10.1046/j.0266-4909.2001.00198.x

Liou, H. C., & Peng, C. Y. (2009). Training effects on computer-mediated peer review. *System*, *37*(3), 514–525. doi:10.1016/j.system.2009.01.005

Liou, H. C., & Peng, Z. Y. (2009). Training effects on computer-mediated peer review. *System*, *37*(3), 514–525. doi:10.1016/j.system.2009.01.005

Liu, J., & Hansen, J. (2002). *Peer response in second language writing classrooms*. University of Michigan Press. doi:10.3998/mpub.8952

Liu, J., & Sadler, R. W. (2003). The effect and affect of peer review in electronic versus traditional modes on L2 writing. *Journal of English for Academic Purposes*, *2*(3), 193–227. doi:10.1016/S1475-1585(03)00025-0

Liu, N., & Carless, D. (2006). Peer feedback: The learning element of peer assessment. *Teaching in Higher Education*, *11*(3), 279–290. doi:10.1080/13562510600680582

Lockhart, C., & Ng, P. (1995). Analyzing talk in ESL peer response groups: Stances, functions, and content. *Language Learning*, *45*(4), 605–655. doi:10.1111/j.1467-1770.1995.tb00456.x

Long, M., & Porter, P. (1985). Group work, interlanguage talk and second language *acquisition. TESOL Quarterly*, *19*(2), 305–325. doi:10.2307/3586827

Lu, R., & Bol, L. (2007). A comparison of anonymous versus identifiable e-Peer review on college student writing performance and the extent of critical feedback. *Journal of Interactive Online Learning*, *6*(2), 100–115.

Lundstrom, K., & Baker, W. (2009). To give is better than to receive: The benefits of peer review to the reviewer's own writing. *Journal of Second Language Writing*, *18*(1), 30–43. doi:10.1016/j.jslw.2008.06.002

Lynch, M. M. (2002). *The online educator: A guide to creating the virtual classroom.* Routledge Falmer. doi:10.4324/9780203458556

Manchón, R. M. (2012). Multiple perspectives on L2 writing development. In R. M. Manchón (Ed.), *L2 writing development: Multiple perspectives* (pp. 17–46). Walter de Gruyter. doi:10.1515/9781934078303.1

Mangelsdorf, K. (1992). Peer reviews in the ESL composition classroom: What do the students think? *ELT Journal, 46*(3), 274–285. doi:10.1093/elt/46.3.274

Mangelsdorf, K., & Schlumberger, A. (1992). ESL student response stances in a peer-review task. *Journal of Second Language Writing, 1*(3), 235–254. doi:10.1016/1060-3743(92)90005-A

Maykut, P., & Morehouse, R. (1994). *Beginning qualitative research: A philosophic and practical guide.* The Falmer Press.

McConnell, D. (2002). The experience of collaborative assessment in e-learning. *Studies in Continuing Education, 24*(1), 73–92. doi:10.1080/01580370220130459

McGroarty, M. E., & Zhu, W. (1997). Triangulation in classroom research: A study of peer revision. *Language Learning, 47*(1), 1–43. doi:10.1111/0023-8333.11997001

Mendonca, C. O., Johnson, K. E., & Mendonca, C. O. (1994). Peer review negotiations: Revision activities in ESL writing instruction. *TESOL Quarterly, 28*(4), 745–769. doi:10.2307/3587558

Min, H. T. (2005). Training students to become successful peer reviewers. *System, 33*(2), 293–308. doi:10.1016/j.system.2004.11.003

Min, H. T. (2006). The effects of trained peer review on EFL students' revision types and writing quality. *Journal of Second Language Writing, 15*(2), 118–141. doi:10.1016/j.jslw.2006.01.003

Min, H. T. (2016). Effect of teacher modeling and feedback on EFL students' peer review skills in peer review training. *Journal of Second Language Writing, 31*, 43–57. doi:10.1016/j.jslw.2016.01.004

Mory, E. H. (2004). Feedback research review. In D. H. Jonassen (Ed.), *Handbook of research on educational communications and technology* (pp. 745–783). Lawrence Erlbaum.

Narciss, S. (2008). Feedback strategies for interactive learning tasks. In J. M. Spector, M. D. Merrill, J. G. van Merrienboer, & M. P. Driscoll (Eds.), *Handbook of research on educational communications and technology* (3rd ed., pp. 125–144). Lawrence Erlbaum Associates.

Narciss, S. (2012). Feedback in instructional contexts. In N. M. Seel (Ed.), *Encyclopedia of the sciences of learning* (pp. 1285–1289). Springer. doi:10.1007/978-1-4419-1428-6_282

Narciss, S. (2013). Designing and evaluating tutoring feedback strategies for digital learning environments on the basis of the interactive tutoring Feedback Model. *Digital Education Review, 23*, 7-26. Retrieved from http://greav.ub.edu/der

Nelson, G. I., & Carson, J. G. (2006). Cultural issues in peer response: 'revisiting culture. In K. Hyland & F. Hyland (Eds.), *Feedback in second language writing: Context and issues* (pp. 42–59). Cambridge University Press. doi:10.1017/CBO9781139524742.005

Nelson, G. L., & Carson, J. G. (1998). ESL students' perceptions of effectiveness in peer feedback groups. *Journal of Second Language Writing*, *7*(2), 113–131. doi:10.1016/S1060-3743(98)90010-8

Nelson, G. L., & Murphy, J. M. (1992). An L2 writing group: Task and social dimensions. *Journal of Second Language Writing*, *1*(3), 171–193. doi:10.1016/1060-3743(92)90002-7

Nicol, D. J., & Macfarlane-Dick, D. (2006). Formative assessment and self-regulated learning: A model and seven principles of good feedback practice. *Studies in Higher Education*, *31*(2), 199–218. doi:10.1080/03075070600572090

Notar, C. E., Wilson, J. D., & Ross, K. G. (2002). Distant learning for the development of higher-level cognitive skills. *Education*, *122*, 642–650.

Palloff, R. M., & Pratt, K. (2013). *Lessons from the virtual classroom: The realities of online teaching* (2nd ed.). Jossey-Bass.

Paulus, T. M. (1999). The effect of peer and student feedback on student writing. *Journal of Second Language Writing*, *8*(3), 265–289. doi:10.1016/S1060-3743(99)80117-9

Pham, T. H. T., & Gillies, R. (2010). Designing a culturally appropriate format of formative peer assessment for Asian students: The case of Vietnamese students. *International Journal of Educational Reform*, *19*(2), 72–85. doi:10.1177/105678791001900201

Rahimi, M. (2013). Is training students' reviewers worth its while? A study of how training influences the quality of students' feedback and writing. *Language Teaching Research*, *17*(1), 67–89. doi:10.1177/1362168812459151

Razak, N. A., & Saeed, M. A. (2014). Collaborative writing revision process among learners of English as a foreign language (EFL) in an online community of practice (CoP). *Australasian Journal of Educational Technology*, *30*(5), 580–599. doi:10.14742/ajet.786

Razı, S. (2014). *Plagiarism detectors in undergraduate academic writing assignments*. Paper presented at the 14th International Languge, Literature and Stylistics Symposium, İzmir, Turkey.

Razı, S. (2015). Cross-checked problems in undergraduate academic writing. In K. Dikilitaş, R. Smith, & W. Trotman (Eds.), *Teacher-researchers in action* (pp. 147–161). IATEFL.

Razı, S. (2016). Open and anonymous peer review in a digital online environment compared in academic writing context. In C. Goria, O. Speicher, & S. Stollhans (Eds.), Innovative language teaching and learning at university: enhancing participation and collaboration (pp. 49-56). Dublin, Ireland: Research-publishing.net. doi:10.14705/rpnet.2016.000404

Richardson, J., & Swan, K. (2003). Examining social presence in online courses in relation to students' perceived learning and satisfaction. *Journal of Asynchronous Learning Networks*, *7*, 68–88.

Rinehart, D., & Chen, S. J. (2012). *The benefits of a cycle of corrective feedback on L2 writing: improving Taiwanese technological university students' English writing proficiency*. Lambert Academic Publishing.

Robinson, J. M. (2002). In search of fairness: An application of multi-reviewer anonymous peer review in a large class. *Journal of Further and Higher Education*, *26*(2), 183–192. doi:10.1080/03098770220129451

Rollinson, P. (1998). *Peer response and revision in an ESL writing group: a case study* (Unpublished PhD thesis). Universidad Autónoma de Madrid.

Rollinson, P. (2005). Using peer feedback in ESL writing class. *ELT Journal*, *59*(1), 23–30. doi:10.1093/elt/cci003

Ruecker, T. (2011). The potential of dual-language cross-cultural peer review. *ELT Journal*, *65*(4), 398–407. doi:10.1093/elt/ccq067

Ruegg, R. (2015). The relative effects of peer and teacher feedback on improvement in EFL students' writing ability. *Linguistics and Education*, *29*, 73–82. doi:10.1016/j.linged.2014.12.001

Saeed, M. A., Ghazali, K., & Aljaberi, M. A. (2018). A review of previous studies on ESL/EFL learners' interactional feedback exchanges in face-to face and computer-assisted peer review of writing. *International Journal of Educational Technology in Higher Education*, *15*(6), 1–25. doi:10.118641239-017-0084-8

Saito, H., & Fujita, T. (2004). Characteristics and user acceptance of peer rating in EFL writing classrooms. *Language Teaching Research*, *8*(1), 31–54. doi:10.1191/1362168804lr133oa

Sato, M. (2013). Beliefs about peer interaction and peer corrective feedback: Efficacy of classroom intervention. *Modern Language Journal*, *97*(3), 611–630. doi:10.1111/j.1540-4781.2013.12035.x

Savignon, S. J., & Roithmeier, W. (2004). Computer-mediated communication: Texts and strategies. *CALICO Journal*, *21*(2), 265–290. doi:10.1558/cj.v21i2.265-290

Sengupta, S. (1998). From text revision to text improvement: A story of secondary school composition. *RELC Journal*, *29*(1), 110–137. doi:10.1177/003368829802900107

Shrum, J. L., & Glisan, E. (2005). *Teacher's handbook: Contextualized language instruction* (3rd ed.). Heinle.

Shute, V. J. (2008). Focus on formative feedback. *Review of Educational Research*, *78*(1), 153–189. doi:10.3102/0034654307313795

Smits, M. S., Boon, J., Sluijsmans, D. M., & Van Gogh, T. (2006). Content and timing of feedback in a web-based learning environment: Effects on learning as a function of prior knowledge. *Interactive Learning Environments*, *16*(2), 183–193. doi:10.1080/10494820701365952

Sommers, E., & Lawrence, S. (1992). Women's ways of talking in teacher-directed and student-directed peer response groups. *Linguistics and Education*, *4*(1), 1–36. doi:10.1016/0898-5898(92)90018-R

Speck, B. W. (2000). Grading students' classroom writing: Issues and strategies (ASHE-ERIC Higher Education Report Vol. 27, No. 3). The George Washington University, Graduate School of Education and Human Development.

Stanley, J. (1992). Coaching student writers to be effective peer evaluators. *Journal of Second Language Writing*, *1*(3), 217–233. doi:10.1016/1060-3743(92)90004-9

Topping, K. (1998). Peer assessment between students in colleges and universities. *Review of Educational Research*, *68*(3), 249–276. doi:10.3102/00346543068003249

Tsui, A. B. M., & Ng, M. (2000). Do secondary L2 writers benefit from peer comments. *Journal of Second Language Writing*, *9*(2), 147–170. doi:10.1016/S1060-3743(00)00022-9

Tunison, S., & Noonan, B. (2001). On-line learning: Secondary students' first experience. *Canadian Journal of Education*, *26*(4), 495–514. doi:10.2307/1602179

Van den Berg, B. A. M., Admiraal, W. F., & Pilot, A. (2006). Designing student peer assessment in higher education: Analysis of written and oral peer feedback. *Teaching in Higher Education*, *11*(2), 135–147. doi:10.1080/13562510500527685

Van der Pol, J., Van den Berg, B. A. M., Admiral, W. F., & Simons, P. R. J. (2008). The nature, reception, and use of online peer feedback in higher education. *Computers & Education*, *51*(4), 1804–1817. doi:10.1016/j.compedu.2008.06.001

Van Popta, E., Kral, M., Camp, G., Martens, R. L., & Simons, P. R. J. (2017). Exploring the value of peer feedback in online learning for the provider. *Educational Research Review*, *20*, 24–34. doi:10.1016/j.edurev.2016.10.003

Villamil, O. S., & De Guerrero, M. C. M. (1996). Peer revision in the L2 classroom: Social-cognitive activities, mediating strategies, and aspects of social behaviour. *Journal of Second Language Writing*, *5*(1), 51–75. doi:10.1016/S1060-3743(96)90015-6

Villamil, O. S., & De Guerrero, M. C. M. (2006). Sociocultural theory: A framework for understanding of socio-cognitive aspects of social behavior. In K. Hyland & F. Hyland (Eds.), *Feedback in second language writing: Context and issues* (pp. 23–41). Cambridge University Press. doi:10.1017/CBO9781139524742.004

Vygotsky, L. S. (1978). *Mind in society: The development of higher psychological processes*. Harvard University Press.

Wang, W. Q. (2014). Students' perceptions of rubric-referenced feedback on EFL writing: A longitudinal inquiry. *Assessing Writing*, *19*, 80–96. doi:10.1016/j.asw.2013.11.008

Warshauer, M., & Ware, P. (2006). Automated writing evaluation: Defining the classroom research agenda. *Language Teaching Research*, *10*(2), 157–180. doi:10.1191/1362168806lr190oa

Weissberg, R. (2006). Scaffolded feedback: theoretical conversations with advanced L2 writers. In K. Hyland & F. Hyland (Eds.), *Feedback in second language writing: Contexts and issues* (pp. 246–265). Cambridge University Press. doi:10.1017/CBO9781139524742.015

Wette, R. (2015). Teacher-led collaborative modelling in academic L2 writing courses. *ELT Journal*, *69*(1), 71–80. doi:10.1093/elt/ccu043

Wu, W. C. V., Petit, E., & Chen, C. H. (2015). EFL writing revision with blind expert and peer review using a CMC open forum. *Computer Assisted Language Learning*, *28*(1), 58–80. doi:10.1080/09588221.2014.937442

Yang, M., Badger, R., & Yu, Z. (2006). A comparative and teacher feedback in a Chinese EFL writing class. *Journal of Second Language Writing*, *15*(3), 179–200. doi:10.1016/j.jslw.2006.09.004

Yin, R. K. (2003). *Case study research: Design and methods* (3rd ed.). Sage.

Yu, S., & Lee, I. (2015). Understanding EFL students' participation in group peer feedback of L2 writing: A case study from an activity theory perspective. *Language Teaching Research, 19*(5), 572–593. doi:10.1177/1362168814541714

Zamel, V. (1995). College composition and communication. *National Council of Teachers of English, 46*(4), 506-521.

Zhan, L. (2016). Written teacher feedback: Student perceptions, teacher perceptions, and actual teacher performance. *English Language Teaching*, *9*(8), 73–84. doi:10.5539/elt.v9n8p73

Zhang, S. (1995). Reexamining the affective advantage of peer feedback in the ESL writing class. *Journal of Second Language Writing*, *4*(3), 209–222. doi:10.1016/1060-3743(95)90010-1

Zhao, H. (2010). Investigating use and understanding of peer and teacher feedback on writing: A comparative study in a Chinese English writing classroom. *Assessing Writing*, *15*(1), 3–17. doi:10.1016/j.asw.2010.01.002

Zhao, H. (2014). Investigating teacher-supported peer assessment for EFL writing. *ELT Journal*, *68*(2), 155–168. doi:10.1093/elt/cct068

Zhu, W. (2001). Interaction and feedback in mixed peer response groups. *Journal of Second Language Writing*, *10*(4), 251–276. doi:10.1016/S1060-3743(01)00043-1

Zhu, W., & Mitchell, D. (2012). Participation in peer response as activity: An examination of peer response stances from an activity theory perspective. *TESOL Quarterly*, *46*(2), 362–386. doi:10.1002/tesq.22

KEY TERMS AND DEFINITIONS

Face-to-Face Feedback: The provision of oral or written feedback in response to students' oral or written work.

Global Concerns in Writing: Content-related issues in writing such as idea development, textual organization, coherence, unity, clarity of purpose and awareness of audience.

Intra-Feedback: A procedure where students are initially engaged in writing individual comments and then form small groups to discuss the comments on the same written work in a collaborative fashion.

Local Concerns in Writing: Language-related issues in writing such as spelling, word choice, and grammatical accuracy.

Non-Revision-Oriented Feedback: Synchronous or asynchronous comments related to the establishment of a friendly learning enviroment and good classroom dynamics.

Online Feedback: Asychronous or synchronous oral or written feedback provision in response to students' oral or written work on a virtual platform such as weblogs and wikis in a delayed time frame.

Peer Feedback: A collaborative activity where learners provide one another information interactively by commenting on and critiquing each other's work in both written and oral formats during the writing process

Process Approach to Teaching Writing: A non-linear, exploratory, and a bottom-up pedagogical approach to teaching writing that involves a series of stages such as planning, drafting and revising.

APPENDIX A. THE PEER RESPONSE SHEET

Name of the feedback provider:
Name of the writer:

1. Does the paragraph have a proper topic sentence? If so, what is it? If not, what needs to be changed to form a proper topic sentence?
2. Is it clear from the topic sentence what the writer is going to discuss? From what aspects do the writer would like to discuss the topic sentence? If not, what needs to be changed to make it clear?
3. Are all the aspets mentioned in the topic sentence (i.e., controlling ideas) well-supported? If not, would you like to add more details and support?
4. Is the order of the supports logical and effective? If not, how can you reorganize the supports to make them more effective?
5. Can you follow the flow of ideas easily through the transition words or phrases that the writer has used? If not, what transition words or phrases are needed to be added (or omitted) to help you follow the flow of ideas better?
6. Is the paragraph logically developed? Are there any parts in the paragraph that need to be developed further?
7. Do you think the paragraph has a proper summary sentence? If so, what is it? If not, how can you improve it?
8. Have you detected any inappropriate use of language /vocabulary in the paragraph? If so, how can you improve such inaccurate linguistic/lexical forms?
9. What did you like best about the paragraph?
10. Are there any parts in the paragraph that are unclear or confusing for you? If so, underline those parts and ask your partner to explain them to you.

APPENDIX B. THE HOLISTIC SCORING SCALE (ADAPTED FROM GILLES & SNOOK, 1990)

The Holistic Scoring Scale:

A 5 essay: is outstanding in all areas of the composition. It is particularly distinguished in content, containing an interesting, original and sometimes even risky thesis. The essay is fully developed and organized in a way that underscores and builds upon the main idea. A 5 essay addresses the assignment thoroughly, incorporates fresh and precise language, while usage and mechanics show few if any deviations from the conventions of standard edited English. What distinguishes a 5 essay is the ease with which mature ideas flow from one to the other in support of a central thesis.

A 4 essay: is solid in all areas of the composition. The topic is interesting and worthwhile to an educated reader, and the ideas are fully developed and logically organized. A 4 essay addresses the assignment sufficiently. The paper contains few if any deviations from standard usage and incorporates precise and appropriate diction.

A 3 essay: is a sound composition that may lack the focus and development of a 5 or 4 essay. It is characterized by a clear thesis given sufficient treatment to validate it; however, a 3 essay often fails to intrigue an educated reader. A 3 essay addresses the assignment, but not in the same way a 5 or 4 paper does. The essay is orderly and contains few deviations from standard English, but may be marked by a rigid or mechanical organization that draws attention away from the ideas themselves.

A2 essay: is weak in one or more areas of the composition. It is difficult to understand for some reason: it may have a vague, poorly stated, or simplistic thesis; it may lack a cohesive organization that leads the reader from one point to the next; it may fail to develop ideas in a manner satisfactory to an educated reader; it may address the assignment superficially, or it may have significant problems with grammar, spelling, or language usage. Any number of weaknesses may characterize a 2 essay.

A1 essay: is unacceptable in one or more of the areas of the composition. It may fail to address the assignment, containing numerous serious grammatical errors, unclear or misleading expressions, illogical organization, or negligible content. In short, it does not succeed in communicating ideas to an educated reader.

APPENDIX C

Items	Strongly Agree	Agree	Not sure	Disagree	Strongly Disagree
1. I found the peer feedback training (both face-to-face and online) useful.	31.25 %	34.37%	25%	9%	-
2. I found the peer feedback activities (providing oral and written peer feedback) useful in general.	28%	37.50%	25.62 %	3.12 %	3.12 %
3. I found face-to-face oral peer feedback activities useful.	28.62%	40.62%	18.75 %	15%	-
4. I found online written peer feedback activities useful.	25%	35%	10%	15%	5%
5. I had difficulty providing face-to-face oral feedback.	20%	15%	5%	30%	20%
6. I had difficulty providing online written feedback.	15.13%	34.47%	25%	35%	15%
7. I had difficulty understanding face-to-face oral feedback.	10%	15%	5%	40%	30%
8. I had difficulty understanding online written feedback.	20%	15%	20%	35%	20%
9. I consulted my peers for confusing face-to-face oral feedback.	15%	20%	10%	30%	25%
10. I consulted my peers for confusing online written feedback.	38%	15%	12%	30%	20%

Chapter 8
Significance of Transcript Availability for Viewing TV Coverage of US Elections

Buğra Zengin
Namık Kemal University, Turkey

ABSTRACT

With its engaging spectacle and discussion of key issues, the US presidential election debates draw high viewership of not only Americans but also people of other nationalities through not only US-based television channels but also other mainstream channels around the world. Since many of the viewers are likely to be the learners of English as the most widely used Lingua Franca, who listen for as well as to the content, they should be helped with their processing of what they listen to, given the listening comprehension challenges. The demanding factors of speech rate and vocabulary necessitate the use of transcripts as an aid for listening comprehension. Transcripts facilitate the required vocabulary buildup and allow for semantic mapping through frequency counts of recurrent key words or word strings uncovering linguistic patterns and laying foundation for the enriching listening practices. Therefore, this chapter focuses on using transcripts accordingly.

INTRODUCTION

Through the 21st century skills map (P21), American Council on Teaching of Foreign Languages has shared its vision of integrating the teaching of 21st skills and that of world languages (ACTFL, 2011). A parallel work has been envisioned with Cambridge University Press the only difference being the English language replaced with other world languages aimed in P21 (The Cambridge Framework for Life Competencies, 2020). A similar mentality has been observed to emerge in a book titled "English for 21st Century Skills" (Eds: Mavrid & Xerri) in the field of English language teaching published by Express Publishing. As described in the publishing company's website, "well-known ELT experts as well as teacher educators, practitioners and leaders from around the world explore how creativity, collaboration, critical thinking, inclusion, wellbeing, leadership, and other key new literacies can be

DOI: 10.4018/978-1-7998-7876-6.ch008

developed in both formal and informal learning contexts". The potential beneficiaries of the up-to-date developments in this field of considerable importance range from seasoned and novice teachers, and trainers to researchers and policy makers.

Despite the need for their integration in the learning/teaching practices, the curricular claims regarding the 21st century skill are not seen as actually realised as frequently as talked about. Therefore this chapter shares the author's idea for the actualisation of these aspirations. Proposing a practically solid proposition this chapter shares the author's share of walk the walk with an illustrative example. The chapter argues for the utilization of the public availability advantage of transcription as a means to help with the viewing of presidential election debate broadcasts for contents listened to for their social social significance and relevance aimed to communicate through candidates' rhetorical strategies.

BACKGROUND

Offering a cognitive perspective Goh (2000) listed ten problems of real-time second language listening comprehension during such concurrrently occuring cognitive processing phases as perception, parsing and utilisation, all three of which are interrelated and recursive. In Goh's work, half of the real-time cognitive constraints were reported to occur during perceptual processing such as word recognition and attention failure. The listeners were also found to be inefficient in parsing and unable to utilise a mental representation even if the input is parsed.

A number of related factors considered to possibly cause the problems were listed: "sound-script and word-referent processes were not automatised, poor sound representations of familiar words, failure to use appropriate comprehension tactics, a lack of appropriate schematic knowledge, insuficient prior knowledge, preoccupation with knowing the meaning of certain content words, limited processing capacity in short-term memory and shallow processing." Low ability listeners' low-level perception problems were reported to appear higher in number compared to the high ability listeners (Goh, 2000, p. 69).

Goh (1998a, b: as cited in 2000) provides strategies such as perception practice as well as listening strategy practice (cognitive, metacognitivee, social-affective tactics). One of the perception practice items is to follow along with the transcript of recording. On the other hand, according to Renandya and Farrel (2010), simply there is not enough evidence to endorse the implementation of listening strategy training in English as a foreign language classrooms.

On the other hand, arguing for a text-based approach over a strategy-based one, Ridgway (2000: 179) states that "in listening, working from the text, or from texts in general, may be a more productive way of approaching comprehension than working from the notion of strategies." For, the development of skills and automaticity in processing spoken discourse, the most needed for learners is ample practice of actual listening. Renandya and Farrel (2010, p.55) "also believe that an overemphasis on strategy training may undermine the value of practice."

Among the materials with facilitation potential for practice procedures transcripts have great but untapped potential. However, machine only solutions more often than not fall short of the desired accuracy level. Pan, Jiang, Yao, Picheny and Qin (2010) state that real-time transcriptions generated by automated speech recognition (ASR) systems assist those lacking any help with listening comprehension but only if the accuracy rate is high. The 20% word error rate (WER) has been found to be the most likely critical threshod for a transcript to be accepted and considered useful compared to no transcript condition. However, it is 10% WER that makes a significant increase in comprehension performance.

Therefore, the transcription services providing high level accuracy should be preferred. For instance, the services of the rev.com have this kind of quality using human plus artifical intelligence approach. The rev.com website includes audio visual component accompanied by the transcription text, which is not erroneous since they are the product of both aritifial intelligence and humans. The technology is available to decode the utterance through on-screen automated speech recognition (ASR) but it is highly likely to come across errors in many video posts. Cao, Yamashita and Ishida (2016) analysed two different conditions (audio only and audio+ASR transcripts) in an experiment participated by non-native speakers (NNSs). ASL transcripts were found to hold potential for facilitating listening comprehension like solving word recognition problems except in the case of confusing, imperfect ASR (e.g., errors and no punctuation). It was also difficult to shift between modal contents. Though supplemental information is provided through ASR transcripts may well turn into overwhelming excessive information given the inadequate time for their full exploitation. Of all problem occurences identified in three patterns, the immediate listening comprehension problems surpass the other two patterns with 76% most of which consist of word or phrase-related problems (Cao, Yamashita, Ishida, 2016).

Practical solutions for most of these word or phrase-related problems need to be designed. Word clouds are one of the key factors significant in the efficiency of these solutions. Besides eye-catching compact memory-enhancing visualisations representing the text content, word clouds provide information on frequency of words. Baker (2006, p.47) highlights frequency as "one of the most central concepts underpinning the analysis of corpora" and seeing word cloud generators as efficient learning tools for analysis, Filatova (2016) argues for their benefits as free, user-friendly online tools of corpus pedagogy which would can be effectively used in reading, writing, vocabulary comprehension, reading, writing and rhetoric skills for all proficiency levels.

This study aims to show how to make transcripts part of the solution. Its account is shared hoping that it can be adapted for other contexts. Facilitating vocabulary build-up as a foundation prior to listening-intensive learning processes, the transcripts with a human plus machine approach can function as a reliable aid along the road.

METHODOLOGY

This study has a practical methodology. It is an account of writing Google search queries with Google operators (inurl, allinurl, intext, allintext, filetype, site etc.) to find out websites which provide accurate transcripts. Two sources were determined. One is rev.com ["presidential election" site:rev.com] and the other is a section of cnn.com [http://transcripts.cnn.com/TRANSCRIPTS/]. Once transcript URL or PDFs are accessed, frequency lists were made of high-frequency word/sord strings related to recurrent patterns of social significance from the spoken discourse transcripts as a preparation laying foundation for listening authentic content. The content chosen was the first two debates of Democratic primaries in the 2020 United States presidential election which has dominated the news cycle and online search trends (only second to the coronavirus around globe). US election campaigns are highly televised by US-based international TV news networks and mainstream channels around the world. Using counts of high-frequency words through wordcloud.com and word sequences through Microsoft word for vocabulary build-up purposes as the right warm-up for follow-up viewing/listening.

Besides, Youtube streams commenting on US election have been analysed. Since there is no transcript service like rev.com, the best option was listening attentively, taking notes and making summaries including comments or analyses. These summaries demonstrate that the high quality of the coverages deserve transcription services which would make it possible for viewers to make comparisons let alone their potential benefits for language learning.

Instrument

Google search queries have been used to find out transcripts. To narrow down searches search operators/commands are used. For instance, phrases or other word strings are enclosed with the quotation mark-opcrator (couble quotes, "…") to find the exact string as it is an operator for exact match objectives. The minus-operator (-) is applied to delete a specific word or word strings. In a query intitle:term, the intitle operator limits the search results to documents containing term in the title. The site operator restricts search to a parcitular website, .edu domains, which indicates academic websites, education), (or second level domain .ac short for academia), or country domains, military or government domains, etc. It is also possible to narrow down results to paricular file types (.doc, .docx, .pdf, .ppt, .pptx, .txt, or .srt file extensions). With .txt or .srt file extensions it is also possible It is possible to construct such a query as ["presidential election" site:rev.com] to reach (rev.com) or [allinurl:cnn transcripts] to reach (http://transcripts.cnn.com/TRANSCRIPTS/).

As for the videos in the section of the chapter with brief accounts from some Turkish sources related to US democracy and election, Language Learning with Youtube extension (LLY) can be used to have captions on screen. The extension LLY is another instrument as an aid, a Google chrome extension allowing two subtitle sets (original language of the content and the language translated to, if there is any available. I remember a colleague of mine from a non-language major informing me that it is possible to add our translations to the content. The only video with transcription is Atatürk's democracy-themed address to Americans and its transcription was accessed with Google search queries.

RESULTS

Wordcloud Frequency Counts

Below are the categories and coding units selected out of the transcripts of the first 2020. Democractic presidential debate (night 1 and night 2) (see a detailed version in the appendix).

Night 1: People (total frequency / tf: 1059-most direct ones being 407- out of 44 themes); health (tf: 114 codes out of 9 themes); environment (45 codes out of 7 themes); economy (tf: 211 out of 16 themes); education (tf: 80 out of 9 themes); military (tf: 28 out of 6 themes); crime (tf: 128 out of 18 themes); law (tf: 105 out of 16 themes); policy/politics (tf: 647 out of 76 themes); and others

Night2: People (total frequency / tf: 300 out of 29 themes); health (tf: 132 out of 10 themes); environment (tf: 30 out of 2 themes); economy (tf: 203 out of 29 themes); education (tf: 46 out of 8 themes); military (tf: 46 out of 6 themes); crime (tf: 54 out of 12 themes); law (tf: 63 out of 10 themes); policy (tf: 317 out of 37 themes); emotion/feelings (tf: 16 out of 4 themes); belief (tf: 9 out of 2 themes); others

Details of the Night 1 of the First Democratic Debate

People: people (112), going (100), can (86), will (82), need (75), country (70), I'm (64), we're (58), want (57), sure (33), I've (31), we've (28), We're (26), community (19), kids (19), communities (18), children (17), human (11), race (11), immigration (8), universal (8), national (8), folks (8), immigrants (7), language (7), workers (7), parents (7), women (7), we'll (7), young (6), woman (6), black (6), Latino (5), nations (5), lad (5), Can (5), individual (4), people's (4), daughters (4), farmers (4), father (4), gay (4), woman's (3), child (3)

Health: health (28), healthcare (28), insurance (20), Medicare-for-all (7), pharmaceutical (7), medicare (7), premiums (6), clean (6), abortion (5)

Environment: climate (19 + Climate 2 = 21), carbon (8), green (5), energy (4), planet (4), water (3)

Economy: work (28), deal (26), economy (21), jobs (21), companies (19), money (19), tax (12), corporations (9), corporate (9), crisis (9), income (8), job (8), cost (7), wage (7), industrial (5), capital (3)

Education: school (22), college (18), reason, future (7), potential (6), education (6), sense (6), students (5), research (4)

Military: nuclear (9), military (6), battle (4), deployed (3), weapons (3), allies (3)

Crime: gun (20), border (14), guns (14), threat (10), police (7), violence (6), criminal (6), safe (6), security (5), shooting (5), genocide (5), offense (5), crimes (5), drugs (5), shootings (4), Assault (4), crime (4), safety (3)

Law: law (25), rights (10), justice (9), Act (9), civil (8), asylum (7), citizens (5), order (5), legislation (4), inequality (4), court (4), impeachment (3), citizenship (3), provisions (3), Supreme (3), repeal (3)

Policy/politics: care (36), America (33), know (33), issue (31), change (27), government (21), Democrats (20), policy (17), families (15), democratic (14), foreign (14), home (14), family (13), public (13), democracy (11), campaign (11), votes (11), world (11), Washington (10), federal (10), issues (10), voters (9), promise (8), reform (8), build (8), bills (8), vote (8), administration (7), Republican (7), electoral (7), services (7), values (7), works (7), plans (7), executive (6), Political (6), problems (6), option (6), power (6), consolidation (5), geopolitical (5), opportunity (5), political (5), solutions (5), priority (5), program (5), matters (5), leader (5), solve (5), LGBTQ (5), goal (5), accountability (4), responsibility (4), contributions (4), negotiate (4), activists (4), politics (4), elected (4), voting (4), polls (4), politically (3), leadership (3), bipartisan (3), referendum (3), elections (3), technical (3), emergency (3), caucuses (3), policies (3), midterms (3), mobilize (3), mission (3), union (3), equal (3), voter (3), fix (3)

Others: system (14), free (11), protect (6), ensure (6), China (6), forgotten (5), access (5), chance (5), reason (5), rating (5), fear (5), challenges (4), qualified (4), challenge (4), airplane (4), urgency (4), record (4), report (4), dream (4), wrong (4), rest (4), commitment (3), serving (3), process (3), career (3), fights (3), worry (3)

Detaileds of the Night 2 of the First Democratic Debate

People: People (85), children (27), families (26), family (16), student (13), parents (12), public (12), child (12), voters (11), communities (8), women (7), black (7), diversity (6), immigration (5), immigrants (5), kids (5), citizenship (4), women's (4), People (4), folks (4), humanitarian (3), generational (3), parent's (3), popular (3), father (3), Social (3), human (3), local (3), woman (3)

Health: healthcare (45), insurance (30), Medicare (22), health (11), healthier (5), abortion (5), prescription (4), emergency (4), cancer (3), drugs (3)

Environment: climate (27), environmental (3)

Economy: tax (20), economy (17), money (15), jobs (15), debt (14), taxes (12), work (12), industry (8), economic (7), income (7), loan (6), trade (5), greed (5), manufacturers (4), corporations (4), businesses (4), spending (4), business (4), tariffs (4), profits (4), wealth (4), fuel (4), gas (4), oil (4), cut (4), industrial (3), affordable (3), capitalism (3), corporate (3)

Education: college (11), education (8), schools (7), school (6), science (4), reason (4), intellectual (3), tuition (3)

Military: war (15), weapons (14), military (6), troops (4), allies (4), intelligence (3)

Crime: crime (8), poliçe (7), violence (6), security (6), guns (6), trafficking (3), deportation (3), Security (3), criminal (3), shooting (3), buyback (3), combat (3)

Law: rights (16), court (7), civil (7), justice (6), freedom (6), Supreme (5), Court (5), inequality (4), Rights (4), prosecutor (3)

Policy/politics: will (104), issue (33), issues (13), policy (12), corruption (11), democracy (9), politics (9), job (9), foreign (7), progressive (5), leadership (5), coalition (5), solution (5), reality (5), streets (5), racist (5), polls (5), left (5), accountability (4), interests (4), voting (4), values (4), reform (4), crisis (4), votes (4), view (4), responsibility (3), alternative (3), socialism (3), demonize (3), measures (3), caucuses (3), leaders (3), deliver (3), urgency (3), racism (3), Let's (3)

Emotion/feelings: love (5), fear (4), guts (4), hope (3)

Belief: God (6), religion (3)

Others: technology (5), documents (4), checks (4), attack (3)

MS Word Word String Counts

Counts of the Themes in Word Sequences Numbers Representing Frequences in the Transcript:

Foreign policy 9, industrial policy 3

Climate change 29, climate crisis2, Green economy 3 (+clean economy 1), green energy 1, green technology 1, green products 1, green new deal 3,

Free college 7 + free public college 1 + make community college free 1, Public education 2, Pre-K 4

student loan 8 (including student loan debt 5), student debt 8,

Private prisons 2, private detention centers 2, private insurance 12, private health insurance 4,

Special interests 3 (+ big money interests), corporate consolidation 4, Government benefits 2,

Human right 9, rule of law 2, above the law 5, undocumented immigrants 5, undocumented people 2, undocumented children 1, income inequality 7, Low income 2, structural change/reform/problems 1, institutional racism 2, systemic racism 1

Tax cut(s) 8, Tax code 2, tax bill 4, hard earned taxpayer dollars 2, minimum wage 5, paid family leave 2, national paid leave day 1, paid sick days 1, minimum wage 5

Healthcare system 7 (including broken healthcare system 2), Sickness care system 1; Health care 5,

Healthcare plan 4, universal healthcare coverage 3, healthcare option / provider / issue / services / coverage / guarantee / as a basic human right 1, healthcare as a right 3, Government healthcare 4, Government run plan 3

Health insurance 13, mental health 3 (+ mentally healthier), big farma 2

Reproductive health / healthcare 2, reproductive freedom / rights 3, reproductive justice 1

pharmaceutical companies 5, pharmaceutical prices 2, opioid crisis 2, opioid addiction 1 insurance compan(y/ies) 15, drug compan(y/ies) 4, tech companies 3, oil companies 1

Pharmaceutical industry 1, solar industry 1, oil and gas industry 2, fossil fuel industry 2, military industrial complex 1

working people 10, working class 6 (including working class people 2, working class family/families 1+1), Middle class 8, African american 11, Young people 4

Small businesses 3 (+small business owner 1), Small-to-midsize businesses 1, startup businesses 1

White House 8, United States 65, day one 16 (on day one = 10)

DISCUSSION AND CONCLUSION

With the publicly available means of analysis, the publicly available transcripts are searched for words or word strings related to socially relevant content, which then are grouped into larger categories detecting thematic connections. It can be considered as an activity aiming at vocabulary build up for follow-up listening as well as execise for analysis. The transcribed streams of the debates show signs of speechwriters candidates work with. Speechwriters can be googled to find their audio files together with their subtitle files (which can be googled with adding filetype:srt to the search query). They may benefit those who listen for rhetoric examples to model. Besdies debates, speeches can be found among the presidential documents of the website of The American Presidency Project (non-profit, non-partisan) available at https://www.presidency.ucsb.edu/ Speechwriters' work can be modelled." Another rich source is americanrhetoric.org. Benefits of listening to this kind of content increase if the debaters are those political leaders, who are supposed to be efficient listeners, aware of the importance of listening to constituents, people.

The teachers and learners of English language can easily access these transcripts free of charge from the rev.com. In an interdisciplinary study (Saka and Zengin, 2020), the transcripts of televised programs serving Coronavirus awareness were sampled from rev.com and cnn transcript website and suggested for use in content and language integrated learning approach. Some other recommendations for their use in teaching activities would be using transcripts in accompaniment with ausio-visual broadcasts (optionally with bimodal subtitles) of the debates, which would facilitate the listening comprehension. Listening challenges may be facilitated with use of transcripts as reading materials. Some activities can develop vocabulary skills like semantic mapping activities or discourse analysis activities using wordclouds, or asking students to pick their favorites from the utterances of the candidates participating in the debates, list a sample with their reasons for choosing them as their favorites. There can also be writing tasks. Picking some quotes, formulaic language examples and specialised language examples out of the debate content as a pre-task. Several creative and imaginative election debate dialogues can be written modelling the debates viewed and read incorporating what they picked the transcripts into these dialogues to enrich their dialogue regarding lexical items. Following the written work, recommendations for speaking activities doing role play perfomances of their dialogue scripts, recording them and viewing, voicing critical apprasials of what is viewed. The task topic may include scriptwriting or role play performances containing political analysts' or journalists' analyses or commentaries about debate performances and the work of pollsters. Most of the above-mentioned activities were carried out during the process leading to the production of this paper with language skills course given to English preparation year and first year students.

The transcripts of audio/visual streams can be evaluated for materials development. Conceding the adequacy of EFL/ESL teachers' knowledge and creativity for the undertaking of their own materials development, Pardo and Tellez (2009, p. 172) motivate them to make use of their potential and take on thc task of developing their instructional materials not confining this to text developers. "Teacher-developed materials boost not only effective learning settings and outcomes, but also teachers' pedagogical practice/performance" (Pardo and Tellez 2009, p. 184). Teaching materials, as Tomlinson (2009 p. 4) states, "can be anything which is deliberately used to increase the learners' knowledge and / or experience of the language".

Aided with accurate transcripts, teachers and materials development staff can make the most noticing effect for the input to convert into intake and uptake for perceptive and productive skills respectively to provide benefits in terms of language as well as content of interest and value to people. The content focused here is the first two debates of the Democratic primary in the highly televised 2020 US election. Use of the transcripts representative of the US election discourse can be evaluated in the framework of content-based instruction (CBI) or content and language integrated learning (CLIL). The fact that election content is interesting can be shown with the Googling trends of the year 2020. The election results was the second most trending Google search after Coronavirus. The peak of the search was between November 1 and November 7, 2020.

Studying the students' perceptions on how CBI supports learning of a foreign language, Corrales and Maloof (2011) highlight the increased motivation felt with the topics, materials, and activities which are relevant, meaningful, interesting, and useful to them. Writing on tertiary Content Based ESL instruction, Oxford (1993: 75) underlines the primary goal being "communicative competence in the target language, and an associated aim is content knowledge". Gupta (2020) defines CLIL as "an educational approach teaching content through an additional language of students with a dual focus on content and language learning" underlining its wide recognition "due to its nature of dual learning goals". Stoller (2004: 261) expresses the basic feature of CBI as "distinguished by its dual commitment to language and content-learning objectives". Thompson and McKinley (2018) report CBI and CLIL fostering "additive bilingualism through a dual focus on both content and language learning" along with other approacches to content-centered second language learning (bilingual education, immersion, English medium instruction).

The chapter also reflects the author's summary of the coverage of US election in Turkey demonstrating the awareness raising quality of the comments about the US election as well as content demosntrating connections between Turkish and American democracies. The non-availability of transcripts for the coverage is criticised for falling short of the long lasting effect of the written mode. Actually what would be the ideal is the availability of the transcripts of the coverage and their translations into English as well as the availability of English subtitles displayed on-screen along with the Turkish soundtrack. Given the content, the following examples are:

The earliest dating back to centuries ago. The year 2020 has been termed as the Patara Year by the Republic of Turkey Ministry of Culture and Tourism. The Ancient Patara City is located in the southern province of Antalya in Southwest Turkey. The Turkish coalition of America (TCA) has published a brochure providing the significance of Patara, "which was the capital of Lycia, an ancient civilisation of democratic principles that inspired framers of the U.S. Constituion and, ultimately, influenced the formation of the United States. TCA's website summarize the contributions of the Lycian government, known as the Lycian League to the American constituion. "First, it was a model of a federal union the strength of whose parts in the national councils is proportionate to their size. Second, it showed the possibility of popular government that was representative. Third, it offered the example of a strong national government with its own strong officers and the power to make laws that applied directly to individual citizens." (TCA Issue paper-9, 2007, June 8). The TCA's brochure titled (Patara: Its role in U.S. History, n.d.) shows that among Anatolian civilisations, the Lycians held a distinctive place with "a fierce love of freedom and independence, and resisted all attempts at outside domination". Also, Patara is home to the world's first democratic parliament building, the semicircular rows of which gave rise to the seating arrangement in the chambers of the US. Congress today. Built in 2nd century BC, the building was restored in 2012 by the Grand National Assembly of Turkey.

An independent think tank publishing their website content both in Turkish and English, Istanbul Political Research Institute (IstanPol) can be benefited from by Turkish learners of English who want to read reports about a broad range democratic topics including U.S. elections. IstanPol "works towards the production of scientific knowledge on democratization, human and social development and foreign policy and produces research projects and policy proposals on these issues to share with the public." (n.d. en.istanpol.org/hakkimizda). Its co-founder, Seren Selvin Korkmaz often shares her views in the national and international media. As stated in Istanpol's website, her "research interests include populism, political parties, voter perception, election strategies and political economy of exclusion" (n. d. en.istanpol.org/blank).

As a university student, Yunus Emre Erdölen (then a university senior student majoring in law, commentator invited to TV programs covering US election) provided updates with his Tweets in Turkish then. Later he appeared in Youtube programs and wrote articles for daktilo1984 as a journalist (recently, he has made appearances as one of those commentators in the national TV channels about US election results). In Elis Gjevori's (TRT World producer) article titled 'The View from Turkey regarding US elections' Erdölen has been quoted as saying "Campaign methods, slogans used in elections are under close scrutiny by content producers who are in search of new ideas".

The 2020 presidential election has been covered with several TV shows in Turkey. Some viewers criticized the amount and way of coverage but the program moderators and the analysts' comments provided justification of the importance attributed to the US election for awareness of both good and bad examples that should be noticed. The coverage of the US election in the TV news channels and the Youtube programs confirmed the conclusions of the current paper. Pertinent connections were made between the elections in both countries with comparisons and contrasts implied or explicitly underlined.

In a Youtube series of daktilo1984, Üsküdar Motoru (2020, November 8) the program host Yunus Emre Erdölen interviewed the two time pollie winner political communication expert Gülfem Saydan Sanver who highlighted the main aspects of election. The first th duration of the process: an election takes a long time (two years including primaries). Secondly, they take the form of shows. Thirdly the amount of money involved in the financial aspect of elections is huge (Sanver, 2020). The first two aspects are suitable for the use of elections in language learning. The fact that many events taking the forms of shows motivates learners bringing fun to the activities. Most of the democratic debates took place before the Coronavirus pandemic, which lead to digital campaigns. The lengthy process is also beneficial to the learning due to the enriching and lengthy input exposure and the possibility of spreading the events to the syllabus (Sanver, 2020).

Evaluation of election polls is beneficial empathywise and emotionwise, which is significant for language and literature majors with their focus on emotions. In HT 360 (hosted by Afşin Yurdakul) of (Haberturk TV, Novermber, 6) a Turkish national TV channel, Asked about the errors of polls in swing states in the last US election, Sanver (MA & PhD on political communication, strategist) cited the German political scientist Elisabeth Noelle-Neumann's spiral of silence theory, which states that as social beings, human beings keep silent if they think they would not be socially acceptable. Sanver states that the reason for the 'shy' constituents is the witch-hunting, smearing counter-campaigns causing them to feel themselves guilty due to their choices, which leads inaccurate data. Sanver argues for the need to combine emotion analyses (in-depth interviews, neuro-marketing) with quantitative analysis for the quality of precision to understand the 'shyness' and take measure against the use of demagogy in the campaigns which are run not only on issue but on emotions. At another program on special coverage of the US election (Haberturk TV, 2020, November 4), making comparisons among populist (authoritarian) leaders, Sanver demonstrates similarities between them stating they copy each other's campaign strategies communication strategies they even use the same slogans. She says the resulting discourse normalizes what is not normal. Those seeing that arbitrary use of power works owns this use.

In the same program (Haberturk TV, HT 360 November 6, 2020), Prof.Dr. Emre Erdoğan criticises the misleading and unreal presentation by some politicians, and somepeople in the media of vote counting processes as a horse race, and of the election results as a matter of life and death and exagerrate the identity politics, this tenses up people. And comes the anger and disappointment. This is especially the case in regions with ethnic tensions. Erdoğan stresses that given the possibility of some people to hide their choices in the polls, who is in front is no so interesting in the election forecasts. More interesting in polling are the reasons for who is not/voting for who, or what the motivations are what the arguments are. Emre Erdogan states that instead of understanding people's reasons for voting, we have started to listen the way they justify their reasons for voting. We have quit discussing the background such as class, ethnicity, rural/urban distinction, consumption patterns, future anxiety, coronavirus in the background and the discussion has turned to a simple "who did you vote for" (also seen in Turkey). And, ranging from field researchers to academics, so many people profit from this wrong direction of the discussions (HT 360 November 6, 2020). Actually, an image or an impression of illegality can have manipulative effects on the pricings of the political developments, which would result in nobody's victory but the greed's. Political literacy should be evaluated in connection with financial literacy. This may reveal patterns between volatilities in the financial domain and deviations from the legal norms and from the expectations set by the social contract.

It is also necessary to highlight that in some of the programs in the Turkish media have presented US electon with so much detail that the amount of detail has been enough to suffocate the purpose. The meaning is lost when the information is too much, or when loaded with too much information or fed too much informaton (inforbesity), or when intoxicated with too much information (infoxication) or exposed to information bombardment (information explosion). The detail gets so crazy that we say what is the point? So this condition makes it difficult to understand issues, and making decisions about these issues.

Professor of political science, Ersin Kalaycıoğlu (Haberturk TV, HT 360, November 5) emphasizes that it would be overgeneralisation to overrate the repercussions of the current election in US emphasizing that currently US is classified in the category of flawed democracy, not full democracy citing the Economist intelligence unit (EIU). According to EIU's Democracy Index 2019 "a regression and/or stagnation of democracy has been reflected in the declining average scores for the advanced democracies of the US and Europe over many years." Listing the main manifestations fo the recession, the report ascertains that the "regressive trends in the mature democracies developed from the 1990s onwards, accelerated in the 2000s and reached their apogee in the decade that closed in 2019 (p. 6)." The report (2019, p. 4) warns that "the average global score for democracy fell from 5.48 in 2018 to 5.44 (on a scale of 0-10)."

I would like to finish this paper with the leader of Turkey Mustafa Kemal Atatürk's address to the American people (Atatürk VideoKanalı, January 12, 2007):

"Honorable Americans,

Expressing my wish to say a few remarks, the natural origins of the affection and sincerity between Turkish and American nations, mutuality of which I am confident of. I would like to mention a few words about the natural origion of undeniable sincere relationship between the people of Turkey and those of the United States of America. Turks are democrats in the first place. If it is not completely understood by th civilised world, its reasons have been expressed very well by the US ambassador pointing to the last days of the Ottoman Empire. On the other hand it is democracy Americans have relied on as of the moment of their inception. It is through this blessing that Americans has presented their existence in the civilised humankind. They have built a great national unity. It is from this viewpoint Turkish nation feels deep and strong affection about American nation. I hope that this observation will root the affecton between two nations. Howevever, nor will this be all, this will help all the humankind to like/care for each other, to eradicate all the past superstitions that prevent this mutual affection, and put the world at rest and at peace.

Honorable Americans,

This is the human goal of the Turkish nation, the new republic of Turkey, of which I am the leading representative. I have no doubt that American nation, who have excelled in this great goal in the first place, are together with the Turkish nation".

As these remarks prove, Atatürk has been his nation's voice for the vision of shareing leadership with the USA to enhance the progress of democracy. The aim of this chapter has been written with same spirit.

FUTURE RESEARCH DIRECTIONS

Hopefully, in the near future, transcription services will be available everywhere and provided with parallel texts (in many languages) as well as subtitling multiplicity (including inter-lingual, intra-lingual and reversed subtitling). Variety of text visualisation and learning tools for corpus pegagogy and text visualisation can be in future studies such as Voyant tools, Antconc, Wordle, or TagCrowd. The global availability of transcripts with policy content and their translations would facilitate exchange of insights for solutions to our problems across nations. The means to access spoken discourse on record including the textual records would serve the expansion of transparency and accountability as well. The treatment of the election debate transcripts in this paper may hopefully encourage academics to engage in policy making processes. Text/content mining can be applied in this direction. As Ayan and Demirel (2017) "the studies using text/content mining are beneficial for all followers including researchers and learners inside and outside the field of language education and help people focus less frequently studied topics and areas."

REFERENCES

American Council on Teaching of Foreign Languages. (2011). *21st Century Skills map*. Retrieved from https://www.actfl.org/sites/default/files/pdfs/21stCenturySkillsMap/p21_worldlanguagesmap.pdf

Ayan, E., & Demirel, E. (2017). Analyzing Research Tendencies of ELT Researchers and Trajectory of English Language Teaching and Learning in the last Five Years. *International Journal of Curriculum and Instruction*, *9*(2), 89–106.

Baker, P. (2006). *Using Corpora in Discourse Analysis*. Continuum.

Cao, X., Yamashita, N., & Ishida, T. (2016). How non-native speakers perceive listening comprehension problems: Implications for adaptive support technologies. In T. Yoshino, G. Zurita, T. Yuizono, N. Baloian, G-D. Chen, & T. Inoue (Eds.), *Collaboration Technologies and Social Computing - 8th International Conference, CollabTech 2016, Proceedings* (pp. 89-104). Springer Verlag. 10.1007/978-981-10-2618-8_8

Corrales, K., & Maloof, C. (2011). Student perceptions on how content based instruction supports learner development in a foreign language context. *Zona Próxima*, *15*, 40–53.

Erdölen, Y. E. (2020, November 8). *ABD Seçim ve Kampanya Değerlendirmesi|Gülfem Saydan Sanver & Yunus Emre Erdölen | Üsküdar Motoru #3* [daktilo1984 Youtube Channel]. Retrieved from https://www.youtube.com/watch?v=2LMbpvI1Xfg&feature=youtu.be

Filatova, O. (2016). More than a word cloud. *TESOL Journal*, *7*(2), 438–448. doi:10.1002/tesj.251

Gjevori, E. (2020, November 3). *The View from Turkey regarding US elections*. Retrieved from https://www.trtworld.com/magazine/the-view-from-turkey-regarding-us-elections-41143

Goh, C. C. (2000). A cognitive perspective on language learners' listening comprehension prob lems. *System*, *28*(1), 55–75. doi:10.1016/S0346-251X(99)00060-3

Gupta, K. C.-L. (2020). Researcher-teacher collaboration in adopting critical Content and Language Integrated Learning (CLIL): Processes, challenges and outcomes. *Trabalhos em Lingüística Aplicada*, *59*(1), 42–77. doi:10.1590/010318136014125912020

Habertürk. (2020, November 5). *HT 360, Donald Trump görevi Joe Biden'a devretmeyi reddedebilir mi?* [Habertürk TV Youtube Channel]. Retrieved from https://www.youtube.com/watch?v=3DbOt5f79Jk

Habertürk. (2020, November 6). *HT 360, Kim bu mahcup Trumpçılar?* [Habertürk TV Youtube Channel]. Retrieved from https://www.youtube.com/watch?v=1Pgy12jjyrg

Habertürk. (2020, November 4). *Türkiye'nin Nabzı, ABD Seçim Özel*. Habertürk TV Youtube Channel.

Mavridi, S., & Xerri, D. (Eds.). (2020). *English for 21st century skills*. Express Publishing. https://www.expresspublishing.co.uk/en/content/english-21st-century-skills-0

Oxford, R. L. (1993). Progress in Tertiary Content-Based ESL Instruction. *TESL Canada Journal*, *11*(1), 75–97. doi:10.18806/tesl.v11i1.626

Pan, Y.-X., Jiang, D.-N., Yao, L., Picheny, M., & Qin, Y. (2010). *Effects of automated transcription quality on non-native speakers' comprehension in real-time computer-mediated communication.* Paper presented at CHI 2010: Sound and Speech, Atlanta, GA.

Pardo, N. A., & Téllez, M. F. (2009). ELT materials: The key to fostering effective teaching and learning settings. *PROFILE Issues in Teachers Professional Deveopment*, *11*(2), 171–186.

Renandya, W. A., & Farrell, T. S. C. (2011). Teacher, the tape is too fast: Extensive listening in ELT. *ELT Journal*, *65*(1), 52–59. doi:10.1093/elt/ccq015

Ridgway, T. (2000). Listening strategies—I beg your pardon'. *ELT Journal*, *54*(2), 179–185. doi:10.1093/elt/54.2.179

Saka, N. E., & Zengin, B. (2020). Use of Video/Listening Content Transcripts in English for Coronavirus Awareness and Information, *Virology & Immunological Journal*, *4*(1), 1–9. doi:10.23880/vij-16000230

Stoller, F. L. (2004). Content-based instructions: Perspectives on curriculum planning. *Annual Review of Applied Linguistics*, *24*(1), 261–283. doi:10.1017/S0267190504000108

The Cambridge Framework for Life Competences, Introductory Guide for Teachers and Educational Managers. (n.d.). Cambridge University Press. Retrieved from: https://www.cambridge.org/tr/files/6615/9741/7863/CambridgeLifeCompetencies_Introduction_Booklet_digitalPDF_082020.pdf

The Turkish Coalition of America. (2007, June 8). *Patara: Its role in US history.* Retrieved from https://www.tc-america.org/files/Patara-Brochure.pdf

The Turkish coalition of America. (2007, June 8). *The Lycian confederation and the American constitution* TCA Issues & Information – TCA Issue Paper-9. Retrieved from https://www.tc-america.org/issues-information/us-turkey-relations/the-lycian-confederation-and-the-american-constitution-153.htm

Thompson, G., & McKinley, J. (2018). Integration of content and language learning. The TESOL Encyclopedia of English language, 1-13. doi:10.1002/9781118784235.eelt0634

ADDITIONAL READING

Commision on the practice of democratic citizenship (June 2020). Our common purpose: Reinventing American Democracy for the 21st century. https://www.amacad.org/ourcommonpurpose/report

Kasimi, Y. (2020). Democracy in EFL classrooms. *Journal of Language and Linguistic Studies*, *16*(1), 126–136. doi:10.17263/jlls.712666

Kjeldsen, J. E., Kock, C., & Vigsø, O. (2021). Political rhetoric in Scandinavia. In E. Skogerbø, Ø. Ihlen, N. N. Kristensen, & L. Nord (Eds.), *Power, communication, and politics in the Nordic countries* (pp. 365–383). Nordicom, University of Gothenburg., doi:10.48335/9789188855299-18

Muryasov, R., Samigullina, A., & Gazizov, R. (2018). The Category of Evaluation in Barack Obama's Election Discourse: An Empirical Account. *Tarih Kültür ve Sanat Arastirmalari Dergisi*, *7*(1), 1–11. doi:10.7596/taksad.v7i1.1463

Packer, G. (July/August 2021). How America fractures into four parts. *The Atlantic Daily.*

Serrano, J. R. (2008). ELT and citizenship: Basic principles to raise social awareness through language teaching. *HOW, 15*(1), 63-82. https://www.turkishnews.com/tr/content/2014/10/28/ataturks-address-to-the-american-people-1925-ankara/

KEY TERMS AND DEFINITIONS

Intralingual/Bimodal Subtitling: The kind of subtitling in which both soundtrack and the subtitle are the same language (in this case subtitles are captions).

Language Learning With YouTube: Appreciated for its potential fo language learning, this Google Chrome extension provides two subtitles sets (same language subtitles plus possibly its translation.

Reversed Subtitling: The kind of subtitling in which the soundtrack/audio language and the subtitle language are different, but in this case the former is the L1 and the latter is L2. For instance, the soundtrack language is Turkish and the subtitle language is English for viewers whose L1 is Turkish.

Rhetoric: The science and art of persuasive language use. Its study has implications for benefits like communication or creative thinking.

Speech Transcription: Whether it is audio transcribing or video transcribing, transcription of speech is written/textual account of speech based on audio/visual records.

Chapter 9
Benefits of Film Adaptations of Literary Works for English Language Learning

Sunay Öztürk
Fahreddin Kerim Gökay Anatolian High School, Turkey

Buğra Zengin
Tekirdağ Namık Kemal University, Turkey

ABSTRACT

Despite their potential in language learning, films have not been covered adequately in academic research except some attempts to develop strategies. One of these strategies has been to compare film scripts and their translations, to identify chunks, technical terms, and creative language examples and to list them with their translations, and to do creative dialogue writing activities especially by modeling these scripts where chunks are used extensively. Believing in the importance of evaluating the film scripts with their translations, this study aims to investigate the processes the researcher/teacher's high school students went through in learning and using the target language starting with the use of the film adaptations of the literary works, writing their scripts modeling the film scripts, acting their own scripts, and shooting their performances. The mixed method research design was used with the triangulation of the qualitative and quantitative data. The 68 high school students' views were investigated with Likert scale (1-5) questionnaire along with their comments about the items.

INTRODUCTION

In English as a foreign langauge setting, the problem of exposure to input is remarkable. Emphasizing the classroom limits for a quantitatively and qualitatively satisfactory exposure to English, Thurnbury (2002, pp. 20-21) emphasizes that the average input of L2 learner's is, "to say the least, impoverished". Speaking for the effect of the learning atmosphere on communicative claims and learning opportunities, Nyikos and Fan (2007) emphasize the fundamental difference between classroom-limited EFL and

DOI: 10.4018/978-1-7998-7876-6.ch009

socially-enclosed ESL settings. One of the ways to solve the input problem is to tap the input-rich sources of the movies on the Internet. As stated by Zengin and Çubukçu (2013), particularly in low exposure settings of English as a foreign language, series or movies can be a rich material of input. Furthermore, the limited settings of EFL and limited class hours make it essential to benefit from social and fun activities of movies as learning materials.

It is important to try to increase the amount and quality of exposure to authentic English out of the formal settings. In order to do it, we can use movies as resources of genuine English in input-poor learning environments. One of the solutions provided by Zengin (2019) is constructing lists of chunks, jargons and creative language use examples by the comparative study of film scripts and the Turkish translations and productive use of these lists in creative writing dialogue activities as well as watching films using audio and subtitle combinations (Zengin, 2019), especially the integration of Turkish dubbed English captioned combination for language learning purposes (Zengin and Aker, 2016).

Aiming to build on this work and seeing an untapped area of movie adaptations of literary works, this study aimed to focus on the scripts of movie adaptations of novels in English and covered the process from reading novels mostly simplified simplified versions graded for preintermediate level and scripts (both the original and translated versions of these scripts), their comparative study, makings lists of chunks, jargons and creative language use together with their Turkish equivalents, to writing their own scenarios (scripts), acting their scripts and shooting their performances. Therefore, this study demonstrated that use of literature is not limited to canonised classics.

Background

For literary texts, inspite of the potential advantage of Computer Assisted Language Learning (CALL) for the young learners' motivation, this potency is not realized enough. Perceptions of literature restrict its use to the canonized literature and other imaginative and creative fiction productions are marginalised. Interrogating the misunderstanding, the study of Zengin, Altuntaş and Toraman (2016) demostrated that in English Language Teaching (ELT), the use of literature is not restricted to the canonized literature reflecting the students' opinions as to the connection between Voscreen© (an online CALL product which benefits from short clips from English movies and the translation/paraphrase questions of the movie lines) and literature.

However, listening to unsimplified, authentic natural English has challenges. Renandya and Farrell (2011) point features which make speech especially hard for learners of English as a foreign language and speech rate is one of them. There is an inverse correlation between speech rate and comprehension, that becomes nearly impossible with speech rate reaching at a critical level, the level that "is normally much lower than that for the more advanced learners" for many EFL learners (Renandya & Farrell, 2011, p. 53). Another feature, that makes it hard for EFL learners to recognise words, with sounds modified, dropped, or added when woven together (e.g. reduction of _What is up?' to _Sup?'). The next feature is the blurry boundaries between words which are tending to blend with each other. The last but not the least feature is "the real-time processing requirements of speech" with little control over what is said and the speed it is said.

In a study with EFL college students in China listing top ten listening comprehension difficulties of EFL college students Zeng's (2007: cited in Renandya & Farrell) has found the fast rate of speech to be the most important problem of his students. Unable to catch the words, they can get distracted, and they can not even recognize the words they knew. It is not surprising to see eighty-five percent desribing new vocabulary as the fourth problem. One of the sources might be the shortage of knowledge of a low-frequency vocabulary like the technical vocabulary of a profession as well as collocations or other mutli-word units.

Together with missing subsequent input (%85) and sentence complexity (60%) the sources of listening problems show the importance of the comprehension of formulaic language along with the imputed knowledge of its components. Processing language through multi-word lexical units can save time and also reduce the nervousness, a factor reported by the seventy percent, and anxiety and frustration reported by the forty-five percent. Processing through multi-word units can also prevent waste of limited sources of working memory and can save them for higher levels of language so that they can use background knowledge reported by fifty-five percent.

Furthermore, the matter of unfamiliar pronunciation has a lexico-grammatical aspect, too. For example, Celce-Murcia (1995, p. 377) points lexicogrammatical signals (e.g. discourse markers, lexical chunks, and word order) as items, that are "critical in signaling information to the listener." According to Kremmel, Brunfaut, and Alderson (2015), although the recognition of the formulaic nature of language increases, the critical role of phraseological knowledge is still underestimated. On the other hand, warning against memorisation recall, Brown (1987, p. 185) warns against the likely errors in the process of keeping prefabricated patterns due to the rottenest being a factor of "a lack of knowledge of the structural rules for forming the particular utterance."

As Hughes-Trappes Lomax (2004, p. 15) points out, in language teaching,"if grammar and lexis are presented as more separate than they are, thus obscuring the inter-connectedness in lexico-grammar", this may bring about potential disadvantages. Likewise, if the part between four skills is perceived and applied in a way that can cause the failure to attend to the common field between listening and reading and what is common to speaking and writing.

Similar seperation might be seen in the conceptualization of spoken and written media"as discrete types rather than points on a continuum" or in the arbitrary part of text-making features between spoken and written forms, "thus obscuring those text-making features that are common to discourse of all kinds." Renandya and Farrell (2011) take the side of those who benefit from the common ground between reading and listening. Criticizing the insufficient attention given to the learners' perceptions and emphasis on strategy-based approach, Renandya and Farrell (2011, p. 56) think that applying the extensive reading approach to the listening skill in the form of an extensive listening approach can be the answer which can help with the listening problems. They point to the strong evidence for extensive reading, that results in the improvement of "word recognition skills, vocabulary, reading comprehension, fluency, and general language proficiency" as well as "more positive attitudes towards reading and language learning in general." According to Renandya and Farrell (2011, p. 56), the students read anything they find pleasurable in this approach. The only condition is that "they should choose reading materials that they can understand on their own and that they read a lot of these materials." They believe that the same result can be obtained in the listening skill, if lots of practice is done with extensive listening materials which are comprehensible and enjoyable. According to them, use of listening strategies is boring and difficult for foreign language learners due to the challenges in the nature of listening especially for the beginner level students.

TV series and movies might be accepted as enjoyable and comprehensible listening materials. They also let working memory save space for higher levels of second language use leading to improving all competencies in the framework of intercultural communicative competence, for example sociolinguistic competence. Several solutions have been reported. Zengin and Aker (2016) recommend integration of

Turkish dubbed English subbed combination to the audio-subtitle variations. This way, the larner whose previous foreign language learning culture is reading oriented comprehends with more ease and it is also easy for him/her to combine the subtitle read with the listening language being his/her native language. However, dubbing quality is not a guarantee and the negative associations with dubbing cause the learner to tend to avoid this combination despite its potential advantages.

Zengin and Aker's study (2016) tried including a new combination of audio and subtitle languages by reversing the direction of L1 and L2 in the traditionally applied interlingual subtitling. The subtitle is the target language and audio language is the EFL learner's native language. As the cognitive skill of reading is easier than listening. In order to facilitate the preparation of the foundation for productive skills, reading can be used.

One of the ways of doing this is reading English subtitles when enjoying the full comprehension of the audio language being the native language of the learner. EFL learners, who may have problems with jargon (e.g. medical), can notice them more easily in the subtitle, which facilitates their learning. If they do not want to learn the technical terminology of a professional domain, they can shift their attention to other components of the text, instead of being distracted by unrelated input not losing connections between other components of the co-text. Accelerating the vocabulary learning procedure might pave the way to gradual facilitation of listening comprehension with which learners can gain what is unusual to listening (stress, rhythm, intonation and variety of accents).

However, due to the learners' various needs and preferrences, alternative solutions need to be developed and the use of movies in teaching English is not limited to audio and subtitle /caption preferences. Also solutions should take into account the differences in learning cultures. Given that the first phases of formal education is reading oriented, the film scripts have potential to build a foundation for the facilitation of listening and speech production problems.

RESEARCH AIM

Believing the importance of evaluating the film scripts with their translations, this study aims to investigate the processes the researcher/teacher's high school students went through in learning and using the target language starting with the use of the film adaptations of the literary works, writing their scripts modeling the film scripts, acting their own scripts and shooting their performances.

METHODOLOGY

The mixed method research design was used with the triangulation of the qualitative and quantitative data. The students were given an extracurricular activity plan according to which were responsible for reading the literary works, watching the film adaptations of the literary Works, comparing the language used in the literary Works and their film adaptations, writing and acting their own scripts and performing them by shooting videos. These activities were conducted for 2 to 3 months. The opinions of the participants have been investigated with Likert scale (1-5) questionnaire with 34 items. and their comments about the question items.

Participants

The participants are 68 high school students of researcher/teacher from a high school in Istanbul. Fourty-two (42) of the participants are 11th grade and the twenty-six (26) of them are 10th grade pre-intermediate students. The sixty-eight (68) of the participants consist of 24 boys 44 girls.

Research Design

By benefiting from triangulation of the qualitative and quantitative data, the mixed method research was prefered to use. In order to get further insights, a quantitative Likert scale (1-5) questionnaire was made in Turkish. In the process of constructing the questionnaire, all the previous experience accumulated in the previous publications of the supervisor was made use of. The supervisor completed several studies on the use of films for the teaching and learning of English as a foreign language. The data collection tools included subtitles, captions (bimodal subtitling), audio language and their combination, as well as questions on the inclusion of literature in the content of VoScreen.com, which applies the use of films successfully through its website or application. Throughout the process, ideas of practitioners (teachers of English as a foreign language in the Enlish preparation school) and focus groups consisting of English philology students and English preparation year students English were asked to contribute to the quality of the questionnaires in all these works. So the supervisor was experienced and shared it with the researcher/teacher in this study, who benefited from his observations in the field, and mainly focus groups consisting of his students combined with the supervisor's recommendations. Since the originality of this study was the use of literature based movies, the above-mentioned previous experiences were adapted to servet his direction. The focus group was the students of Fahreddin Kerim Gökay Anatolian High School.

The reason for utilizing a mixed design was complementarity, that "searches detail, enlargement, illustration, the clarification of the results from one method with the results from the other method" (Greene, Caracelli, & Graham, 1989; cited in Christensen & Johnson, 2004, p. 423). As for quantitative component of the research design, an academician in the mathematics department of the same university the supervisor worked at assisted in the quantitative data analyses, which was conducted with SPSS. As she taught statistics at university level, her feedback assured that the procedure was on the right track. For the qualitative data, the students were required to writen their comments related to the questionnaire items. Almost all of them completed the task. Out of their reports, only four of them was selected. The selected student accounts were those participants (four females, one male) who were considered to reflect the most realistic account of the process and who took the time to write elaborate evaluations regarding the aspects of the items with the utmost care. These accounts and field observations both verified, complemented, served to clarify and explain and justify the conclusions that can be made out of the quantitative component of the research. Besides high level of focus and involvement, these students presented their comments in detail and with much clarity. The interperetations and conclusions of the researcher were discussed with the real participants for verification.

RESULTS

Quantitative Findings

For the item 1, about two thirds of the participants (45 out of 68 students) agreed to the statement that we don't see literature limited to the classical works of the great writers (canonical) such as Shakespeare, Voltaire etc. For the item 2, %73,5 of the participants (50 out of 68 students) agreed to the statement I think the pop culture products such as Harry Potter etc. are literary works too.For the item 3, a little more than half of the participants (37 out of 68 students) agreed to the statement Scenario texts based on literary works are also literary products.

For the item 4, % 61,8 of the participants (42 out of 68 students) agreed to the statement I support using literature in learning English.For the item 5, % 22,15 of the participants (16 out of 68 students) agreed to the statement novels I read are rather shortened versions.For the item 6, % 23,6 of the participants (16 out of 68 students) agreed to the statement one of the reasons I prefer simplified versions is that I can't make time for the original. For the item 7, %35,17 of the participants (25 out of 68 students) agreed to the statement one of the reasons I prefer the simplified versions is that it is more suitable for my language level. For the item 8, % 61,8 of the participants (42 out of 68 students) agreed to the statement film scripts based on literary works provide an alternative source of quality in terms of using literature in learning English.

For the item 9, %37,13 of the participants (26 out of 68 students) agreed to the statement among the scenarios based on literary works, the ones I prefer to read are those based on the literary works I have read before.For the item 10, %30,9 of the participants (21 out of 68 students) agreed to the statement reading scripts based on literary works is advantageous in terms of time.

For the item 11, about half of the participants (35 out of 68 students) agreed to the statement scripts based on literary works are suitable for me in terms of language level.For the item 12, %27,8 of the participants (19 out of 68 students) agreed to the statement I enjoy reading scripts based on literary works. For the item 13, nearly quarter of the participants (20 out of 68 students) agreed to the statement I make comparisons between the literary works I have read before and the script texts based on these works.

For the item 14, approximately two thirds of the participants (46 out of 68 students) agreed to the statement scenario texts based on literary works also have features that encourage reading and wondering about the literary work on which they are based.For the item 15, %56,9 of of the participants (38 out of 68 students) agreed to the statement most scripts based on literary works are film scripts.For the item 16, more than quarter of the participants (22 out of 68 students) agreed to the statement most scripts based on literary works are Tv Series scenarios.For the item 17, %31,13 participants (22 out of 68 students) agreed to the statement the script texts based on literary works increased my motivation to read.

For the item 18, more than half of the participants (37 out of 68 students) agreed to the statement film scripts based on literary works became a good source for finding chunks.For the item 19, %58,8 of the participants (40 out of 68 students) agreed to the statement film scripts became a good source for finding examples of technical language usage in the special field (military, police, law, medicine) from different types of scripts based on literary works, such as military, police, law, medicine, etc.

For the item 20, more than two thirds of the participants (57 out of 68 students) agreed to the statement scripts based on literary works have become a good source of spoken English. For the item 21, a little less than two thirds of the participants (48 out of 68 students) agreed to the statement comparing scripts based on literary works with texts translated into Turkish helped my English learning.

Table 1. Reliability Statistics

Cronbach's Alpha	Cronbach's Alpha Based on Standardized Items	N of Items
,883	,889	34

For the item 22, %60,3 of the participants (41 out of 68 students) agreed to the statement comparing scripts based on literary works with texts translated into Turkish improved my translation skills.For the item 23, %63,3 of the participants (43 out of 68 students) agreed to the statement reading literary works and writing script tasks based on these scripts improved my writing skills.For the item 24, more than half of the participants (42 out of 68 students) agreed to the statement the scenarios we wrote in this way helped our collaboration ability.

For the item 25, %85,17 of the participants (59 out of 68 students) agreed to the statement acting the scripts we wrote improved my speaking skills.For the item 26, almost all the participants (65 out of 68 students) agreed to the statement when watching movies, I also pay attention to non-verbal dimensions such as body language.

For the item 27, most of the participants (61 out of 68 students) agreed to the statement I took care of non-verbal communication elements such as gestures while shooting and playing the script we created. For the item 28, over two thirds of the participants (55 out of 68 students) agreed to the statement we enjoyed shooting and playing our own script.

For the item 29, almost two thirds of the participants (50 out of 68 students) agreed to the statement I used Turkish subtitles to understand the film. For the item 30, %68,11 of the participants (47 out of 68 students) agreed to the statement I used English subtitles to improve my pronunciation.For the item 31, a little more than half of the participants (38 out of 68 students) agreed to the statement it helped to motivate my English learning out of class.

For the item 32, %84,13 of the participants (58 out of 68 students) agreed to the statement while writing our own scenarios, we took the example of language usage in books and film subtitles.For the item 33, more than two thirds of the participants (55 out of 68 students) agreed to the statement I realized that scripts contain more examples of spoken language than literary works.For the item 34, about two thirds of the participants (51 out of 68 students) agreed to the statement in the course of these studies, my English to Turkish translation skills improved as well as my translation skills from Turkish to English.

The Cronbach alpha of the scales for participants in two grades as a whole is 0.883, which is very high since normally 0.70 is accepted as an adequate level.

Qualitative Findings

For the item 1, according to F1 nowadays in the publishing market, different kinds of authors and kinds of books can be found, M1 thinks that participants may have thought nowadays literary works can be a classic in the future and there are a lot of unsung literary works. For F2 literature is a huge combination that cannot consist of only "classics" and for F3 recently many different types of books are being written and these constitute the current popular culture, and mostly students prefer to read them.

For the item 2, F1 thinks these books were top of the bestseller for a long time and it counted as literature books. According to M1 Participants may have thought they have literary technics and they have a genuine style. Writing these works needs literary talent and F2 thinks that the participants who agreed to this statement (%73,5) may have thought that it is a fact because we are at an age we can easily find a novel, short story etc. or scripts and as for F3 the author has created different genres and worlds to appeal to readers, and most people have read these books so they have become the trendy books.

For the item 3, according to F1 producers might have shot a movie benefiting from a book whose subject was different from its original and the students don't like it. For M1 participants may have thought that writing a scenario has different aims. That's why half of participant disagreed. F2 thinks that the participants (a little more than half) consider that it is possible to say adaptations of literary works are also literary products so that might be the reason why most of them agreed to this one. According to F3 the films that are adaptations of the books may not be aimed at everyone's style and therefore cannot provide the audience that the book provides. This is why people may not count scripts as literary products.

For the item 4, F1 thinks reading English books is a really good way to improve their English. For M1, participants may have thought literature has very rich vocabulary and a lot of books were written in daily language. According to F2 utilizing literature is just one of the opportunity we have to learn English so 61,8% of the participants may have found it true by agreeing. As for F3, we have translated many books in English as an international language and we can improve our English grammar and vocabulary by reading these books.

For the item 5, F1 thinks that high school students can read the whole book. They don`t prefer the short versions or summaries of books. For M1, participants may have thought simplified versions of the books dont give same taste. According to F2, the English level has an influence on people to choose the length of a novel and as for F3, this survey is conducted at high school level and it is more useful for them to read the original version of the books for that age group and therefore they might not prefer the short versions.

For the item 6, F1 thinks students choose the original ones to sum ups. According to M1 participants may have thought reading short versions of the books can help you gain time but its not worth it. F2 thinks that %23,6 of the participants could be interested in something else instead of reading a thick novel so that is a possibility. As for F3, most people do not prefer to read the short versions because they think the original is more meaningful.

For the item 7, F1 thinks that %35,17 agreed on this question because they have been educated in English since the fourth class and they have reached good levels of grammar or vocabulary and they have got skills in recent years to read the original ones. M1 thinks that participants may have thought that they can improve themselves more by reading original books thats why they don't prefer short versions of the books. As for F2, the participants think that they choose short version of novels because of their English levels so the rest of them find it true. According to F3, the difference between the original version and the short version is not only the grammar, but also writing events in a different and short way

For the item 8, F1 thinks that when there is an English novel adaptation into movies we have a chance to watch the book. M1 thinks that, participants may have thought scripts have rich vocabulary. Thats why its a good alternative. As for F2, learning with the tools in class instead of just a bunch of theories without practice would be a new way so they may have found it interesting. According to F3, the films adapted from the books provide a better understanding of the books as well as the development of grammar and especially listening.

For the item 9, F1 thinks that; after finishing books that they liked, they start looking for the books them. For M1, participants mostly disagreed because they thought they have already read the original ones. According to F2, participants do not mind reading the scripts of the literary products which they

Table 2. Item Statistics (mean = M) and (standard deviation=SD)

Questions	Mean	Std. Deviation	N
Q1	3.56	1.226	68
Q2	3.82	1.036	68
Q3	3.44	.904	68
Q4	3.62	1.107	68
Q5	2.57	1.111	68
Q6	2.57	1.375	68
Q7	2.79	1.311	68
Q8	3.60	.995	68
Q9	3.19	1.026	68
Q10	2.78	1.091	68
Q11	3.37	.991	68
Q12	2.78	1.157	68
Q13	2.81	1.237	68
Q14	3.56	1.111	68
Q15	3.22	1.131	68
Q16	2.97	1.146	68
Q17	3.07	1.124	68
Q18	3.41	.934	68
Q19	3.37	1.171	68
Q20	3.94	.826	68
Q21	3.76	1.053	68
Q22	3.60	1.081	68
Q23	3.38	1.270	68
Q24	3.82	1.036	68
Q25	4.13	.845	68
Q26	4.29	.600	68
Q27	4.15	.797	68
Q28	4.15	1.149	68
Q29	3.75	1.028	68
Q30	3.75	1.028	68
Q31	3.59	1,136	68
Q32	4.06	.879	68
Q33	4.03	.914	68
Q34	3.97	.992	68

have read before according to the result. As for F3, every person wants to read different kinds of books and improve their general culture. And thinks reading a different kind after reading any kind of book is good in terms of word and subject difference.

For the item 10, F1 thinks %30,9 of students agreed because a script of a movie sometimes will be boring and longer than the book so students prefer the real ones. For M1, less than half of participants agreed because it is hard to find it. While trying to find, it takes a lot of time. According to F2, %30,9 of the participants agreed to this item so I suppose they like reading scripts that take time and they do not complain about that. As for F3, most people prefer to read the original version of the script or watch the filmed version instead of reading the script.

For the item 11, F1 thinks that half of the students agreed to this question because maybe they can see and hear the daily speech that was simple and they understood sentences of the script. For M1, the language level of the scripts is easier than the originals. That's why half of the participants agreed. As for F2, about half of the participants agreed to the statement scripts adapted from literary works are suitable for their English level so they think it is not hard to understand. According to F3, reading these script texts can help them to improve their grammar and vocabulary.

For the item 12, F1 thinks that %27,8 of students agreed that because not everybody enjoyed reading scripts or reading some books. For M1, participants disagreed because scripts are not as aesthetics as much as originals. According to F2, most of the participants do not enjoy reading scripts adapted from literary works. As for F3, each person has their own preferences and some things give him pleasure while other things may not. Most of the students do not agreed to this item because they don't like reading script texts.

For the item 13, F1 thinks most of the students don`t compare script and books. For M1, participants disagreed because they dont prefer to read scripts. According to F2, nearly quarter of the participants agreed to the statement so they compare the books they read their scripts and like it. As for F3, for the students making comparisons between scripts and books might not be very significant.

For the item 14, F1 thinks 46 out 68 of students agreed to this statement because when they watch a movie maybe they get curious about what happened next and decide to buy books original forms of movies. For M1, participants agreed because scripts have less details and short. Thats why they are being curios about details. As for F2, if the film or movie is not an abortive adaptation of its book then it could motivate the audience to read the literary work of it, so two thirds of the participants might have thought in that way. According to F3, a book that they watch the movie may be interesting to them because when you watch a movie you don't know what the characters think, but you can understand what they think when you read the book.

For the item 15, F1 thinks the reasons are the %56,9 of students agreed to this question they have many options for the adaption. They could be movies or tv series. For M1, participants may have thought literary works have very strong plots thats why they are very convenient for film scripts. As for F2%56,9 of the participants agreed that the scripts based on literary products are mostly film scripts so they have watched examples of them and found them successful. According to F3, scripts are often adapted to movies if texts are liked.

For the item 16, F1 22 out 68 of students don`t think the most book adaptions are series because when we look at nowadays adaptions they usually made a movie about a book. For M1, participants thought there were a few of them but not generally that's why they disagreed. According to F2, more than quarter of the participants agreed that most of the scripts of TV series are adaptation of literary products, she thinks they are able to reach some information about it so they answered in this way.

For the item 17, F1-%31,13 of students agreed that because they might be curious about what will happen or they just enjoyed reading scripts For M1, participants disagreed because to them scripst are boring and not exciting. Accorting to F2, %31,13 of the participants agreed that the scripts motivated them so they might be aware of the benefits. As for F3, the reason why some students do not participate is that the scripts do not appeal to each person or not have the same effect on each person.

For the item 18, F1 thinks that more than half of the students agreed to that statement because they might have liked doing it or it`s a simple way to find basic sentences or daily dialogues. For M1, more than half of the participants agreed because scenario texts have daily language. As for F2, more than half of the participants agreed to this statement so they find it accurate to say scripts are good resources to learn chunks from. According to F3, dialogues and chunks about casual life are frequently found in scripts, and therefore some students may think that it is a good resource for finding new chunks with correct grammar.

For the item 19, F1 thinks that %58,8 of students agreed to that because when they read a book about a specific subject or terms they can learn about this topic. For M1, more than half of the participants agreed because scripts can be written about any subject and it has a superficial language so it is easy to learn terminology. As for F2, %58,8 of the participants think that scripts are useful resources about getting the jargon used in different fields so apparently they have experienced that. According to F3, many different topics can be explained in films and scripts. Many different and more technical terms related to these topics can be found in the texts. Students use this kind of scripts to learn more technical terms about medicine, engineering etc.

For the item 20, F1 57 out of 68 students agreed on this question because when they read or watch a script they can see the basic sentences easily or conversations at their English level. For M1, participants agreed because scripts generally have daily language. According to F2, more than two thirds of the participants agreed that scripts are resources for spoken language so I think they might have thought that it would make sense because they may think a written text will benefit their speaking skills. As for F3, everyday dialogues have a great deal of place in scripts, so they are very effective in daily spoken language.

For the item 21, F1 thinks that 48 out of 68 are students agreed to this statement because when they see a word which they don`t know, they look up their dictionary and learn the true form of the vocabulary. For M1, participants agreed because seeing the differences clearly between two language is a very affective learning style. According to F2, a little less than half of the participants agreed to the statement, that does not work for them perhaps. As for F3, learning a language that we don't know very well by comparing it with our own language and seeing the differences between them helps many people to learn that language and it also makes it easier to learn it by comparing it with our own language.

For the item 22, F1 thinks %60,3 of students agreed to this question because when they try to translate sentences which were in English, they try hard to translate in their language and it develops their translating skills. For M1, participants may have thought practising translation improves their translating ability. According to F2, %60,3 of the participants agreed to this one, that means they are happy with the results. As for F3, when translating English into our own language, we see that some words do not correspond in our language, and defining them with some words in our own language improves our translation skills.

For the item 23, F1 thinks that %63,3 of participants agreed to this because when they do their writing tasks in English, their writing skills will be developed. For M1, participants may have thought writing a script that based on a story and think how to write it improve their writing ability. For F2, %63,3 of the participants agreed that reading a book and producing a script based on them improved their writing skills. In her opinion, that is all because they have practised so it has an influence on them. And as for F3, according to the book we read, creating our own scenarios improved our writing skills because we both created our own scripts and tried to be careful about grammar.

For the item 24, F1 thinks that more than half of the participants agreed to this because when they have to make a project like this they have to belong in a team and work with other people and projects like this helped their cooperation skills. For M1, participants may have thought asking questions, making brain storm together, task sharing improve their cooperation skills. As for F2, more than half of the participants agreed that teamwork improved their collaboration skills, that is because a group/team requires collaboration. According to F3, these scenarios were written with a certain task sharing. After reading the books, everyone stated their ideas and stated what topic they wanted to take part in. That's why, our cooperation skills improved.

For the item 25, F1 thinks that %85,17 of students agreed to this statement because when they want to make a movie and use a script they have to talk in front of cameras and it increases their speaking level. For M1, participants may have thought speaking simultaneously and practicing are the most important things to improve their speaking ability. As for F2, %85,17 of the participants apparently experienced that and film is supposed to improve our speaking the most. According to F3, İt had an impact on our speaking skills because we presented the script that we wrote, the words which we chose ourselves, and it had a great impact on the speech because we prepared ourselves.

For the item 26, F1 thinks that almost all of the participants agreed to this statement because when we see a movie or tv series we want to feel the subject and their acting skills so body language is a very important thing. For M1, participants may have thought they are already notice the body language uncounsciously thats why they agreed. As for F2, almost all the participants agreed to the statement because people care about our gestures instead of focusing on our speech most of the time. According to F3, body language is as important as speech language because it is the body language that tells us what the player is thinking at that moment. We can adopt their views on the events, even on their facial expressions.

For the item 27, F1 thinks that most of the participants agreed that because of one of the important things actors do is to make gestures in front of the cameras. For M1, participants may have thought they have to use gestures to express themselves. As for F2, most of the participants agreed to this one and that mentions what I was trying to say in the previous question. According to F3, we must pay attention to mimics because we must show our feelings to those who are watching us.

For the item 28, F1 thinks that over two-thirds of the participants agreed that because she thinks that when they have made a project with their friends they enjoyed it and they had fun. For M1, participants may have thought doing a thing with friends in every way is enjoyable. What they are doing is no matter. For F2, predictable result is; over two thirds of the participants agreed to this. I am not surprised at all, to be honest. In my opinion, stimulating activities will always be hundreds of miles ahead of "classic" education. According to F3, many people had fun shooting their own films because they had spent time with their friends and learned English in a more fun way.

For the item 29, F1 thinks that almost two-thirds of students agreed on this statement because of when they see a movie with subtitles at the same time they hear the pronounce of these words and it helps their project. For M1, some of participants are not very good at English thats why they used subtitles. As for F2, almost two thirds of the participants agreed that they used Turkish subtitles to understand the movie they watched and that means they know how to utilize subtitles. According to F3, students had a hard time while trying to understand what the film character said because we don't use English very much in our daily life. That's why they got help from subtitles.

For the item 30, F1 thinks %68,11 of students agreed to this statement because they have heard the pronunciation and they have read the subtitle at the same time. For M1, more than half of the participants agreed because they thought when they dont see the word how written, it is hard to understand and they can't pronunce. According to F2. %68,11 of the participants agreed and that means they enjoy pronunciation and using the subtitles for that to me. As for F3, most students give importance to pronunciation but some students do not pay attention to pronunciation thinking that they will not use it, but it is easy to learn the pronunciation of difficult words by using subtitles.

For the item 31, F1 thinks that a little more than half of the students agreed on this statement because when they saw how enjoyable learning a different language is they didn't want to give up and they motivated about learning a new language like English. For M1, nearly half of the participants disagreed because they thought that is a complicated thing and not worth it. Nearly half of the participants agreed because they had fun and worthed it. According to F2, a little more than half of the participants agreed to this and surprised me but it is really satisfying to see they enjoy learning English outside the class. That is because shooting a film is a good deal more fun and affectionate than sitting in class. As for F3, most people don't like the English they learn at school because it is compulsory, but this kind of activity motivates the students. It allows them to learn English, not really to memorize, but to learn and have fun.

For the item 32, F1 thinks that %84,13 of students agreed on this statement because when they have written a script of their movie they use the examples in books and movies. For M1, they agreed because there is not much to learn how to write a scenario. And its a good way For.F2, %84,13 of the participants agreed to this statement and I think that is what should happen because we cannot practice without an instance in this case. According to F3, since books and film writings are written by professional people, they are more accurate and understandable in terms of grammar, so the students received help from the grammar books or the subtitles of the films.

For the item 33, F1 thinks that more than two-thirds of participants agreed to this statement because in movie scripts writers use the simple form of verbs and daily language. For M1, more than half of participants agreed because they thought scripts have daily language. As for F2, 55 out of 68 participents agreed to this statement, I think scripts have dialogues more than literary products do so that must be the reason they agreed to the statement. According to F3, scripts have more daily dialogues and events than literary works. That's why scripts are more common in daily life.

For the item 34, F1 thinks that about two-thirds of students agreed to this statement because when students have thought Turkish expressions and translated them into English and it helped their translating skills. For M1, more than half of the participants agreed because they thought when we translate English to Turkish at the same time we can learn to translate Turkish to English. We can see two diffreent ways at the same time. For F2, 51 out of 68 participants agreed, in her opinion, why they agreed to this one is the benefitial interchange between these two languages used for the project. The participants might have been satisfied by some chunks/words that they have learned within the time they spent. According to F3, while making sentences, we first make Turkish sentences and then translate them into English, so our Turkish language skills have improved.

DISCUSSION AND CONCLUSION

Compared to other studies using films in the teaching and learning of English, this study feeatures originality with the use of literature based films by considerably high number (68 = 24 males, 44 females) of high school students. Besides, the students act out their dialogues they produced out of script studies. They also videotaped these drama activities. It was a unique example of students involvement at the out-of-class settings. However, these out-of-class settings were integrated to the process with discipline.

It was concluded in this study that the researcher/teacher's high school students improved their productive use English reading literary works, watching their film adaptations both Turkish subbed and later English captioned, reading the scripts (scenarios) of these films and the Turkish translations of these scripts, noticing and making lists of chunks, jargons and creative language use examples out of the works they studied.

Despite the potential to develop speaking skills, the development of strategies to benefit from films has been neglected and very few studies have been conducted. One of these strategies has been to compare film scripts and their translations, to identify chunks, technical terms and creative language examples and to list them with their translations, and to do creative dialogue writing activities especially by modeling these scripts where chunks are used extensively. Believing in the importance of evaluating the film scripts with their translations, this study aims to investigate the processes the researcher/teacher's high school students went through in learning and using the target language starting with the use of the film adaptations of the literary works, writing their scripts modeling the film scripts, acting their own scripts and shooting their performances. The mixed method research design was used with the triangulation of the qualitative and quantitative data. As the participants, the 68 high school students's views were investigated with Likert scale (1-5) questionnaire and their comments about the question items.

According to the ratings of the Likert scale items, their responses can be summarised as such: Many participants defend using literature in learning English. However, their conceptualisation of literature is not limited to the classical works by major authors like Shakespeare, Voltaire etc. (canonical-literature). According to many, products of popular culture like Harry Potter etc. are literary works as well. The film scripts based on literary works are also accepted as products of literature and serve as an alternative source in terms of literature use in the learning of English. Intriguing their curiousity, these motivate them to read the literary works they are based on. These serve as a good source for spoken English as well. Comparing them with their Turkish translations helps them with the learning of English and develops their translation skills. While watching a movie, they notice nonverbal elements of communication as well as verbal ones. While acting the scripts they wrote they notice such nonverbal communication elements as gesture and mimics. They enjoy acting out their own scripts. They use Turkish subtitles to comprehend the movie, and English captions to develop their pronunciation. In the writing process of their own scripts, they take examples of language use from the literary works and film subtitles. Acting the scripts they wrote develops their speaking skills and collaboration skills. They notice that the scripts include more examples of spoken language than the literary works. In the process of the activities, their translation skills develop in both directions (from Turkish to English as well as from English to Turkish). The activities motivate them to continue learning English in out-of-class settings.

Asked whether the novels they read are abridged versions rather than unabridged versions (item 5), about two thirds of the participants responded that they do not have such a preferrence. It is only about one quarter who said they do (meaning they prefer abridged editions) and their reasons were to do with time and level of English. There are also a few low-rated items about the scripts of movie adaptations of literary works. Asked whether reading these scripts brings an advantage regarding time, only about one third of the responses were positive. It was only 19 (almost 30%) out of 68 participants who stated they took pleasure from reading such scripts in contrast to about 50% responding negatively. Asked whether they make comparisons between the literary works they read and the scripts of their movie adaptations a bit over fourty percent responded negatively whereas the rate was almost thirty percent for those undecided and responding positively.

Mean values for some questionnaire items were within undecided range. Asked whether the scripts they preferred to read were those based on the literary works they had read before, those agreeing, undecided and disagreeing numbered 26, 22 and 20 respectively (m: 3.19). With those disagreeing (f: 24) slightly outnumbering those undecided and agreeing (f: 22), there was a similar distribution of responses to whether the scripts based on literary works developed motivation for reading.

A bit over half of them (f: 35) found the language of the scripts at their level whereas those disagreeing were not even half of those agreeing (f: 13). However, about 30 percent undecided participants caused the mean value (m: 3.37) to be close to agreement in general, however neither very high nor low either. There was a similar level of rating bordering agreement with more than half of the participants responding positively to whether the literary work based scripts were good sources to find chunks (m: 3.41) and specialised language examples (jargons) (m: 3.37) and whether students' tasks of script writing based on the literary work based scripts they read developed their writing skills (m: 3.38).

The fifteenth item of the questionnaire was written to investigate if the participants knew that scripts based on literary works include TV series besides movies. Although many thought the literary work based scripts are only movies almost one quarter disagreed, (and rightly so) suggesting that they knew that there are also TV series. One fifth were undecided. The reason for a possible confusion is likely to be the wording of the question in the item, because film in Turkish means both a cinema film (movie) and TV series (shows).

The implications for teaching is that use of literature in language learning should be conceptualised as to involve the scripts of TV series and movies. In fact many movies are adaptations of literary works. It is likely that activities developed in the use of film scripts improve the creativity, self-confidence and collaboration. The study demonstrated that productive use of films in teaching English develops imagination, creativity, and the literacies regarding information, media, and innovative technology and leadership through the learning and use of English.

The process also has benefits for education in general through the teaching of English using films. The productive activities using films in teaching English developed imagination and creativity, the literacies regarding information, media, and innovative technology and leadership through the learning and use of English. The activities in our study were examples of using imagination in foreign language learning. Imaginative education is of key importance but neglected in teaching. Zengin, Başal and Yükselir (2019) who suggest that creativity and imagination is not given their due value in the teaching.

Use of films scripts in this study provided benefits for the researcher/teacher's high school students' sociolinguistic competence. Film scripts provide beneficial examples for the variety of language use regarding genre and register. Webb and Rogers (1995) believes science fiction programs to be one of the most demanding genres, including more technical vocabulary than other genres. Arguing for the interaction between foreign language teaching (FLT) and sociolinguistics, Bayyurt (2013) argues that the teaching of successful communication in L2 is through the use of correct register or reasonable variety, where to use the proper register or variety. She clearly criticizes FLT research for its neglect for sociolinguistics.

The participants in this study argues that the reverse translation (from Turkish to English) improves the productive language skills like Zengin and Aker (2016). The conclusions of Zengin and Aker (2016) support the idea that there are various ways to make translation a source of effective language learning. The application recommended in their study parallels reverse translation, that is translation out of the first language into the target language, that refers a contribution to productive skills. Reverse translation facilitates the recognition and production of vocabulary, a key source challenging the listening comprehension. Accelerating the vocabulary learning process can pave the way to gradual assistance of listening comprehension with which learners can gain what is odd to listening (variety of accents, stress, rhythm, intonation). With reference to the genre, familiarity with vocabulary bases on genre difference too.

Translation has a strong potential if evaluated as part of the multidisciplinary natüre of English language teaching. Arguing against thc overemphasis on interdisciplinary and intradisciplinary separatism but not separation for heuristic reasons, that facilitates a study), Cook and Barbara Seidlhofer (1995, p. 7) state that the areas of translation and lexicography, while sometimes by convention considered separate, might equally well be encompassed by the definition of applied linguistics. According to Hall and Cook (2012), "the continued use of translation activities in language classrooms around the world has also been increasingly acknowledged in recent years." As they state that the field has started to enjoy the beginning of "a re-evaluation of teaching that relates the language being taught to the students' own language," this study argues that English as a foreign language can be taught with the help of the students' own language. Köksal (2005) argued for the incorporation of translation skill into the skills taxonomy of foreign language learning and teaching.

The researcher/teacher's high school students were self-reported to have fun and enjoyment. As stated in Zengin and Aker (2016), in Turkey, lots of young people might watch English language series through websites of streaming companies. Benefiting from this motivation, most probably they could be guided on how best to utilize this rich source of input and strategies need to be improved to infuse education into entertainment.

Our thesis is a powerful application of CALL demonstrated by the high school students' information, multi-media and technology literacies. They performed the activies in informal settings outside the formal setting of classroom. Hubbard and Levy (2006) argue that interpretation of computer-assisted language learning (CALL) in all major CALL journals is not limited to the language teaching in the classroom, on the contrary CALL is interpreted way beyond the classroom. There are a variety of this interpretation in some articles. As the studies of Zengin and Çucukçu (2013) and Zengin, Doğan and Çubukçu (2015) have attempted to call attention to the need to equip backseat TV and cinema systems (that are costly but inefficient given its potential) of intercity coaches for high-frequency student commuter, there is a probablity to connect the potential overlaps between education and entertainment. The results of this study has inferences for foreign language teaching anxiety, a neglected area according to Aydın (2016) as well since foreign language teaching majors are not immune to input problem specially concerning chunks.

FUTURE RESEARCH DIRECTIONS

Investment on vocabulary can rise the possibility of payoff from investments like 3-D virtual learning environments, use of which Zengin (2015) argues for to improve listening skills. The results of this study may have an effect on policy makers to argue providing the spectators with some other ways of combining subtitle andlanguage /captions so that the viewers, who want to improve their language skills, can use them to their advantage.

AUTHORS NOTE

This book chapter is based on the Masters thesis of the first author conducted in collaboration with and under the supervision of the second author.

REFERENCES

Aydın, S. (2016). A Qualitative Research on Foreign Language Teaching Anxiety. *The Qualitative Report, 21*(4), 629-642. https://nsuworks.nova.edu/tqr/vol21/iss4/2

Bayyurt, Y. (2013). Current Perspectives on Sociolinguistics and English Language Education. *The Journal of Language Teaching and Learning, 3*(1), 69–78.

Brewster, J., Ellis, G., & Girard, D. (1992). *The Primary English teacher"s guide*. Penguin.

Brown, H. D. (1987). *Principles of language learning and teaching* (2nd ed.). Prentice-Hall.

Celce-Murcia, M. (1995). Discourse analysis and the teaching of listening. In G. Cook & B. Seidlhofer (Eds.), *Principle and practice in applied linguistics* (pp. 363–377). Oxford University Press.

Christensen, L., & Johnson, B. (2004). *Educational research: quantitative, qualitative and mixed approaches* (2nd ed.). Pearson Education.

Cook, G., & Seidlhofer, B. (1995). An applied linguist in principle and practice. In G. Cook & B. Seidlhofer (Eds.), *Principle and practice in applied linguistics* (pp. 1–26). Oxford University Press.

Cook, G., & Seidlhofer, B. (Eds.), *Principle and practice in applied linguistics* (pp. 363–377). Oxford University Press.

Hall, G., & Cook, G. (2012). Own-language use in language teaching and learning. *Language Teaching, 45*(3), 271–308. doi:10.1017/S0261444812000067

Hubbard, P., & Levy, M. (2006). The scope of CALL education. In P. Hubbard & M. Levy (Eds.), *Teacher Education in CALL* (pp. 3–20). John Benjamins., doi:10.1075/lllt.14.04hub

Kırkgöz, Y. (2009). Globalization and English Language Policy in Turkey. *Educational Policy, 23*(5), 663–684. doi:10.1177/0895904808316319

Köksal, D. (2005). *Çeviri Eğitimi: Kuram ve Uygulama*. Nobel.

Kremmel, B., Brunfaut, T., & Alderson, J. C. (2015). Exploring the role of phraseological knowledge in foreign language reading. *Applied Linguistics, 1*(24), amv070. Advance online publication. doi:10.1093/applin/amv070

Renandya, W., & Farrel, T. S. C. (2011). Teacher, the tape is too fast! Extensive listening in ELT. *ELT Journal, 65*(1), 52–58. doi:10.1093/elt/ccq015

Trappes-Lomax, H. (2004). Discourse analysis. In A. Davies & C. Elder (Eds.), *The Handbook of Applied Linguistics* (pp. 133–164). Blackwell Publishing. doi:10.1002/9780470757000.ch5

Zengin, B., Başal, A., & Yükselir, C. (2019). Investigation into the perceptions of English teachers and instructors on the use of literature in English language teaching. *The Reading Matrix: An International Online Journal, 19*(1), 155–166.

Zengin, B., & Cubukcu, F. (2013). English language learners' attitudes to backseat TV systems and subtitled movies. *Turkish Studies – International Periodical for the Languages. Literature and History of Turkish or Turkic*, *8*(8), 1561–1573. doi:10.7827/TurkishStudies.5353

Zengin, B., Doğan, D., & Çubukçu, F. (2015). English major students' attitudes towards movies and series as language resources. In V. Turel (Ed.), *Intelligent Design of Interactive Multimedia Listening Software* (pp. 276–295). IGI Global. doi:10.4018/978-1-4666-8499-7.ch011

Zengin, B., & Yetkin Aker, D. (2016). Education Policy and the Effect of Subtitle Options in Learning Process: A Case of Turkish Students as English Language Learners. *Humanitas*, *4*(7), 485–498.

Zengin, B., & Yetkin Aker, D. (2016). Education Policy and the Effect of Subtitle Options in Learning Process: A Case of Turkish Students as English Language Learners. *Humanitas*, *4*(7), 485–498.

ADDITIONAL READING

Aksu Ataç, B., & Köprülü-Günay, S. (2018). The role of subtitles in foreign language teaching. *International Online Journal of Education & Teaching*, *5*(3), 525–533.

Eslamieh, R., & Javankhah, N. (2018). Domestication and Foreignisation in Dubbing andSubtitling of Duncan Jones" English Movie *Warcraft* into Persian. *International Journal of English Language &Translation Studies*, *6*(1), 162–170.

Karakaş, A., & Sarıçoban, A. (2012). The impact of watching subtitled animated cartoons on incidental vocabulary learning of ELT students. *Teaching English with Technology*, *12*(4), 3–15.

Metruk, R. (2019). Using English movies and TV programs for developing listening skills of EFL learners. *Information Technologies and Learning Tools*, *70*(2), 227. doi:10.33407/itlt.v70i2.2488

Murshidi, G. A. (2020). Effectiveness of movies in teaching and learning English as a foreign language at universities in UAE. *Psychology and Education*, *57*(6), 442–450.

Okyayuz, S. (2016). Subtitling Historical DramaTV Series: Constraints and Considerations *International Journal of English Language&. Translation Studies*, *4*(1), 45–59.

Pujadas, G., & Muñoz, C. (2019). Extensive viewing of captioned and subtitled TV series: A study of L2 vocabulary learning by adolescents. *Language Learning Journal*, *47*(4), 479–496. doi:10.1080/09571736.2019.1616806

Rao, P. S. (2019). The Impact of English Movies on Learning English in ESL/EFL Classrooms. *Research Journal of English Language and Literature*, *7*(4), 430–438.

Seferoğlu, G. (2008). Using feature films in language classes. *Educational Studies*, *34*(1), 1–9. doi:10.1080/03055690701785202

Thammineni, H. B. (2016). Movies supplement English classroom to be effective in improving students' listening and speaking skills–a review. *International Journal of Studies in English Language and Literature*, *4*, 35–37.

Winke, P., Gass, S., & Sydorenko, T. (2010). The effects of captioning videos used for foreign language listening activities. *Language Learning & Technology*, *14*(1), 65–86.

Yuksel, D., & Tanriverdi, B. (2009). Effects of watching captioned movie clip on vocabulary development of EFL learners. *The Turkish Online Journal of Educational Technology – TOJET, 8*(2), 48–54.

Zatlin, P. (2005). *Theoretical Translation and Film Adaption: A Practitioner"s View*. Multilingual Matters Ltd.

KEY TERMS AND DEFINITIONS

Script: Interlingual/Standard Subtitling: Screenplay written by screenwriters for a film, television show, or a video game. Scripts may be adaptations of literary works.

Subtitles: The lines of text from the original or translated script. There are different combinations. Standard subtitles are interlingual and demonstrate the translations of the spoken dialogue. Captioning is the intralingual kind (soundtrack and subtitle are the same lanaguage. There is also reversed subtitling, which can simply be defined as the reversed version of the standard subtitling.

APPENDIX

The Questionnaire

1. We don't see literature limited to the classical works of the great writers (canonical) such as Shakespeare, Voltaire etc.
2. I think the pop culture products such as Harry Potter etc. are literary works too.
3. Movie scripts based on literary works are also literary products.
4. I support using literature in learning English.
5. Novels I read are rather shortened versions.
6. One of the reasons I prefer simplified versions is that I can't make time for the original.
7. One of the reasons I prefer the simplified versions is that it is more suitable for my language level.
8. Scenario texts based on literary works provide an alternative source of quality in terms of using literature in learning English.
9. Among the scenarios based on literary works, the ones I prefer to read are those based on the literary works I have read before.
10. Reading scripts based on literary works is advantageous in terms of time.
11. Scripts based on literary works are suitable for me in terms of language level.
12. I enjoy reading scripts based on literary works.
13. I make comparisons between the literary works I have read before and the script texts based on these works.
14. Scripts based on literary works also have features that encourage reading and wondering about the literary work on which they are based.
15. Most scripts based on literary works are film scripts.
16. Most scripts based on literary works are Tv Series scenarios.
17. the script texts based on literary works increased my motivation to read.
18. Scripts based on literary works became a good source for finding chunks.
19. Scripts became a good source for finding examples of technical language usage in the special field (military, police, law, medicine) from different types of scripts based on literary works, such as military, police, law, medicine, etc.20) Scripts texts based on literary works have become a good source of spoken English.
20. Comparing scripts based on literary works with texts translated into Turkish helped my English learning.
21. Comparing scripts based on literary works with texts translated into Turkish improved my translation skills.
22. Reading literary works and writing script tasks based on these scenarios improved my writing skills.
23. The scenarios we wrote in this way helped our collaboration ability.
24. Acting the scenarios we wrote improved my speaking skills.
25. When watching movies, I also pay attention to non-verbal dimensions such as body language.
26. I took care of non-verbal communication elements such as gestures while shooting and playing the script we created.
27. We enjoyed shooting and playing our own scenario.

28. I used Turkish subtitles to understand the film.
29. I used English subtitles to improve my pronunciation.
30. It helped to motivate my English learning out of class.
31. While writing our own scenarios, we took the example of language usage in books and film subtitles.
32. I realized that script texts contain more examples of spoken language than literary works.
33. In the course of these studies, my English to Turkish translation skills improved as well as my translation skills from Turkish to English.

Chapter 10
Exploring Teachers' Beliefs About Listening in a Foreign Language

Pelin Irgin
TED University, Turkey

ABSTRACT

There is a substantial amount of research in the field of language teacher cognition reiterating that teachers' beliefs influence their classroom practices; however, teacher beliefs and practices do not always correspond because of the differences in both researchers' and teachers' conceptualization of beliefs and practices. This study aims to investigate the beliefs and practices of 96 language teachers in Turkey. A mixed-method research design was employed to answer the questions to what extent teachers' beliefs and practices reflect the issues stated in the literature on second language listening, what their beliefs and listening instructional practices are, whether their beliefs and practices converge, and if so, what factors underpin them.

INTRODUCTION

The study of language teacher cognition and teacher beliefs has been pivotal to understand how language teachers learn, what they do and what reasons stand behind their actions while teaching. Professionals such as language teacher educators, researchers, teachers and policy makers in many previous research (Borg, 2015; Borg, 2003; Farrell, 2003; Kumaravadivelu, 2012) interested in exploring the nature of language teacher cognition, the relationship between language teachers' cognitions and classroom practices, the impact of context on language teachers' cognitions and practices, and the processes of pre-service teacher learning. By the nature of language teacher cognition, the rationalization of cognition, knowledge, beliefs, attitudes, teacher efficacy, theories, assumptions, principles, thinking and decision making is a way to understand the ultimate goal of research on teachers and their practices. The "teacher cognition" is a term that seems to embrace the knowledge, thoughts, beliefs, and actions which are used interchangeably in the literature (Goa & Ma, 2011). In this chapter, the term teacher beliefs is

DOI: 10.4018/978-1-7998-7876-6.ch010

used interchangeably referring to understanding how teachers' thinking and conceptions influence their behaviors and practices in the classroom (Zheng, 2015).

In field of teacher cognition, what teachers do is underpinned and influenced by a set of pre-active, inter-active and post-active of cognitions they have (Borg, 2015). However, as teaching is a dynamic process requiring effective teaching and communication, the relation between teacher cognition and their practices in the classroom is not linear and unidirectional. Teachers who elicit greater achievement gains spend a great deal of time actively instructing their students. Their classrooms feature more time spent in interactive lessons featuring teacher-student discourse and less time spent in independent seatwork. Further, teachers who are the most effective spend "a minimum amount of time in lecture-presentations" (Eggen & Kauchak, 2016, p. 535). At this point, to explore teachers' beliefs, Barcelos (2003) outlined three approaches; (1) normative approach, (2) metacognitive approach, and (3) contextual approach. First, the normative approach is focused on exploring by appraising or establishing the values and norms rather than referring to the actual practices and implantations. The second is the metacognitive approach, where the aim is on teachers' thinking and explaining the process of teaching and learning through self-discovery and self-reporting. Finally, the third is the contextual approach referring to the teaching and learning with contextual integrity, more specifically external and environmental factors. Overall, by the integration of these three approaches, it might be possible to draw a frame to understand the teacher beliefs and the sources of teacher behaviors in their classroom practices.

BACKGROUND

Numerous studies (Barcelos, 2003; Borg, 2003; Borg, 2006; Borg & Burns, 2008) on teacher beliefs and language teacher beliefs (Graham et al., 2014; Kumaravadivelu, 2012; Phipps & Borg, 2009; Sah & Shah, 2020; Utami, 2016; Zeng, 2018) followed the combined approach to explore teacher beliefs. Language teachers' beliefs were represented in a belief system, which governed their thoughts, words, and actions in classroom practices (Kumaravadivelu, 2012). Language teachers reflected their beliefs as the driving force behind decisions they made throughout their professional career. Hence one might assume that teachers follow practice based approach, in which they think, say and do as a reflection of their beliefs. The belief system of language teachers can be explained with a set of fundamental assumptions. First of all, language teacher beliefs are formed early and tend to self-perpetuate, persevering even against contradictions caused by reason, time, schooling, or experience. The earlier a belief is incorporated into the belief structure, the more difficult it is to alter. Second, teachers tend to hold on to beliefs based on incorrect or incomplete knowledge, even after scientifically correct explanations are presented to them. Beliefs about teaching are well established by the time a student teacher enters a teacher education program. Lastly, teacher beliefs affect teaching behavior, and guide teachers in defining, selecting, organizing knowledge and information presented to students. By their nature, some language teacher beliefs are more incontrovertible than other beliefs (Borg, 2015).

Moreover, it was found that TESOL teachers had different belief systems about teaching and learning, and different approaches in the classroom. According to Kumaravadivelu (2012), concepts such as best method and good teaching should be abandoned in favour of recognition of diversity in language teachers. The main argument at this point was that the best teaching was individually the best step for each teacher. This may correspond to the combined approach. On the other hand, the findings of other research in the field (Ellis, 2018) reflected a practice based approach to understand language teachers' beliefs. For example, Ellis (2018) showed that there were inconsistencies between what was believed and what was practiced. Language teachers paid primary attention to meaning rather than grammar (the incidental focus on form). Similarly, Phipss and Borg (2009) examined the disparity in the grammar teaching beliefs and practices of three teachers of English by observing and interviewing them over a period of eighteen months. Their conclusion suggested that, in teaching grammar, the beliefs of the teachers were not always aligned with their practices. Disparity was less in the case of core beliefs while disparity was more in the case of peripheral beliefs. Core beliefs were more influential in shaping teachers' instructional strategies, and peripheral beliefs might cause inconsistencies in teacher behavior. However there are many studies that still argue the existence of tension between teachers' beliefs and their practices.

TEACHER BELIEFS AND PRACTICES IN ESL/EFL PEDAGOGY

Numerous studies have been conducted in teacher beliefs and their practices in both ESL and EFL pedagogy. Some of the relevant studies were on grammar teaching (Farrell, 2003; Phipps & Borg, 2009) while the others were in certain language skills teaching and approaches in various context (Borg, 2019).

Zacharias (2005) investigated teachers' beliefs about internationally-published materials by conducting a survey of tertiary English teachers in Indonesian context. In the study it was explored to find out the mismatches between the teachers' beliefs and their practices in actual classroom setting. It was revealed that the majority of the Indonesian teachers participating to the study believed the internationally-published materials were preferable to the locally published ones. Most of the teachers showed that the materials chosen from English-speaking countries were preferred. Their beliefs were on the perfect use of English language in those materials while their beliefs for the locally-published materials were that the locally-published ones were not always available. The findings of this research emphasized that there was an immediate need for Indonesian teachers to be empowered on developing their own teaching materials that would facilitate their particular teaching and learning practices.

In a different research in an ESL context conducted by Barrot (2016), language teacher beliefs and practices were examined. Also, the previous studies on teacher beliefs in recent 20 years were reviewed. While some studies confirmed that there was a consistency between teachers' beliefs and their practices, some of them confirmed that teacher beliefs were not always reflected in their classrooms. It was seen that teachers' beliefs and practices might not always reflect the principles in second language teaching. Hence, from a socio-cognitive transformative perspective, it was aimed to explore five experienced ESL teachers' beliefs for teaching and to determine whether their beliefs and teaching practices were consistent with the principles in ESL pedagogy. The findings of the study showed that there were consistencies between teachers' beliefs and practices, and teachers attributed to some factors such as the abstractness of principles and their attitudes towards this divergence to their students and education system.

TEACHER BELIEFS ABOUT LISTENING AND PRACTICES

Previous research into teachers and student teachers' beliefs shed further light on beliefs about listening. Language teachers' beliefs, more specifically about listening and their instructional practices, have been gaining considerable attention recently (Li, 2020; Borg, 2019).

In Chinese context, Li and Renandya (2012) examined the effective approaches applied by EFL teachers to teaching listening. From Chinese EFL teachers' perspectives, their students had some major difficulties in L2 listening, and they were supposed to their students' problems in listening. 10 Chinese EFL teachers were interviewed on their beliefs and their classroom practices to deal with their students' difficulties in listening. During the interviews, the teachers identified the main reasons of their students' listening difficulties, and the approaches they preferred in dealing with their problems. Teachers were also expected to explain the teaching procedures they followed in their language classes, especially in listening tasks. The findings revealed that Chinese EFL teachers preferred the bottom-up approach in L2 listening rather than the integration of both bottom-up and top-down approaches in teaching L2 listening. For example, students' coping with fast speed and recognizing words in speech were some of the bottom-up approaches. However, it was seen that there were differences in teachers' beliefs that might be identified as regarding the importance of both approaches and listening strategy instruction to the students. Overall, it seems that teachers need to involve macro and micro skills in listening by following an integrated approach to address their difficulties in L2 listening. Similarly, Gao and Liu (2013) investigated 325 college English teachers' beliefs in teaching listening and their relationships with teaching practices. The results of the study showed that college English teachers emphasized the critical role of listening skill and teaching listening. Also, they highlighted both micro and macro levels in listening. However, it was observed that Chinese teachers did not follow their beliefs in teaching practices faithfully. What they applied in practices and what they believed were disconnected, which was in line with the results of Li and Renandya's (2012) research.

In Schultz et al.'s (2015) study, the main focus was on how four prospective teachers attempted to adopt a listening stance in their practices, and how they responded to the constraints of the standardized curriculum of their district. They reported a two-year ethnographic study to investigate how prospective teachers enacted a listening stance in teaching at their preparation program which implied entering a classroom with questions besides answers, and knowledge as well as limitations. It was concluded that the teachers' beliefs on negotiating and their practices was quite critical for teachers to learn. They suggested to support teachers in the current climate of high-stakes testing and mandated curriculum, and to foster explicit teaching of negotiation skills.

Other studies underpinned language teachers' beliefs and practices toward authentic materials in teaching listening. As a relevant study, Mandasari (2016) conducted a research on teachers' beliefs toward authentic materials in teaching listening as an important issue including teachers' beliefs, their world views and mental models about the materials, and how those shaped their learning and teaching practices. Two English teachers teaching listening by using authentic materials were observed in their classrooms. After the observations a questionnaire and some interviews were conducted. The findings revealed that "teachers believed:

1. Authentic materials are materials produced by a native speaker of the target language not for a teaching purpose;
2. Authentic materials must introduce English in the real context used by the native speaker, improve students' knowledge, improve students' English competence in both speaking and writing, improve students' vocabulary, introduce the culture of native speaker, improve students' listening ability, motivate students to learn, motivate students to learn autonomously;
3. Authentic materials are important to use since it is motivating students and providing some aspect of English that students can learn;

4. Authentic materials are considered to be carefully selected before being taught to the students;
5. Using song attracts students' interest" Mandasari (2016, p.19).

In short, it might be considered that teachers' beliefs about teaching listening were shaped by their concepts toward authentic materials. Teachers may believe in a certain approach in teaching listening however, classroom dynamics and teaching materials might inhibit them to reflect their beliefs on their instruction.

Moreover, EFL teachers' beliefs and the realization of their beliefs in classroom practices have recently gained momentum in Iranian context, as well. Karimi and Nazari (2017) conducted a research to investigate Iranian EFL teachers' beliefs about listening and their beliefs-driven instructional practices. A listening beliefs questionnaire was administered to 85 teachers teaching at BA and MA levels and 12 of them were observed in their classroom practices. It was found that there was no significant difference between BA and MA teachers regarding their listening beliefs and beliefs-driven practices. The results of the Phi coefficient of correlation indicated that there was no significant relationship between teachers' beliefs about listening instruction and their listening instructional practices. In their interviews, teachers stated that they had obstacles to actualize their listening beliefs besides some other impediments, which might suggest that teacher beliefs and practices do not always correspond.

More recently, Emerick (2019) examined university language teachers' beliefs about explicit teaching and authentic materials in L2 listening instruction. 60 language teachers teaching at the departments of Spanish and English at eight universities in the US participated to an online questionnaire and semi-structured interviews were conducted to six of the participants. The results of the research indicated that language teachers had positive beliefs on the explicit listening instruction which developed learners' competences in L2 listening. Very few of the university teachers reported the implementation of metacognitive strategy instruction. Indeed, most of them believed that authenticity was one of the most valuable constructs for listening teaching; however, "instructors considered either linguistic features of the listening text or features of the L2 learning context when selecting materials, not both, illustrating differing conceptualizations of authenticity and varied approaches to authentic material selection" (Emerick, 2019, p.107). In other words, the emphasis is on paying more attention to listening pedagogy in L2 teaching setting. This might be also seen in the instruction of listening strategies both implicitly and explicitly (Chen, 2015; Irgin & Erten, 2020) and the use of authentic listening materials.

The review above suggests, on the one hand, that there is an evidence of divergence between language teachers' beliefs about listening and their practices in their classrooms. The findings of the previous research outlined the practical areas that refer to how they believe listening should be taught (Borg & Burns, 2008) and the formal areas what academic studies suggest in theory.

MAIN FOCUS OF THE CHAPTER

In the field of language teacher cognition, Borg and Burns (2008) explain teachers' beliefs as the study of teachers' understanding, thoughts, and actions they do. The definition of teachers' cognition and teachers' beliefs in the literature are interchangeably embraced. However, in the area of second language listening, there is a limited research on teacher beliefs about listening in foreign languages (Borg, 2019; Emerick, 2019; Karimi & Nazari, 2017). Arguably, there is some debate on the operational definition of teacher beliefs and its influence on how teachers behave in their classroom practices. Graham et al.

(2014) investigated whether teacher beliefs about listening and their listening pedagogy converged, and what components might corroborate them. They had found out the mismatch between teachers' belief on the instruction of listening to their students and their listening teaching practices. In a similar vein, Liao and Yeldham (2015) reported on their study investigating EFL teachers' perceptions of their listening instruction that Taiwanese EFL teachers used a comprehension based approach in their teaching (listening classes), commonly had affective problems among their students, and often dealt with their students' comprehension and affective challenges. These studies address an important gap in the field of language teacher cognition as they suggest implications for future research to understand the complex relationship between teacher beliefs and practices, more specifically teaching of listening. Thus, this study aims to investigate the beliefs and teaching of listening practices of 96 foreign language teachers in state secondary schools in the context of Turkey.

To address this gap the following research questions guided current study:

1. What are the EFL teachers' beliefs about listening in a foreign language?
2. What are the EFL teachers' beliefs about instruction of foreign language listening
3. How do the EFL teachers' stated beliefs compare with their stated in-class listening practices?

METHOD

A mixed-method research design was employed to answer the questions to what extent language teachers' beliefs and practices reflect the issues stated in the literature on second language listening, what their beliefs and listening instructional practices are, whether their beliefs and practices converge, and if so, what factors underpin them.

96 foreign language teachers in state secondary schools in the context of Turkey participated to this study. The researcher contacted with 20 state secondary schools randomly selected from the capital of Turkey. The foreign language teachers teaching English in the schools were informed about the aim of the research but only the volunteers participated to the study. Also, they were non-native speakers of the language they taught. Their year of professional teaching experience was ranged from 5 to 20. Very limited number of teachers had master degrees in the fields of English language education, and English language and literature.

The data was collected through a background questionnaire for foreign language teachers (see Appendix A) and a questionnaire entitled teachers' beliefs about listening in a foreign language designed by Graham et al. (2014). The questionnaire included both open-ended items for listening procedures and justifications, and a five-point scale (see Appendix B). The quantitative data was collected by the questionnaire "teachers' beliefs about listening in a foreign language"

Both qualitative and quantitative analysis methods were performed. Descriptive statistics was run to get the ratios of teacher beliefs and their practices in teaching listening. The data gathered from the open-ended items in the questionnaire was coded by two raters and the inter-rater reliability calculated using Cohen's kappa coefficient score was .95 (Miles & Huberman, 1994). To analyze the qualitative data including procedus (what they do/or what students do in L2 listening process in the classroom) and justification (why they do this and/or why this is done) on the teachers' beliefs and their practices, the coding approach was followed (Stipek et al., 2001). This qualitative analysis was conducted to find some possible reasons from language teachers' beliefs about listening and their instruction of listening.

Table 1. Teachers' profile

Age	Varied between 25 and 60
Mother tongue	Turkish, Kurdish, Arabic, German, Dutch
Highest educational attainment (Graduation)	BA (75), MA (15), PhD (6)
Specialization	English Language Education, Linguistics, English Language and Literature, Translation Studies, Educational Sciences
Gender	Female (64) Male (32)
Year of experience	Varied between 3 and 27
Type of students (target population in teaching)	EFL
Languages they usually speak in class	Turkish and English
Languages they usually speak out of class	Turkish

FINDINGS AND DISCUSSION

Teachers' beliefs about listening gathered from the most important procedures and justifications outlined as what they do and why they do, revealed the main categories of procedures and justification Turkish EFL teachers considered to be important. The findings were listed in the order of frequency mentioned. The following Table 2 and Table 3 indicated how many procedures language teachers do in listening and justifications that language teachers follow in listening practices. As Table 2 shows, language teachers seem to be claiming that giving general information about the topic, explaining the rules of listening task, asking students their opinions/ideas or experiences about the topic, and revising key words in pre-listening were recommendations they gave to their students. Most commonly, they attempted to give the general idea of the listening text. Then, the information about the listening task requirements were clearly explained by the teachers to their students. This might prepare students to the listening process as it lets them to be aware of what they will do in listening process. Also, the teachers coded that they asked their students to revise the key words shared in pre-listening tasks in the textbooks. By focusing on the words, teachers might draw a picture about the listening text. It might encourage the students predict the topic of the listening text and make their minds ready for listening cognitively. Regarding the procedures followed by the teachers, Table 2 shows that explaining the structures they learnt in pre-listening process, increasing students' self-confidence, reading aloud the new words in the text before listening, and asking questions about what they understand from the listening text had little resonance among the language teachers.

In Table 3 below, the justifications that language teachers follow in their practices were ordered. It was found that checking what their students understand from the listening text, letting them listen to the text carefully, taking their attention to important parts, and making them know which steps they need to follow were the most important aspects of the language teachers' beliefs in the teaching of listening. The most frequently cited justification was to check what the students understand from the listening text. It shows that language teachers have attempts to learn whether the students grasp the gist of the listening text or not. This reflects their attempts to understand the effectiveness of the listening process. On the other hand, the language teachers stated that they reminded new words, which was the least desirable justification. However, they had reported that they had given importance revising key words in pre-listening in the procedures part, which reflects the inconsistency in their beliefs and practices. Similarly, there was a little evidence of the language teachers' statement that was "they did brainstorming". They had reported that they believed in asking students their opinions/ideas or experiences about the topic with a high importance rank in the procedures (in Table 2).

Table 2. Procedures that language teachers do in listening (The most important procedures 1 to 4)

Procedures (What they do)	Procedure 1	Procedure 2	Procedure 3	Procedure 4	TOTAL
Giving general information about the topic	44	23	12	11	90
Revise key words in pre-listening	26	29	13	4	72
Asking students their opinions/ideas or experiences about the topic	40	26	13	0	79
Explaining the rules of listening task	11	34	29	10	84
Explaining what they are doing to the students	13	2	21	5	41
Explaining the meaning of new words	17	9	3	1	30
Taking listeners' attention with warm-up	15	1	6	4	26
Talking about listening (pre-listening)	42	8	3	1	54
Guessing what they are doing (pre-listening)	28	16	2	0	46
Focusing on listening (during-listening)	6	5	30	2	43
Expressing their ideas about text (post-listening)	0	0	21	24	43
Explaining the structures they learnt (pre-listening)	3	0	0	0	3
Listening to the text twice	16	22	0	0	38
Listening to the text till the end	0	2	0	4	6
Checking the answers of the questions	0	12	10	0	22
Focusing on content words while listening	0	5	3	0	8
Increasing students' self-confidence (let them believe they can do)	0	0	2	0	2
Directing students' attention on new words in the listening text	11	26	10	2	49
Studying new words	1	2	2	4	9
Read the pre-listening questions	10	25	0	4	39
Underlining some words in the text and write them to the board before listening	8	3	2	0	13
Read aloud the new words in the text before listening	0	2	0	0	2
Asking questions about what they understand from the listening text	0	0	0	4	4

Note: The numbers of times procedures reported by teachers

Table 3. Justifications that language teachers follow in listening practices (The most important procedures 1 to 4)

Justifications (Why they do)	Justification 1	Justification 2	Justification 3	Justification 4	TOTAL
To make students familiar with the subject	17	15	10	5	47
To take their attention to important parts	18	22	9	6	55
To do brainstorming	2	0	3	8	13
To make them know which steps the students need to follow	3	16	2	34	55
To check their knowledge	2	20	8	0	30
To prepare students for listening	3	12	13	1	29
To motivate students for listening task	0	0	0	4	4
To encourage students to express their ideas about the listening text	11	6	4	3	24
To encourage them for prediction	5	28	10	2	45
To make students familiar with vocabulary and structures	3	6	5	14	28
To make students concentrate on the listening	23	14	2	0	39
To let students listen to the text carefully	42	29	11	5	82
To check what they understand from the listening text	46	23	15	7	91
To remind new words	11	0	0	1	12

As seen in Table 4, the language teachers' responses for pre-, during-, post-listening process were ordered by the percentages that were ranked from 1 (never) to 4 (always). It was found that language teachers sometimes give pupils vocabulary items that will be used in the passage (% 57). Before listening, they always remind learners of vocabulary linked to the topic (% 46). They sometimes ask pupils to predict vocabulary they might hear (e.g. verbs, nouns) (% 46). They sometimes ask pupils to discuss possible answers to the questions (% 58). It is seen that there was a disagreement in the teachers' beliefs stated in the procedure part (in Table 2) and their responses for pre-listening process.

The findings also revealed that the language teachers sometimes ask pupils to verify their predictions while listening (% 55). They frequently ask pupils to focus on key words while listening (% 59). However, they did not write down any statement and/or claim about their beliefs on verifying the prediction and focusing on key word while listening in the procedures and justification part. They stated their beliefs and practices about focusing on key words only in pre-listening process.

Additionally, during the listening, language teachers stated that they frequently pause the listening track/CD when the passage is played for the first time (% 65) and for the 2nd time (% 77). However, they did not share their beliefs about what they do in listening and why they do so in the procedures and justification part even though they categorized both statements as frequently done during the listening. Furthermore, they stated when they pause the listening track/CD, they never try to pause it at the end of each speaker (% 88), to pause it at the end of each question (% 90), to pause it at the end of natural speech boundaries (% 95). At this point, it was seen that there was a consistency as these statements were never applied in teaching of listening, more specifically during-listening process.

The findings of this research revealed the divergence between language teachers' beliefs about listening (procedures, what they do) and their practices in their classrooms (justification, why they follow). While the teachers' beliefs reflected the effective completion of listening tasks referring to the pre-, during and post-listening procedures; however, their practices do not emphasize on the use of metacognitive listening strategies and metacognitive awareness to make learners autonomous. The findings of the previous research outlined that there should be a meaningful relation between the practical areas that refer to how they believe listening should be taught (Borg & Burns, 2008) and the formal areas what academic studies suggest in theory.

As Likert-type items listed in Table 5 below, it was found that there was a high level of agreement at the item which was the language teachers' teaching learners how to listen more effectively. This agreement highly reflected language teachers' beliefs about carrying out the listening activities in their instructions. Also, with the items referring to how learners should cope with the new words and where the language teachers' observe major difficulties in the listening, the issues about bottom-up and to-down processing in listening were explored. For example, the use of context as a top-down strategy received a high level support while using context and linguistic knowledge which are bottom-up strategies was at a

Table 4. Teachers' responses for pre-, during-, post-listening process (percentages)

Statements followed by the language teachers in listening process	1	2	3	4
Before listening, I remind learners of the vocabulary linked to the topic.	1	20	33	46
Before listening, I give pupils the vocabulary items that will be used in the passage.	14	57	25	4
Before listening, I ask pupils to predict the vocabulary they might hear (e.g. verbs, nouns)	13	46	28	13
Before listening, I ask pupils to think of the ideas/facts etc. That might be discussed in the passage.	11	33	40	16
Before listening, I ask pupils to discuss the possible answers to the questions.	27	58	8	6
While listening, I ask pupils to verify their predictions.	2	55	30	13
While listening, I ask pupils to focus on key words.	0	27	59	14
During the listening, I pause the listening track/CD when the passage is played for the first time.	4	12	65	19
During the listening, I pause the listening track/CD only when the passage is played for the 2nd time.	0	5	77	23
When I pause the listening track/CD, I try to pause it at the end of each speaker.	88	10	2	0
When I pause the listening track/CD, I try to pause it at the end of each question.	90	6	4	0
When I pause the listening track/CD, I try to pause it at the end of natural speech boundaries.	95	5	0	0
After listening, I tell pupils what the answers are (e.g. Picture A, Picture B, etc.)	30	44	16	10
After listening, I ask pupils what answers they put (e.g. Picture A, Picture B, etc.)	2	10	57	30
After listening, I ask pupils to answer using the target language words/phrases (e.g. "He went to the cinema")-(in target language)	4	45	42	9
After listening, I ask pupils how they felt about the task.	15	40	33	12
After listening, I ask pupils what they did to complete the task.	26	56	17	1
After listening, I advise pupils how to deal with difficulties next time.	22	48	22	8
After listening, I ask pupils to use language/structures used in the passage in a productive follow-up task.	9	49	39	3

Note: 1 (Never), 2 (Sometimes), 3 (Frequently), 4 (Always)

level of agreement over 3. This shows that the language teachers did not have high agreement in the use of context, and they preferred to take the attention on the linguistic issues such as (focusing on) vocabulary and grammar. It can be argued that teachers' beliefs about teaching listening were shaped by their concepts toward authentic matcrials. Teachers may believe in a certain approach in teaching listening; however, classroom dynamics and teaching materials might inhibit them to reflect their beliefs into their instruction. Lastly, the language teachers had remarkably low level of agreement in classroom discussions after listening. This reveals that they did not give importance to the metacognitive awareness, the use of metacognitive strategies, critical discussions after listening. However, when students critically discuss on listening comprehension questions with their classmates after listening, they internalize the topic (Field, 2008; Graham et al., 2010; Irgin & Erten, 2020). Also, Chen (2015) suggested the instruction of listening strategies both implicitly and explicitly and the use of authentic listening materials in EFL classes.

To explore whether the language teachers use some specific listening activities in their classes, a set of items was shared with the teachers and they were expected to respond each item regarding they do or not. It was found that the language teachers asked their students to listen out for key words, to listen out for the gist of the passage, to complete gap-fills, to follow a transcription while listening, and to think about how to work out/deal with unknown words with high frequency ratios (over % 80). All of these activities carried out in listening require the use of metacognitive strategies in listening.

On the other hand, the areas less-frequently applied by the teachers were to transcribe (i.e. write down in the target language) everything the students hear, to listen cooperatively (in pairs), to keep a listening

Table 5. Teachers' beliefs about listening (frequencies)

Statements	1	2	3	4	5
When learners do not understand a word, they should work out its meaning from the context.	20	58	10	5	0
When learners do not understand a word, they should work out its meaning from the words/phrases that precede or follow the unknown word.	12	63	17	4	0
When learners do not understand a word, they should work out its meaning from their linguistic knowledge (e.g. knowledge of L2/L1 (first/second language) vocabulary, grammar).	11	48	27	7	3
It is more important for pupils to use the context of the passage to understand than to listen carefully to what is actually said.	10	41	28	11	6
The main difficulties for pupils in listening arise from their lack of vocabulary.	5	22	43	24	2
Pupils' main problems lie in the difficulty they have in identifying where word/phrase/sentence boundaries are.	11	47	29	6	3
The main difficulties for pupils in listening arise from lack of grammatical knowledge.	3	10	39	43	1
The main difficulties for pupils in listening arise from lack of background knowledge about the topic of the passage.	0	14	29	44	9
After listening, pupils should discuss how they completed the listening activity.	5	25	42	22	2
After listening, pupils should discuss how they felt about the listening activity.	8	45	23	17	3
I introduce new vocabulary to pupils orally as individual items.	25	47	14	8	2
I introduce new vocabulary to pupils orally in connected speech.	11	39	26	17	3
It is possible to teach learners how to listen more effectively.	65	23	6	2	0
Listening should be taught differently depending on whether pupils are in their first or subsequent years of learning.	60	28	4	3	1

Note: From agree strongly (1) to disagree strongly (5)

log about how they approach listening tasks (i.e. what they do), to keep a listening log about how they feel about listening, and to use peer-designed listening activities, with high frequencies (over % 88). The listening activities carried out with high-frequency ratios are important listening task requirements to assess learners' listening comprehension (Brown & Abeywickrama, 2019; Irgin, 2021). Similar to the findings of this research, Mandasari (2016) claimed that language teachers believed in the importance to select and use listening materials and activities carefully. Also, their beliefs were on introducing the English language in the real context, and providing authentic materials to motivate student for listening. The results of Emerick's (2019) research indicated that language teachers had positive beliefs on the explicit listening strategy instruction which developed learners' competences in L2 listening. However, very few of the teachers reported the implementation of metacognitive strategy instruction.

Table 6. The listening activities teachers use with their students

I ask pupils to:	Yes, I do this	No, I never do this
Listen out for words they predict they may hear.	73	23
Transcribe (i.e. write down in the target language) everything they hear.	8	88
Listen out for specific details.	93	3
Listen out for the gist of the passage.	92	2
Listen out for key words.	95	1
Complete gap-fills.	90	5
Match what is heard to a written paraphrase.	59	37
Recognize groups of words that occur together.	28	68
Listen out for marker phrases, e.g. "For example", "First of all" (in the target language).	73	23
Listen out for/distinguish between individual sounds.	48	48
Recognize/listen out for words from different word classes (e.g. verbs, adjectives).	61	35
Listen for verb endings.	60	36
Listen out for how individual words change in connected speech.	32	64
Identify the tone of voice/emotion.	58	38
Distinguish one speaker from another.	49	47
Listen to a text read out by me.	79	17
Focus on intonation patterns.	55	41
Make sound-spelling links.	56	40
Follow a transcription while listening.	86	10
Think about how to work out/deal with unknown words.	83	13
Identify word boundaries.	65	31
Listen cooperatively (in pairs).	7	89
Keep a listening log about how they approach listening tasks (i.e. what they do).	0	96
Keep a listening log about how they feel about listening.	0	96
Use peer-designed listening activities.	0	96

Note: Frequency

The findings of this research indicated that the language teachers' beliefs about instructional listening in a foreign language were inconsistent with their practices of teaching listening to their students, which is in line with the findings of previous research focusing on teachers' beliefs and their practices in various EFL contexts (Emerick, 2019; Gao & Liu, 2013; Li & Renandya, 2012). Karimi and Nazari (2017) indicated that there was no significant relationship between Iranian EFL teachers' beliefs about listening instruction and their listening instructional practices. In their interviews, teachers stated that they had obstacles to actualize their listening beliefs besides some other impediments, which might suggest that teacher beliefs and practices do not always correspond. The findings of Li and Renandya's (2012) research revealed that Chinese EFL teachers preferred the bottom-up approach in L2 listening rather than the integration of both bottom-up and top-down approaches in teaching L2 listening, which is similar to the findings of this research. Also, Gao and Liu (2013) found that college English teachers emphasized the critical role of listening skill and teaching listening. Also, they highlighted both micro and macro levels in listening, but they did not follow their beliefs in teaching practices faithfully. What they applied in practices and what they believed in differentiated from each other.

CONCLUSION

The results of the descriptive statistics and coding approach indicate that the teachers' beliefs about instructional listening in a foreign language are inconsistent with their practices of teaching listening to their pupils. The teachers' beliefs reflect the effective completion of listening tasks referring to the pre-, during and post-listening procedures; however, their practices do not emphasize on the use of metacognitive listening strategies and metacognitive awareness to make learners autonomous. The implication of the study is on the necessity of professional development for teachers to develop their students' listening skill.

This research has presented a wide-range of teachers' beliefs and teaching of listening practices in Turkey. It is significant as it addresses a gap in the field of language teachers' cognition, and the view on the complex relationship between EFL teachers' beliefs and listening pedagogies. This study provides evidence of following the comprehension-based approach in teaching of foreign language listening as discussed by Field (2008), and Liao and Yeldham (2015). The foreign language teachers have concerns on the performance of the listening tasks presented the course-books and following up the curriculum in a very limited class duration. At this point, the majority of EFL teachers report that they are in the middle of the decisions; implementing the listening tasks stated in their course plans, which might be named as task completion, or focusing on the listening task comprehension in a deeper sense. However, Borg (2010)'s statement to the teachers in such complexity, "make deeper sense of their work" (p.414) might be enlightening for teacher development and quality of language education.

This current research is limited in three key respects. First, the primary concern was language teachers' beliefs about listening in Turkish context, there is a need for further research designed in a large-scale to examine how language teachers' beliefs can contribute to their teaching listening. Second, since it was limited to language teachers teaching listening at secondary schools in Turkey, further investigations into other school setting such as primary schools and high schools will help to gain aa better understanding of listening instruction. Finally, this study focused on language teachers having various years of experiences. Future research might place more emphasis on what experienced language teachers' beliefs about listening are and how they teach listening in comparison with novice teachers.

REFERENCES

Barcelos, A. (2003). Researching beliefs about SLA: A critical review. In P. Kalaja & A. M. F. Barcelos (Eds.), *Beliefs about SLA: New research approaches*. Kluwer Academic Publishers. doi:10.1007/978-1-4020-4751-0_1

Barrot, J. S. (2016). Examining the teaching beliefs and practices of experienced ESL teachers: A sociocognitive-transformative perspective. *The Southeast Asian Journal of English Language Studies*, *22*(1), 153–163. doi:10.17576/3L-2016-2201-12

Borg, S. (2003). Teacher cognition in language teaching: A review of research on what language teachers think, know, believe, and do. *Language Teaching*, *36*(2), 81–109. doi:10.1017/S0261444803001903

Borg, S. (2006). *Teacher cognition and language education*. Continuum.

Borg, S. (2007). Research engagement in English Language Teaching. *Teaching and Teacher Education*, *23*(5), 731–747. doi:10.1016/j.tate.2006.03.012

Borg, S. (2015). *Teacher cognition and language education*. Bloomsbury Academic.

Borg, S. (2019). *Language teacher cognition: Perspectives and debates*. Springer.

Borg, S., & Burns, A. (2008). Integrating grammar in adult TESOL classrooms. *Applied Linguistics*, *29*(3), 456–482. doi:10.1093/applin/amn020

Brown, H. D., & Abeywickrama, P. (2019). *Language assessment: Principles and classroom practices*. Pearson.

Chen, A. (2015). The impacts of listening strategy instruction on strategy use and listening performance of EFL learners. *International Journal on Studies in English Language and Literature*, *3*(5), 75–87.

Eggen, P. D., & Kauchak, D. P. (2016). *Educational psychology: Windows on Classrooms*. Pearson Education.

Ellis, R. (2008). *The study of Second Language Acquisition*. Oxford University Press.

Emerick, M. R. (2019). Explicit teaching and authenticity in L2 listening instruction: University language teachers' beliefs. *System*, *80*, 107–119. doi:10.1016/j.system.2018.11.004

Field, J. (2008). *Listening in the language classroom*. Cambridge University Press.

Gao, Q., & Liu, L. (2013). Investigating college English teachers' beliefs in listening teaching and their relationships with teaching practices. *Foreign language word*. Retrieved 1 May, 2021 from https://en.cnki.com.cn/Article_en/CJFDTotal-WYJY201302004.htm

Gao, X., & Ma, Q. (2011). Vocabulary learning and teaching beliefs of pre-service and in-service teachers in Hong Kong and mainland China. *Language Awareness*, *20*(4), 327–342. doi:10.1080/09658416.2011.579977

Graham, S., Santos, D., & Francis-Brophy, E. (2014). Teacher beliefs about listening in a foreign language. *Teaching and Teacher Education*, *40*, 44–60. doi:10.1016/j.tate.2014.01.007

Graham, S., Santos, D., & Vanderplank, R. (2010). Strategy clusters and sources of knowledge in French L2 listening comprehension. *International Journal of Innovation in Language Learning and Teaching*, *4*(1), 1–20. doi:10.1080/17501220802385866

Irgin, P. (2021). Testing and assessment of listening, listening test task types, and sample test items. In S. Çelik, H. Çelik, & C. Coombe (Eds.), *Language assessment and test preparation in English as a foreign language (EFL) education* (pp. 91–115). Vizetek.

Irgin, P., & Erten, I. H. (2020). Exploring the role of strategy instruction: Young learners' listening performance and strategy use. *Eurasian Journal of Applied Linguistics*, *6*(3), 415–441.

Karimi, M. N., & Nazari, M. (2017). The congruity/incongruity of EFL teachers' beliefs about listening instruction and their listening instructional practices. *The Australian Journal of Teacher Education*, *42*(2), 62–80. doi:10.14221/ajte.2017v42n2.5

Kumaravadivelu, B. (2012). *Language Teacher Education for a Global Society: A Modular Model for Knowing, Analyzing, Recognizing, Doing and Seeing*. Routledge. doi:10.4324/9780203832530

Li, L. (2020). *Language teacher cognition a sociocultural perspective*. Palgrave Macmillan. doi:10.1057/978-1-137-51134-8

Li, W., & Renandya, W. A. (2012). The effective approaches to teaching listening: Chinese EFL teachers' perspectives. *The Journal of Asia TEFL*, *9*(4), 79–111.

Liao, C. Y., & Yeldham, M. (2015). Taiwanese high school EFL teachers' perceptions of listening instruction. *The Asian Journal of Applied Linguistics*, *2*(2), 92–101.

Macaro, E., Graham, S., & Woore, R. (2015). *Improving foreign language teaching: Towards a research-based curriculum and pedagogy*. Routledge. doi:10.4324/9781315771229

Mandasari, B. (2016). An analysis of teachers' beliefs toward authentic materials in teaching listening. *Teknosastik*, *14*(1), 19–25. doi:10.33365/ts.v14i1.82

Miles, M. B., & Huberman, A. M. (1994). *Qualitative data analysis: An expanded sourcebook*. Sage.

Phipps, S., & Borg, B. (2009). Exploring tensions between teachers' grammar teaching beliefs and practices. *System*, *37*(3), 380–390. doi:10.1016/j.system.2009.03.002

Sah, F. M., & Shah, P. M. (2020). Teachers' beliefs and practices in teaching listening. *Creative Education*, *11*(02), 182–195. doi:10.4236/ce.2020.112013

Schultz, K., Jones-Walker, C. E., & Chikkatur, A. P. (2015). Listening to students, negotiating beliefs: Preparing teachers for urban classrooms. *Curriculum Inquiry*, *38*(2), 155–187. doi:10.1111/j.1467-873X.2007.00404.x

Stipek, D. J., Givvin, K. B., Salmon, J. M., & MacGyvers, V. L. (2001). Teachers' beliefs and practices related to mathematics instruction. *Teaching and Teacher Education*, *17*(2), 213–226. doi:10.1016/S0742-051X(00)00052-4

Utami, D. N. (2016). The EFL teachers' beliefs and their teaching practices. *Journal of Languages and Literature*, *10*(2), 135–144. doi:10.19105/ojbs.v10i2.974

Vandergrift, L., & Tafaghodtari, M. H. (2010). Teaching L2 learners how to listen does make a difference: An empirical study. *Language Learning*, *60*(2), 470–497. Advance online publication. Retrieved May 5, 2021, from. doi:10.1111/j.1467-9922.2009.00559.x

Wilson, J. J. (2010). *How to teach listening*. Pearson Longman.

Yeldham, M., & Gruba, P. (2014). Toward an instructional approach to developing interactive second language listening. *Language Teaching Research*, *18*(1), 33–53. doi:10.1177/1362168813505395

Zacharias, N. T. (2005). Teachers' beliefs about internationally-published materials: A survey of tertiary English teachers in Indonesia. *RELC Journal*, *36*(1), 23–37. doi:10.1177/0033688205053480

ADDITIONAL READING

Bandura, A. (1994). Self-efficacy. In V. S. Ramachaudran (Ed.), *Encylopedia of human behaviour* (Vol. 4, pp. 71–81). Academic Press. (Reprinted in H. Friedman [Ed.], Encyclopedia of mental health. San Diego: Academic Press, 1998)

Borg, M. (2001). Key concepts in ELT. Teachers' beliefs. *ELT Journal*, *55*(2), 186–188. doi:10.1093/elt/55.2.186

Borg, S. (1998). Teachers' pedagogical systems and grammar teaching: A qualitative study. *TESOL Quarterly*, *32*(1), 9–38. doi:10.2307/3587900

Borg, S. (2009). Language teacher cognition. In J. Richards & A. Burns (Eds.), *Cambridge guide to second language teacher education* (pp. 91–101). Cambridge University Press.

Borg, S. (2011). The impact of in-service teacher education on language teachers' beliefs. *System*, *39*(3), 370–380. doi:10.1016/j.system.2011.07.009

Fishbein, M., & Ajzen, I. (1975). *Belief, attitude, intention, and behavior: An introduction to theory and research*. Addison-Wesley.

Pajares, M. F. (1992). Teachers' beliefs and educational research: Cleaning up a messy construct. *Review of Educational Research*, *62*(3), 307–332. doi:10.3102/00346543062003307

Richards, J. C. (1998). *Beyond training*. Cambridge University Press.

Wenden, A. L. (1999). Metacognitive knowledge and beliefs in language learning. *System, 27*(4), (Speacial Issue).

Woods, D. (1996). *Teacher cognition in language teaching*. Cambridge University Press.

Zheng, H. (2009). A review of research on EFL pre-service teachers' beliefs and practices. *Journal of Cambridge Studies*, *4*(1), 73–81.

Zheng, H. (2015). *Teacher beliefs as a complex system: English language teachers in China*. Springer International Publishing. doi:10.1007/978-3-319-23009-2

KEY TERMS AND DEFINITIONS

Attitude: It is "a learned predisposition to respond to an object in a consistently favourable or unfavourable manner" (Fishbein & Ajzen, 1975, p.336).

Belief: It is "a preposition which may be consciously or unconsciously held, is evaluative in that it is accepted as true by the individual, and is therefore imbued with emotive commitment; it serves as a guide to thought and behaviour" (Borg, 2001, p.186).

English as a Foreign Language (EFL): It is the context where English has no official status, it is taught in classroom setting, and the role of English in communication inside the country is very limited such as in Turkey.

Self-Efficacy: It is one's beliefs on his/her potential or capacity to cope with the prospective situations. Self-efficacy beliefs determine how people feel, think, motivate themselves and behave (Bandura, 1994).

Teacher Beliefs: It is defined as "Teachers' psychologically held thinking, conceptions and understanding about EFL teaching and learning" (Zheng, 2015, p. 14).

APPENDIX A

Background Questionnaire For Foreign Language Teachers
Name (or Pseudonym):
Age:
Sex:
Graduation (Department/University): BA_ MA/MEd_ PhD_
Mother tongue:
How long have you been working as a foreign language teacher?
Language(s) you usually speak at home:
Language(s) you usually speak with your pupils at school:
Language(s) you usually speak with pupils out of the school:
Language(s) you are now learning:
Do you generally enjoy doing L2 listening in your class? (Circle one) Yes No
How important is it for you to have pupils proficient in L2 listening? (Circle one) Very important Important Not so important
Do you do anything with your pupils in the EFL class: Before listening to the recording? (Circle one) Yes No If yes, please list: a) During listening to the recording? (Circle one) Yes No If yes, please list: b) After the listening to the recording? (Circle one) Yes No If yes, please list:
What difficulties do you have when you do/teach/practice listening in English in your class?

APPENDIX B

Questionnaire_ Teachers' Beliefs About Listening in a Foreign Language

In all of the following questions, we are interested in how you teach listening at secondary school level. Please, base your answers on how teach pupils in that key stage. There are no right or wrong answers!

1. In the table below, please list the MOST IMPORTANT procedures (3 or 4) that you usually follow when you ask learners to listen to a recording in the classroom (Note: There is no need to rank the procedures in any way).

PROCEDURES (What you do and/or what students do)	JUSTIFICATION (Why do you do this? Why is this done?)

2. How often do you work on listening in class, on average? Please tick one option.
   Less than once a week ______
   Once a week ______
   Two or three times a week ______
   More than three times a week ______
3. In your opinion how does listening at secondary school level compare with the teaching of other skills in terms of difficulty? Number the options below to answer this question (1- the least difficult; 4- the most difficult)
   Listening______ Speaking ______ Reading______ Writing______
4. In your teaching, how much emphasis do you place on the following skills? Number the options below to answer this question (1- the least emphasis; 4- the most emphasis)
   Listening______ Speaking ______ Reading______ Writing______
5. When you ask learners to listen, how often do you do the following?

	Never	Sometimes	Frequently	Always
Before listening, I remind learners of vocabulary linked to the topic.				
Before listening, I give pupils vocabulary items that will be used in the passage.				
Before listening, I ask pupils to predict vocabulary they might hear (e.g. verbs, nouns)				
Before listening, I ask pupils to think of ideas/facts etc. That might be discussed in the passage.				
Before listening, I ask pupils to discuss possible answers to the questions.				

	Never	Sometimes	Frequently	Always
While listening, I ask pupils to verify their predictions.				
While listening, I ask pupils to focus on key words.				
During the listening, I pause the listening track/CD when the passage is played for the first time.				
During the listening, I pause the listening track/CD only when the passage is played for a 2nd time.				
When I pause the listening track/CD, I try to pause it at the end of each speaker.				
When I pause the listening track/CD, I try to pause it at the end of each question.				
When I pause the listening track/CD, I try to pause it at the end of natural speech boundaries.				

	Never	Sometimes	Frequently	Always
After listening, I tell pupils what the answers are (e.g. Picture A, Picture B, etc.)				
After listening, I ask pupils what answers they put (e.g. Picture A, Picture B, etc.)				
After listening, I ask pupils to answer using target language words/phrases (e.g. "He went to the cinema")-(in target language)				
After listening, I ask pupils how they felt about the task.				
After listening, I ask pupils what they did to complete the task.				
After listening, I advise pupils how to deal with difficulties next time.				
After listening, I ask pupils to use language/structures used in the passage in a productive follow-up task.				

6. How far do you agree with the following statements for the pupils for LISTENING? Please circle <u>one</u> number below to show your level of agreement (1=Agree strongly; 5=Disagree strongly).

When learners do not understand a word they should work out its meaning from the context.	Agree strongly Disagree strongly 1 2 3 4 5
When learners do not understand a word they should work out its meaning from the words/ phrases that precede or follow the unknown word.	1 2 3 4 5
When learners do not understand a word they should work out its meaning from their linguistic knowledge (e.g. knowledge of L2/L1 (first/second language) vocabulary, grammar).	1 2 3 4 5
It is more important for pupils to use the context of the passage to understand than to listen carefully to what is actually said.	1 2 3 4 5
The main difficulties for pupils in listening arise from their lack of vocabulary.	1 2 3 4 5
Pupils' main problems lie in the difficulty they have in identifying where word/phrase/ sentence boundaries are.	1 2 3 4 5
The main difficulties for pupils in listening arise from lack of grammatical knowledge.	1 2 3 4 5
The main difficulties for pupils in listening arise from lack of background knowledge about the topic of the passage.	1 2 3 4 5
After listening, pupils should discuss how they completed the listening activity.	1 2 3 4 5
After listening, pupils should discuss how they felt about the listening activity.	1 2 3 4 5
I introduce new vocabulary to pupils orally as individual items.	1 2 3 4 5
I introduce new vocabulary to pupils orally in connected speech.	1 2 3 4 5
It is possible to teach learners how to listen more effectively.	1 2 3 4 5
Listening should be taught differently depending on whether pupils are in their first or subsequent years of learning.	1 2 3 4 5

7. What, in your view, is the main purpose of carrying out listening tasks? Please rank the following in order of importance, with 1 for the most important reason, 5 the least important.
To extend learners' vocabulary ______
To increase learners' opportunities to practice listening ______

To teach learners how to listen more effectively ______
To assess how well learners can listen ______
To provide learners with a model of pronunciation ______
If you have another reason, please write it here:

8. Which of these listening activities do you use with pupils in different years of foreign language learning? Tick all the boxes that apply.

I ask pupils to:	Yes, in all years of their education	Yes, but only in certain years of education (please specify year)	No, I never do this
Listen out for words they predict they may hear.			
Transcribe (i.e. write down in the target language) everything they hear.			
Listen out for specific details.			
Listen out for the gist of the passage.			
Listen out for key words.			
Complete gap-fills.			
Match what is heard to a written paraphrase.			
Recognise groups of words that occur together.			
Listen out for marker phrases, e.g. "For example", "First of all" (in the target language).			
Listen out for/distinguish between individual sounds.			
Recognise/listen out for words from different word classes (e.g. verbs, adjectives).			
Listen for verb endings.			
Listen out for how individual words change in connected speech.			
Identify tone of voice/emotion.			
Distinguish one speaker from another.			
Listen to a text read out by me.			
Focus on intonation patterns.			
Make sound-spelling links.			
Follow a transcription while listening.			
Think about how to work out/deal with unknown words.			
Identify word boundaries.			
Listen cooperatively (in pairs).			
Keep a listening log about how they approach listening tasks (i.e. what they do).			
Keep a listening log about how they feel about listening.			
Use peer-designed listening activities.			

9. How often do you use listening activities from textbooks? Please tick one option:

Always______ Most of the time______ Sometimes______ Rarely______ Never______

If you have ticked any option other than "Always", please give details to explain your answer: List 2 positive aspects and 2 negative aspects of listening tasks frequently found in textbooks.

Positive aspects
Negative aspects

Chapter 11
Examination of Speaking Test Performance in Structured Group Tasks:
An Interactional Perspective

Elif Tokdemir Demirel
Kırıkkale University, Turkey

Zeynep Baser
Kırıkkale University, Turkey

ABSTRACT

This chapter explores the use of structured group tasks in the testing of speaking skills in the context of a tertiary level first year speaking course. Participants included undergraduate university students from the department of English translation and interpreting at a state university located in Central Turkey. The structured task used in the study guided students into a group discussion during the speaking exam. The discussions in the exam sessions were video-recorded and transcribed. The transcriptions were analyzed from an interactional perspective. The students were also given an open-ended questionnaire to reveal their opinions about the ins and outs of the speaking test task tried out in the study. The questionnaire consisted of two parts. The first part required students to fill out their demographic information, and the second part inquired students' opinions about the group speaking task. The results of the analysis and questionnaire responses revealed additional opportunities provided by the use of structured tasks in speaking exams.

INTRODUCTION

Testing speaking often presents a challenge on the part of teachers as there are several variables affecting students' performance, one of which is the task design. There are a number of test types (reading aloud, picture description, oral presentation, interview, role-play, etc.) which might be used in order to

DOI: 10.4018/978-1-7998-7876-6.ch011

test speaking skills, yet a majority of these tests are criticized because of reliability issues or for being too limited to measure a variety of functional and interactional patterns. Testing speaking is a multi-dimensional process and draws on a diverse set of interrelated factors such as learners' background, language knowledge and skills, general world knowledge, graders' behavior, environmental conditions, etc. The scoring behavior of graders might also vary, and thus lead to differences in learners' speaking test results. To illustrate, Polat (2020) investigated the scoring behavior of 6 expert graders. The graders evaluated 24 English language learners' video recordings including an individual and a pair discussion task for each.

A weakness of many speaking tests is that they focus excessively on the individual performance rather than the student in interaction (Başer, 2011). However, real life speaking is not based on only individual performance. To this end, Bachman & Savignon (1986) criticized the methods for failing to demonstrate the relationships between different skills as well as for failing to recognize the full context of language use and the expanded conception of language proficiency as 'communicative competence' (p.381).

Teaching speaking also entails teaching the dynamics of interaction. Some of the skills involved in using talk as interaction involve knowing how to do the following things:

Opening and closing conversations
Choosing topics
Making small-talk
Joking
Recounting personal incidents and experiences
Turn-taking
Using adjacency pairs
Interrupting
Reacting to others
Using an appropriate style of speaking (Richards, 2016, p.19).

Although speaking skill encompasses many different aspects as pointed out above, the tasks in speaking tests in English departments at universities often test individual performance of students and are usually in the form of a monologue or narration. And the studies investigating the testing of speaking skills point out that this feature of speaking test design is a weakness. Therefore, regarding the speaking tests carried out in English programs at universities, the following questions arise: Why does testing speaking not include the dynamics of interaction and the relationships between different skills? Why are students tested on their individual speaking performance in the form of a monologue or narration rather than communication?

As an alternative to the testing of speaking with monologue, narration tasks, etc., structured tasks facilitate task performance and are associated with greater fluency (Skehan, 2001). Drawing on from these opinions, in this study, students' speaking skills were tested with a structured task in an English speaking course. The purpose of the study was to find out how closely the conversation in the group exam tasks resemble the features of real-life interaction, and also to find out whether this type of testing speaking skills creates positive attitudes towards testing of speaking, and whether it helps reduce anxiety or not.

To this end, the research questions investigated in the present study are as follows:

1. What are students' perceptions about being evaluated based on their performance within groups as compared to individual performance evaluation in speaking?
2. What are the examples of interactional patterns (e.g. turn-taking, sequence organization, turn design and repair) observed with upper-intermediate groups of learners?
3. What strategies (e.g. questioning, challenging, summarizing and scaffolding) do upper-intermediate students often attempt to use to maintain the flow of communication within a group?

The next section presents an overall background of research on speaking tests, challenges the students and the raters have to deal with, and testing speaking through group discussions.

BACKGROUND

Speaking is one of the most important language skills. However, as Foot (1999) asserted, it is also the most difficult skill to test. Different types of speaking tests (e.g. picture story, oral presentation, interview, etc.) exist, yet most of them are often criticized because of reliability issues or for being too limited to measure a variety of functional and interactional patterns.

To illustrate, one of the most oft-used speaking tests, which requires teacher-student interaction, is a direct interview. As Underhill (1987) pointed out, it is challenging to ensure a reliable assessment of students' speaking skills with this method considering the fact that their responses might vary depending on the subject they will speak about, the role of the teacher as the assessor, as well as their background and personality. Instead, the tests where students get prepared in advance might be a better alternative in order to better understand the organization of language and ideas since the content will be more manageable and thus everyone will have something to say (Saville and Hargraves, 1999). Another type of speaking test is the pre-arranged information gap where an information gap between students or between a student and an interviewer is deliberately created, such as learner-learner description and picture story. Even though these methods might demonstrate students' oral proficiency, as Underhill (1987) highlighted, they are limited to a specific communication type, which is 'the transfer of factual information' (p.32). Furthermore, as Weir (2005) indicated, although describing something might be an important operation in some contexts such as oral presentations, talks or lectures, this type of test does not give sufficient information about the students' ability to interact orally or negotiate meaning. Apart from these types of speaking tests, there are so-called "mechanical / entirely predictable tests", the examples of which include reading aloud, sentence completion and transformation, translating / interpreting, etc. (Underhill, 1987, p.33). These tests are quick and students' responses can be scored easily. However, they are not authentic and communicative due to their complete predictability. In many speaking tests, the task is designed as a monologue; however real life speech is rarely in the form of a monologue; in contrast it happens with the collaboration of speakers (Bosker et al., 2014). In this sense, if one of the purposes of a speaking test is to measure fluency, then the setting should involve more than one speaker and should allow for interactional features to take place. As McCarthy (2010) noted, "fluency also involves the ability to create flow and smoothness across turn-boundaries and can be seen as an interactive phenomenon in discourse" (p. 1).

In order to ensure reliable testing of speaking skills, more communicative and interaction-based methods are recommended. To this end, student-student interaction might be considered one of the most effective types of speaking tests because it is easy to measure a number of functional and international patterns in pair or group work. For instance, Saville and Hargraves (1999) listed some of them as follows: turn-taking, exchanging information, justifying opinions, agreeing / disagreeing, suggesting and speculating. Underhill (1987) explained that this type of speaking test would be particularly beneficial for students by stating that "there were no victims, only active agents" (p.44). Weir (2005) emphasized other strong points of student-student interaction as follows: authentic material use, purposeful but unpredictable communication. Furthermore, he added that improvisational skills might be also observed in student-student interaction (e.g. check on understanding, clarify by summarizing, indicate understanding by gestures). Ducasse and Brown (2009) evaluated this type of speaking test from a pragmatic perspective and elaborated on the economic benefits of testing. Accordingly, this type of speaking tests might be time and cost efficient since students are tested together and the raters will evaluate two or more students simultaneously (p.424). Even though student-student interaction either in pair or group work is recommended in the relevant literature, it is essential to be aware of dynamics of this testing method in order to ensure the reliability of the speaking test.

An alternative to individual testing of speaking, group speaking tests allow learners to demonstrate their speaking skills in "appropriate, contextualized and communicative language use" (Bachmann, 1990, p. 84). To start with, Stephenson and Hall (2021) investigated how high-scoring learners managed their talk and performed task requirements in group speaking tests. It is not possible to say that students have interactional competence simply assessing their independent performance as McNamara and Young (1998, as cited in May, 2009) stated, and thus testing speaking through a method requiring student-student interaction is essential, yet it is important to be aware of the dynamics of this testing which might influence students' speaking skills.

MAIN FOCUS OF THE CHAPTER

Testing speaking through interaction between students in pair or group work enables students to demonstrate real life interactional abilities (Gan, Z., Davison, C. & Hamp-Lyons, L., 2009). As Weir (2005) indicated, this type of test allows students to develop the task on their own. Therefore, the power is not in the hands of one person, in other words an assessor, anymore. Instead, each student has the responsibility for managing the conversation. To this end, Jacobs and Ochs (1995, p.177) emphasized "there is a distributed responsibility among interlocutors for the creation of sequential coherence, identities, meaning and events" (as cited in May, 2009, p. 398). Additionally, Ducasse and Brown (2009) reported that students continuously shifted forth and back between the role of listener and speaker in this type of test. In this regard, they said "namely the equal flow of the conversation as the two interactants move equally between speaker and listener role, and participate equally in the management of the interaction" (p.440). The word 'equal' is noteworthy in their quote considering the fact that student-student interaction tests are often criticized for not giving equal opportunity to each student to demonstrate their language skills and having the risk of one student being more dominant. Nevertheless, Brooks (2009) highlighted the fact that sharing the responsibility is vital and students' performances are dependent upon each other in this type of test. To this end, she stated, "the participant may exercise differential responsibility but the process and the product is a joint construction" (p.342). Thus, Saville and Hargreaves (1999) believed that students felt more comfortable and secure regarding the fairness of assessment, as their score did not depend upon only one person. On the other hand, assessors should do still careful (Foot, 1999). As Brooks (2009) warned, "the co-constructed nature of interaction in a test situation presents challenges to language testers in terms of rating and fairness as test-takers' and co-participants' performances are dependent upon each other" (p.343).

Galaczi (2008) studied the interactional patterns of pairs in the Cambridge First Certificate of English. He identified the main interactional patterns as follows: collaborative, parallel, and asymmetric. The first one, collaborative interaction is the one in which both participants have an equal role and complete the task given to them mutually. Fisher (1996) indicated that interaction becomes like real life communication when managed collaboratively (as cited in Gan, Z., Davison, C. & Hamp-Lyons, L., 2009). Most of the other studies also showed that collaborative interaction leads to a higher performance in the oral proficiency tests. To this end, Galaczi (2008) reported that the students received higher scores in collaborative interaction while they received lower scores in parallel interaction, in which the students performed more independently. It is also found out that lower-level students had positive results from a collaborative interaction with higher-level students, and higher-level students were not affected negatively when they worked with lower-level students. Furthermore, Davis (2009) found out that lower proficiency level students produced more language (words) when they worked collaboratively with higher proficiency level students. On the other hand, Davis (2009) argued that proficiency level had no or little influence. He was of the opinion that more language production (words) does not mean improved performance in terms of accuracy, fluency or discourse management. In asymmetric interaction, it is seen that there is a dominant speaker and a passive speaker. Asymmetric dyads are potentially the most problematic from an assessment perspective and student-student interaction tests are mostly criticized because of this problem (Galaczi, 2008). In this regard, Ducasse and Brown (2009) suggested that there were certainly different management abilities depending upon language proficiency level. Weir (2005) says "if there is a large difference in proficiency between the two, this may influence performance and judgments made on it" (p.153). May (2009) touches upon asymmetric interaction in her and indicates "the consequences for a candidate involved in an asymmetric interaction are very real, not simply a matter of rater perception" (p.416).

To sum up, there have been a number of issues on the dynamics of interaction in student-student speaking tests. The present study aims to better understand how this type of speaking test reflects real life speaking skills and whether students have a positive attitude towards structured group tasks in terms of testing environment and scoring. Thus, the main focus of this chapter is to discuss the use of structured tasks for testing students' speaking skills based on the application of such a structured task. The structured task used in the study guided students into a group discussion. Accordingly, students were expected to encourage each other to participate. The discussions were video-recorded and transcribed verbatim using the Jefferson Coding system (see Appendix A) used in Conversational Analysis. Besides, students filled out an open-ended questionnaire and shared their demographic information, language proficiency and experience of speaking tests, and their opinions on being tested individually or in groups.

SOLUTIONS AND RECOMMENDATIONS

In line with the research investigating more interactive ways of testing speaking, we argue that speaking exam tasks should be designed in a way to include all facets of talk as interaction such as knowing how to open and close a conversation, making small-talk joking, turn-taking and reacting to others (Richards, 2016). Research on testing speaking in the last decades indicates that test takers perform better when provided opportunities for collaborative interaction (Galaczi, 2008; Davis 2009). Although collaborative interaction in test situations creates challenges in terms of fair evaluation, a carefully designed task in which all roles and limitations are determined would be helpful in addressing this challenge. Therefore, it is suggested that in order to decrease students' anxiety in the testing of speaking, to improve collaboration between students, and to make the evaluation of speaking multi-faceted, tasks, which create interactional opportunities, are more advantageous compared to individual performance tasks.

MATERIALS AND METHODS

This section is divided into three subheadings: (i) participants, (ii) speaking exam group task, and (iii) open-ended questionnaire. The information about the participants of this study and the instruments used to collect the relevant data as well as the data analysis procedure for each instrument are explained below.

Participants

Participants included undergraduate university students from the department of English translation and interpreting at a state university located in Central Turkey. A total of 17 groups (4 students in each group) participated in the speaking test. The questionnaire was responded to by 33 of the students who took the speaking test. Participants were between the ages of 18 and 23. The mean age was 19. In total, 20 male and 13 female students took part in the study. The students had either completed a compulsory English preparatory program or passed the proficiency exam before they began their education at the department. Thus, they were expected to have at least upper-intermediate level of English, B2 according to CFR. Nonetheless, a self-assessment grid, which is shared by the Council of Europe in Turkish, was also delivered. Accordingly, the number of the students reported their level of English is as follows: 17 C1, 12 B2, 3 C2, and 1 Not Specified. All of the students took part in the study on a voluntary basis.

Speaking Exam Group Task

In the speaking exam, the students in the 1st year speaking class were divided into groups of four. There were a total of 68 students who took the speaking exam. The exam was carried out in a meeting room and groups of four were taken into the exam room one by one. All of the exam sessions were video recorded. Each exam session took about 15 to 20 minutes. The students were first given instructions about the exam, and during the exam the teacher did not interfere with the group discussion. The instructor gave the groups handouts (Appendix A) with a set of questions: four questions for each group member on a given topic. The students were coded as Student A, B,C and D. They were then instructed to ask and answer questions on their group topic, starting with Student A asking his/her questions to Student B and so on. The teacher also instructed students that when questions were being asked and answered, if one of the other group members wanted to make a contribution, they could step in. In the meantime, the teacher just listened and took notes on a speaking exam grading rubric.

Randomly selected portions of the exam sessions, namely 8 sessions, were transcribed for analysis. The transcriptions of the selected exam sessions were analyzed globally in terms of interactional structure to examine the features of turn-taking, sequence organization, turn design and repair (Hoey & Kendrick, 2018). The purpose of the analysis was to determine to what extent the structured task helped students to overcome speaking test anxiety and what extent the spoken production during the exam reflected characteristics of real-life interaction.

Open-Ended Questionnaire

After the speaking exam, the students were given an open-ended questionnaire to reveal their opinions about the ins and outs of the speaking test task tried out in the study. The questionnaire consisted of two parts. The first part required students to fill out their demographic information (e.g. full name, gender, age, languages, etc.) whereas in the second part, the students were asked 9 open-ended questions regarding their previous experience of English language learning, types of speaking tests they had been exposed to before, self-assessment of their speaking skill, their opinions on individual and group speaking tests, and their experience in the last speaking test.

The answers to the questions 1-9 were read meticulously. Questions 1, 3, and 4 required only short answers whilst the students were supposed to share their experiences and opinions in Question 2, 5, 6, 7, 8, and 9 (see Appendix C). While analysing the data obtained through the questionnaire, the students' demographic information and their short answers to Question 1,3 and 4 were entered into an excel file. The responses to the other questions were coded and interpreted, and they were also entered into the same excel file to capture a general understanding.

Below is a detailed analysis of the questionnaire results under three subheadings: (i) perceived usefulness of prior English language learning for speaking, (ii) previous testing methods of speaking and self-assessment, and (iii) general attitude towards being tested individually or in groups.

RESULTS

This section is divided into two subheadings: (i) analysis of the open-ended questionnaire and (ii) analysis of group discussions from an interactional perspective. The results obtained from the data are presented in detail.

Analysis of the Open-Ended Questionnaire

The first part of the questionnaire aimed at collecting information on the demographics and English background of the students. The native language of the students was Turkish with only one student from Kyrgyzstan whose native language is Turkmen language. The mean age of the students was 19 and there were 20 female and 13 male students. In terms of English proficiency, according to the self-reported fluency levels, 17 students had a level of C1 (52%), 12 had a level of B2 (36%) and 3 had a level of C2 (9%) and 1 student did not report a proficiency level. The students reported they have been learning English for approximately 10 years.

Perceived Usefulness of Prior English Language Learning for Speaking

In the questionnaire, the students were also asked about the perceived usefulness of previous English education they received for improving speaking skills. As a response to this question, it was seen that a majority of the students did not perceive their previous English education as useful for improving speaking skills (n=21, 64%) as shown in Figure 1.

Figure 1. Perceived Usefulness of Prior English Education for Improving Speaking

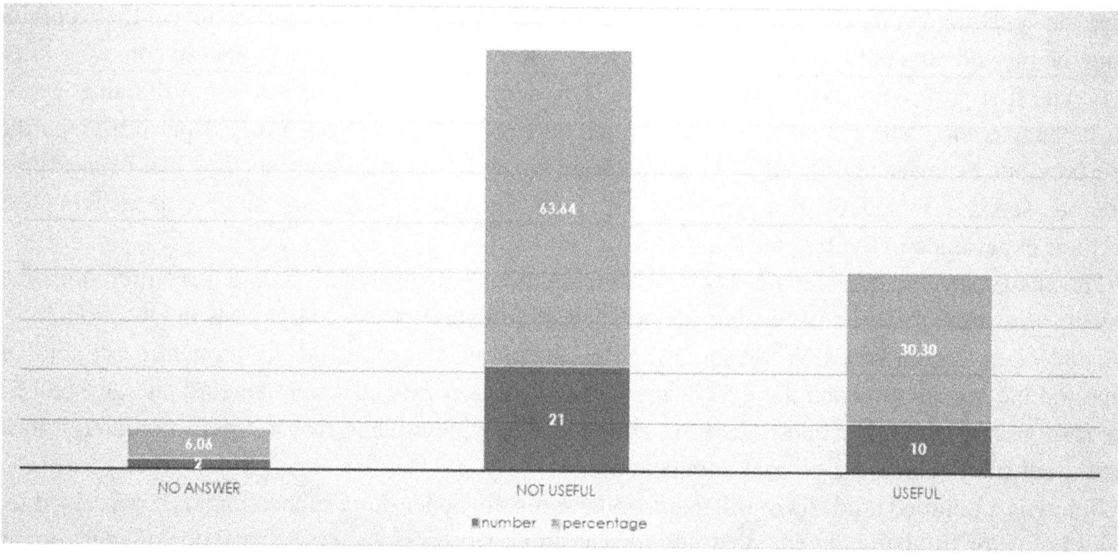

Among the 10 students (30%) who perceived English learning useful for speaking skills were those who took private courses, those who thought that only high school education was useful and 1 student who is from Kyrgyzstan. A majority of the students also reported that they had attended the Intensive English program of the university and that they had taken a separate speaking class of at least 2 hours per week (n=25, 76%).

Previous Testing Methods of Speaking and Self-Assessment

In order to examine how the English speaking skill of the students was tested previously, the students were asked about the methods used in their previous English speaking exams. The answers were mixed but it was seen that the most frequently used method of evaluation was individual question and answer between the student and the teacher. The responses were as follows: 12 students (36%) had their speaking exams in the form of individual question and answer, 2 students had exams both individually and in groups (6%), 5 (15%) students had their exams only in groups, 2 students (6%) gave oral presentations as speaking exams and 1 student (3%) had an exam in the form of a discussion of an English novel. The remaining responses were seen as irrelevant since the students didn't have English speaking courses.

The next question concerned the self-evaluation of English speaking skills of students. The answers revealed that all students thought their English speaking skills were enough to express themselves but that they needed to improve. Figure 2 below illustrates the students' self-reported assessment of their English speaking skills.

Figure 2. Students' self-reported assessment of their English speaking skills

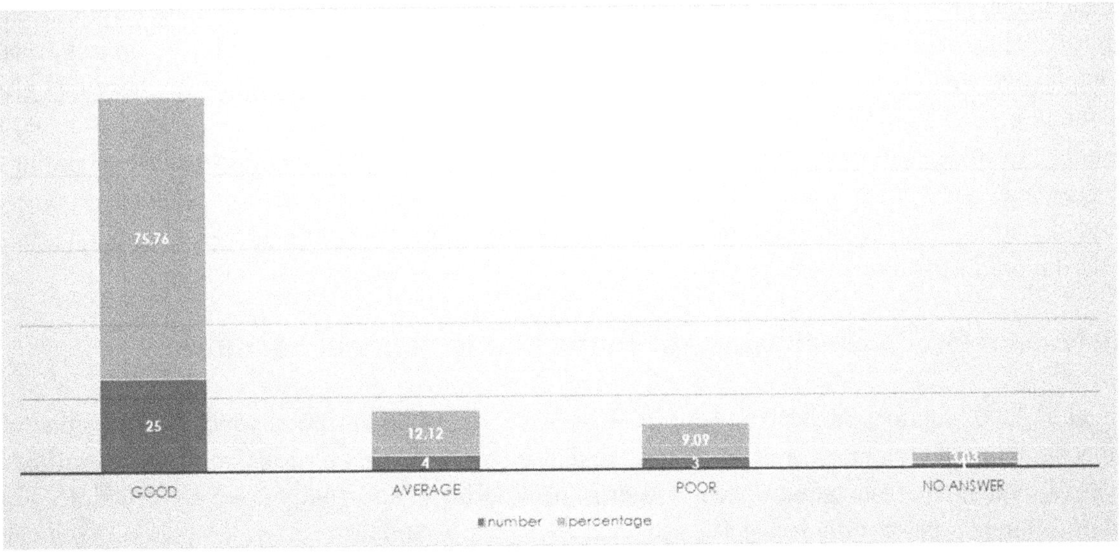

As it can be seen in Figure 2, a great majority of the students (n=25, 75.76%) reported that their English speaking skills were good, whilst others (n=8, 24,24%) reported their level either as average or as poor. Only 1 student did not report the self-assessment of his English speaking skills.

The students were also asked to share their positive or negative thoughts about being evaluated in a group or individually. The positive thoughts about being tested in groups were as follows:

Sample response 1a: It was a fair assessment, both individual and intra-group interaction were evaluated.
Sample response 1b: The recording of the performance made the evaluation more accurate.
Sample response 1c: It was good that the students had the chance to help each other and be evaluated in a more relaxed environment in the mood of a conversation.
Sample response 1d: In general, it is an effective method in terms of evaluating speech ability.
Sample response 1e: The use of different methods was effective.

The students also expressed some negative thoughts about being tested in a group. The negative responses were as follows:

Sample response 2a: The idea that the assessment is unfair.
Sample response 2b: Not getting a chance to explain enough.
Sample response 2c: Getting excited about being filmed / videotaped in front of the camera.
Sample response 2d: Being late for the exam, the exam environment and talking to people you do not know.
Sample response 2e: 4 people is crowded, questions are asked without enough time to think and talk.

In terms of the fairness of the evaluation method, there were responses both in favor of and against group testing but the positive responses focused on the fact that video recording made evaluation more fair and that it also allowed the evaluation of intra-group interaction. Although video-recording created the perception of fair evaluation, it caused some students to get excited. One important positive response was that the group task created a relaxed atmosphere. As negative responses students stated that they felt excited about being videotaped and talking to the people they did not know. These responses give important clues about designing group tasks in the future. For example, although the group task creates an interactive environment, the examiner must make sure that the task is designed in a way that gives equal talk time to all students in the group. And also the rationale for using the videotaping should be explained to students in detail so that they do not feel excited. Another thing to consider, according to the open-ended responses, could be the number of students in each group. In this case, each group consisted of four students. In order to create a group atmosphere this number was seen as optimal since it allowed more inter-group interaction.

Analysis of Group Discussions in Terms of Interactional Features

Schegloff (2007) defines sequence organization as "any kind of organization which concerns the relative positioning of utterances or actions" (p.2) and also points out that turn-taking can be classified as a type of sequential organization since it concerns the ordering of speakers, of turn constructional units and of different types of utterances. The sequence organization and turn design in the exam group task was largely dependent on the instructions given by the examiner in the beginning of the exam session given below:

"Instructions: Your topic is "...". Starting from Student A ask your questions one by one to the next student e.g. Student A to Student B, Student B to Student C... During the question-answer session between two students, if you have something to add, you can step in. You can make clarifications for each other. The Teacher will only be there as a listener. Do not ask for clarifications from the teacher but your group mates. Before we start, you have 3 minutes to go over the questions." (See Appendix B)

It was observed that the students followed the instructions without seeking help from the examiner. The framework of interaction was determined by the instructions; however, during the exam, the students took initiatives to co-construct the interaction by giving contributions to the discussion at relevant points. The contributions which happened randomly by digressing from the planned flow of interaction from Student A to B to C and D happened at relevant points without interrupting the flow of the conversation. This shows that the group task also gave clues about students' interactional competence. Generally, students were capable of managing the sequence organization. One notable observation was that, in the sequence organization, the student who was asking questions was seen as the person managing the talk at that point and students who took turns randomly asked for permission from that student implicitly by making a request or explicitly by using non-verbal communication.

In *Extract 1a* below, we can see an example of an explicit turn taking request from S3. In this extract, S1 is the one who is asking questions and S2 is the one who is answering. However, S3 has something to say about the topic, so he steps in by making a request for a turn from S1, who is seen as the one directing the conversation since it is his turn to ask questions. Turn taking is done by explicitly asking to take the floor. In line 2, S3 says: "Can I say something about it?". Although explicitly asking for the floor seems to be formal compared to real life interaction, it is still remarkable that a student takes the initiative and contributes to the interaction.

Extract 1a-Group 9

(1) S2: (.) uhm >we age really fast because of the air pollution<
(2) S3: Can I say something about it ↑
(3) S1: Yes ((nods))
(4) S3: Uhm (.) I think it affects the environment a::nd uhm (.) when uhm it gets really polluted uhm that affects trees around us and (0.5) that cause (0.5) the trees (.) not working properly I think

Another example of turn taking can be seen in *Extract 1b* below. This extract is from the same group. Here S2 answers a question about the topic: air pollution. S2 has a long turn and in the end she says yes and there is a short pause, S4 who has something to say about the topic sees this as a point where he can take the turn and make a contribution and in line 2, S4 takes the turn and makes an utterance. This also gives a clue about the atmosphere that the exam task has created. It seems as if students are engaged in casual talk and they step in and make a comment and they do this at the correct time without interrupting the normal flow of conversation. They are able to notice the clues in the discourse which mark the end of a turn.

Figure 3. Students in Group 9 engaged in conversation during the exam

Extract 1b-Group 9

(1) S2: <What causes air pollution↑> £ Human causes air pollution £ , I think, a:nd <because we use a lot of>, you know allergic things and ehm, enough to e:hm (1.2) enough to (.) pierce the ozone layer, °you know°, and uhm the oil we use for our cars and what else uhm (.) stuff like that yes (0.8)

(2) S4: Firstly, cars we are [using
(3) S1: [public transportation

In Extract c below, we can see the use of gap fillers and pauses indicating a break in the fluency. The gap fillers and pauses mark a word search by S2 who looks at S3 and uses hand gestures to ask for help. S3 uses the clues from the nonverbal communication and helps S2 by uttering the word "lungs". Although the word S2 is searching is a simple word, S2 may have had trouble because of exam anxiety, but after S3 helps, we see that S2 laughs and seems to be more relaxed and continues her response. Here we can say that the presence of peers in the exam environment helps reduce exam anxiety by providing opportunities to collaborate.

Extract 1c-Group 9

(1) S1: What problems does the air pollution cause?
(2) S2: >OK< The health problems, you know, the u:hm (.) our, which uhm (0.8) which organs ((uses hand gestures and looks at S3)) ?
(3) S3: lungs
(4) S2: £ Ah, yeah, lungs £ ha ha ((laughs)), uhm our lungs are really really sick I think because of (.) you know the bad weather a:nd

Among the skills involved in talk in interaction listed by Richards (2016) are engaging in small talk and recounting personal incidents and experiences. In Extract 2a below, we can see these features. Here S1 is asking the questions and S2 is answering according to the exam instructions. The topic is genes and how they affect our weight. After being asked the question: "What do you think about your weight?" in line 2, S2 gives a personal answer and S1 makes a comment about S2's response, which could be considered as small talk. In line 5, S2 shares a personal experience about how he is trying to lose weight and S3 takes a turn and takes part in the small talk by saying, "you are not fat". The examples in this extract indicate that the students feel that they are not in a threatening environment although they are in an exam and get involved in friendly conversation. The extract also shows that the students in the group have the skills to carry out small talk in English.

Extract 2a- Group 10

(1) S2: We made up from genes
(2) S1: What do you think of £ your weight↑ £
(3) S2: Ehm (.) I think its OK(0.5) e:hm >compared to my age<
(4) S1: You look skinny [but
(5) S2: [not really fat I'm trying to lose some weight ehm about one two kilos (.) but I'm Ok with [my]
(6) S3: [you]are not fat

Fillers in spontaneous speech are approached differently by different researchers. Whereas some consider them as flaws in speech, others stress the positive aspects of speech. For example, Swert (1998) points out that spontaneous speech is a better means of communication than fluent, read speech. Fillers are usually used by speakers when they are thinking what to say next, therefore they reflect a cognitive process in simultaneous speech and they serve a communicative function according to Clark and Fox Tree (2002). Therefore, they should not be regarded as flaws in speech but as a normal part of spontaneous speech. For example, in extract 2b, S1 uses uhm as a filler probably to think about what to say next. It is not regarded as disfluency here but as a reflection of spontaneous speech. These usages also indicate that the students speak naturally in the exam using communicative strategies to keep the conversation flowing with only short interruptions.

Figure 4. Students in Group 10 engaged in conversation during the exam

Extract 2b-Group 10

(1) S1: Do you think its OK to alter genes↑ to change uhm (.) its °nature°
(2) S2: I do not believe we: we need to change(.) we are what we are
(3) S1: What about uhm (.) people with disabilities and down syndrome
(4) S2: Then we can change of course [why not]
(5) S3: [yes maybe]

The group task also allows for students to react to each other (Richards, 2016). In this way, a fuller picture of a student's interactional abilities can be seen. In Extract 3a below S1 reacts to responses of S3 saying yes, (line 2), yeah (line 4) and yes, right (line7).

Figure 5. Students in Group 3 engaged in conversation during the exam

Extract 3a-Group 3

(1) S3: They can use it for you know doing their homework [or]=
(2) S1: [Yes]
(3) S3: =Or any other routines
(4) S1: Yeah.
(5) S1: What is social networking↑
(6) S2: uhm (0.6) social networking <in my opinion> is uhm (.) like social media accounts, olike instagram or the applicationso which we can use to meet new people, isn't it↑
(7) S1: Yes. ((nods)) £ Right £

As the analysis of selected extracts from the transcriptions of the speaking exam shows, the setting of the exam creates a natural environment where components of real life interaction such as turn taking, engaging in small talk, recounting personal experiences, using gap fillers and pauses.

SOLUTIONS AND RECOMMENDATIONS

Although the structured task used in testing speaking in a group has advantages, it also has some limitations to be considered in terms of task design. As a limitation of testing speaking in groups, students felt each of them did not have enough time to respond and that being videotaped caused anxiety. As an implication for future speaking test designers, to overcome these limitations, firstly it is recommended that the students be familiarized with videotaping in the classroom setting, so that when they are videotaped in the exam, they do not have camera anxiety. The second recommendation is that the instructions should clearly state the roles of each student in the exam and enough time should be allocated to the students for them to get ready before the exam session starts and enough time should be given to each student to participate equally.

FUTURE RESEARCH DIRECTIONS

Further research might investigate the influence of changes in task design on the success of group discussion in testing speaking. To illustrate, the present study showed that videotaping students while they were speaking caused them to feel anxious, which might have an impact on their performance since they did not feel comfortable or familiar with this technique.

Future research might also include videotaping group speaking tasks during in-class activities, which might help students feel accustomed to this technique. Additionally, further research might investigate the influence or contributions of preparation before testing their speaking through group discussion, and apart from that, students' speaking skills might be tested with a different task in which students have to collaborate to accomplish (for example, preparing a research proposal or solving a problem as a group), thereby giving ideas about how the dynamics of interaction might change depending on the nature of the structured group speaking task for testing speaking skills of undergraduates.

CONCLUSION

Evaluation of speaking ability is a challenging task which has to take into account several variables which could affect students' performance. Test types used in the evaluation of speaking involve reading aloud, picture description, oral presentation, interview or role play. The different types of evaluation of speaking have been criticized for not being reliable (Polat, 2020) and being limited to certain aspects of speaking while ignoring others (Bachman and Savignon, 1986). As mentioned earlier in the chapter, testing of speaking is usually carried out in a way that does not reflect all components of speaking skills. When students are tested individually through monologues, they are not being tested fully in terms of speaking abilities. Such testing can evaluate vocabulary knowledge, pronunciation and to some extent intonation and fluency, but it does not really capture all aspects of communicative and interactional competence (Başer, 2011). In addition, in individual testing of speaking the anxiety level of students also increases. An alternative method of evaluating speaking skills, which is structured tasks requiring student-student interaction either in pairs or groups, are associated with greater fluency and it is argued that they facilitate task performance (Skehan, 2001).

In this study, a structured group speaking task was implemented in a speaking exam and through analysis of transcriptions and student responses to an open-ended questionnaire, it was aimed at examining the opportunities and limitations such an approach presented. The analysis of transcriptions of exam sessions revealed that when the exam task is designed in a way to include interaction, the evaluation could become more multifaceted to include aspects of real life interactional abilities such as sequence organization, turn design, turn taking, use of fillers and discourse markers, responding to speakers, making small talk and others. Another observation was that compared to individual testing, the affective filter of students was lowered and the atmosphere was more relaxed. It was also seen through the student questionnaire that although most of the students had not experienced being tested in groups in speaking exams, they had a general positive attitude towards being tested this way. They stated that it decreased their anxiety and they had a chance to collaborate with their peers.

As a conclusion, well structured, well-timed and well planned structured speaking tasks used in the evaluation of speaking skills have strengths over other individual testing of speaking and provides a more multifaceted evaluation of speaking ability.

REFERENCES

Bachmann, L. F. (1990). *Fundamental considerations in language testing*. Oxford University Press.

Başer, Z. (October, 2011). *A critical analysis of student to student interaction tests in assessment of speaking*. Paper presented at the 1st International Congress on Curriculum and Instruction, Eskişehir, Turkey.

Bosker, H. R., Quené, H., Sanders, T., & de Jong, N. H. (2014). The Perception of Fluency in Native and Nonnative Speech. *Language Learning*, *64*(3), 579–614. doi:10.1111/lang.12067

Brooks, L. (2009). Interacting in pairs in a test of oral proficiency: Co-constructing a better performance. *Language Testing*, *26*(3), 341–366. doi:10.1177/0265532209104666

Clark, H. H., & Fox Tree, J. E. (2002). Using Uh and Um in spontaneous speaking. *Cognition*, *84*(1), 73–111. doi:10.1016/S0010-0277(02)00017-3 PMID:12062148

Davis, L. (2009). The influence of interlocutor proficiency in a paired oral assessment. *Language Testing*, *26*(3), 367–396. doi:10.1177/0265532209104667

Ducasse, A. M., & Brown, A. (2009). Assessing paired orals: Raters' orientation to interaction. *Language Testing*, *26*(3), 423–443. doi:10.1177/0265532209104669

Foot, M. C. (1999). Relaxing in pairs. *ELT Journal*, *53*(1), 36–41. doi:10.1093/elt/53.1.36

Galaczi, E. D. (2008). Peer-peer interaction in a speaking test: The case of the First Certificate in English Examination. *Language Assessment Quarterly*, *5*(2), 89–119. doi:10.1080/15434300801934702

Gan, Z., Davison, C., & Hamp-Lyons, L. (2009). Topic negotiation in peer group oral assessment situations: A conversation analytic approach. *Applied Linguistics*, *30*(3), 315–334. doi:10.1093/applin/amn035

Hoey, E. M., & Kendrick, K. H. (2018). Conversation analysis. In A. M. B. de Groot & P. Hagoort (Eds.), *Research methods in psycholinguistics and the neurobiology of language: A practical guide* (pp. 151–173). Wiley Blackwell.

Jacobs, A., Fricke, M., & Kroll, J. F. (2016). Cross-Language Activation Begins During Speech Planning and Extends Into Second Language Speech. *Language Learning*, *66*(2), 324–353. doi:10.1111/lang.12148 PMID:27773945

May, L. (2009). Co-constructed interaction in a paired speaking test: The rater's perspective. *Language Testing*, *26*(3), 397–421. doi:10.1177/0265532209104668

McCarthy, M. (2010). Spoken fluency revisited. *English Profile Journal, 1*(1), 1–15. doi:10.1017/S2041536210000012

Polat, M. (2020). A Rasch Analysis of Rater Behaviour in Speaking Assessment. *International Online Journal of Education & Teaching*, *7*(3), 1126–1141.

Richards, J. (2016). *Teaching Speaking #3 – Talk in Interaction* [Blog post]. Retrieved from https://www.cambridge.org/elt/blog/2016/02/03/teaching-speaking-2-talk-interaction-ii/

Saville, N., & Hargreaves, P. (1999). Assessing speaking in the revised FCE. *ELT Journal*, *53*(1), 42–51. doi:10.1093/elt/53.1.42

Schegloff, E. A. (2007). *Sequence Organization in Interaction: A Primer in Conversation Analysis I*. Cambridge University Press. doi:10.1017/CBO9780511791208

Skehan, P. (2001). Tasks and language performance. In M. Bygate, P. Skehan, & M. Swain (Eds.), *Research pedagogic tasks: second language learning, teaching, and testing*. Longman.

Stephenson, M., & Hall, G. (2021). Organizing talk in group speaking tests: Learning from high-scoring students. *ELT Journal*, *75*(1), 42–54. doi:10.1093/elt/ccaa052

Tottie, G. (2011). Uh and Um as sociolinguistic markers in British English. *International Journal of Corpus Linguistics*, *16*(2), 173–197. doi:10.1075/ijcl.16.2.02tot

Underhill, N. (1987). *Testing spoken language: A handbook for oral testing techniques*. Cambridge University Press.

Weir, C. J. (2005). *Language testing and validation*. Palgrave Macmillan. doi:10.1057/9780230514577

KEY TERMS AND DEFINITIONS

Conversational Analysis: It is a method studying human social interactions in verbal or non-verbal conduct.

Jefferson Transcription System: It is a conversation analysis method for annotating speech. It aims to catch not only what is said but also how it is said. For this purpose, it uses familiar keystrokes as symbols in order to indicate the details of speech.

Speaking Anxiety: It is the nervousness that an individual feels during or after speaking, usually while being tested for their oral proficiency.

Structured Group Speaking Task: It is a speaking task in which students are given a list of clear-cut instructions in order to maintain the flow of group discussion.

Student-Student Interaction: It is the interaction between students in the classroom environment in which students ask each other questions, discuss, or reflect on a topic. listen to their peers' opinions, and collaborate to accomplish the task given.

Testing Speaking: It is testing the oral proficiency of learners using different types of techniques/methods such as interviews, oral presentation, reading aloud, picture story, group discussion, etc.

Videotaping: It is recording visual images and sound using a camera.

APPENDIX A

Jefferson Transcription System

Symbol	Definition and use	Key(s)
[yeah] [okay]	Overlapping talk	
=	End of one TCU and beginning of next begin with no gap/pause in between (sometimes a slight overlap if there is speaker change). Can also be used when TCU continues on new line in transcript	
(.)	Brief interval, usually between 0.08 and 0.2 seconds	
(1.4)	Time (in absolute seconds) between end of a word and beginning of next. Alternative method: "none-one-thousand-two-one-thousand…": 0.2, 0.5, 0.7, 1.0 seconds, etc.	
Word Wo:rd	Underlining indicates emphasis Placement indicates which syllable(s) are emphasised Placement within word may also indicate timing/direction of pitch movement (later underlining may indicate location of pitch movement)	
wo::rd	Colon indicates prolonged vowel or consonant One or two colons common, three or more colons only in extreme cases.	
↑word ↓word	Marked shift in pitch, up (↑) or down (↓). Double arrows can be used with extreme pitch shifts.	↑ Wingdings 3 (104) ↓ Wingdings 3 (105) ↑ ALT+24 ↓ ALT+25
. , _ ¿ ?	Markers of final pitch direction at TCU boundary: Final falling intonation (.) Slight rising intonation (,) Level/flat intonation (_) Medium (falling-)rising intonation (¿) (a dip and a rise) Sharp rising intonation (?)	¿ ALT+168
WORD	Upper case indicates syllables or words louder than surrounding speech by the same speaker	
°word°	Degree sign indicate syllables or words distinctly quieter than surrounding speech by the same speaker	° ALT+248
<word	Pre-positioned left carat indicates a hurried start of a word, typically at TCU beginning	
word-	A dash indicates a cut-off. In phonetic terms this is typically a glottal stop	
>word<	Right/left carats indicate increased speaking rate (speeding up)	
<word>	Left/right carats indicate decreased speaking rate (slowing down)	
.hhh	Inbreath. Three letters indicate 'normal' duration. Longer or shorter inbreaths indicated with fewer or more letters.	
hhh whhord	Outbreath. Three letters indicate 'normal' duration. Longer or shorter inbreaths indicated with fewer or more letters. Can also indicate aspiration/breathiness if within a word (not laughter)	
w(h)ord	Indicates abrupt spurts of breathiness, as in laughing while talking	
£word£	Pound sign indicates smiley voice, or suppressed laughter	
#word#	Hash sign indicates creaky voice	
~word~	Tilde sign indicates shaky voice (as in crying)	
(word)	Parentheses indicate uncertain word; no plausible candidate if empty	
(( ))	Double parentheses contain analyst comments or descriptions	

APPENDIX B:

Sample Group Task for Speaking Exam

Instructions: Your topic is "genes". Starting from Student A ask your questions one by one to the next student e.g. Student A to Student B, Student B to Student C… During the question-answer session between two students, if you have something to add, you can step in. You can make clarifications for each other. The Teacher will only be there as a listener. Do not ask for clarifications from the teacher but your groupmates. Before we start, you have 3 minutes to go over the questions.

TOPIC 1

Student A's Questions

1. What images are in your mind when you hear the word 'gene'?
2. What do you know about genes?
3. What do you think of your weight?
4. Do you think it's OK to alter genes?

Student B's Questions

1. Should we alter our genes to keep us slim?
2. How can people keep from being overweight?
3. What is the best way to stay slim?
4. What advice do you have for overweight people?

Student C's Questions

1. Can you eat what you want without piling on the pounds?
2. What do you think of when you hear the word 'slim'?
3. Why are so many people overweight all around the world?
4. Why do people 'rush' to criticize people who are overweight?

Student D's Questions

1. Should we all be vegetarian to be a healthier weight?
2. Will we all be a healthy weight in the future?
3. Why do some researchers suggest thin people might be 'morally superior'?
4. What are your experiences of dieting?

APPENDIX C

Open-Ended Questionnaire

Questionnaire on English Speaking Lessons and Evaluation Methods

Application Date: .... / .... / ....

Personal Information:

First and Last Name: ________________________________

Gender: Female Male

Age: ________________________________

Occupation: ________________________________

I do not work

Languages

Native Language: Turkish Other Please specify: ________

Do you speak one or more foreign languages? Please specify here, if any, which language or languages, along with your level.

Foreign Languages Level Language Exam Score/Certificate (If Applicable)

1. İngilizce __________ ____________________________
2. __________ __________ ____________________________
3. __________ __________ ____________________________

Please answer the following questions:

1. How long have you been learning English?
2. Evaluate your pre-university English courses (primary, secondary, private course, etc.) and their contributions to improving your English speaking skills.
3. Have you participated in the English preparatory program? yes no
4. How many hours of English lessons did you take per week if you attended? Did you have a separate English speaking class? How many hours a week was your English speaking class, if any?
5. What were the previous methods used to assess your English speaking skills? How was the scoring done? Was your ability to speak English evaluated individually or as a group (your interaction with other students)? What are your views on these methods? Explain.
6. Evaluate your English speaking skills.
7. What are your views on individual evaluation of your English speaking skills?
8. What are your views on being evaluated for your English speaking skills as a group?
9. What are your views on the method by which your English speaking skills are evaluated in the latest exam?

Compilation of References

Abdalla Salih, A. R. (2013). Peer response to L2 student writing: Patterns and expectations. *English Language Teaching*, *6*(3), 20–42. doi:10.5539/elt.v6n3p42

Abdous, M., Camarena, M., & Facer, B. (2009). MALL technology: Use of academic podcasting in the foreign language classroom. *ReCALL*, *21*(1), 76–95. doi:10.1017/S0958344009000020

Abraham, L. B. (2008). Computer-mediated glosses in second language reading comprehension and vocabulary learning: A meta-analysis. *Computer Assisted Language Learning*, *21*(3), 199–226. doi:10.1080/09588220802090246

Absalom, M., & Rizzi, A. (2008). Comparing the outcomes of online listening versus online text-based tasks in university level Italian L2 study. *ReCALL*, *20*(1), 55–66. doi:10.1017/S0958344008000517

Ahmadian, M., & Matour, M. (2014). The effects of explicit instruction of connected speech on Iranian EFL learners' listening comprehension skill. *International Journal of Applied Linguistics and English Literature*, *3*(2), 227–236. doi:10.7575/aiac.ijalel.v.3n.2p.227

Alameen, G. (2014). *The effectiveness of linking instruction on NNS speech perception and production*. Graduate Theses and Dissertations. Paper 14004. https://lib.dr.iastate.edu/etd/14004

Alfa, R. R. (2020). Using podcast as authentic materials to develop students' speaking skill. *Journal of English Language Teaching*, *4*, 65–74.

Al-Hazmi, S., & Schofield, P. (2007). Enforced revision with checklist and peer feedback in EFL writing: The example of Saudi university students. *Scientific Journal of King Faisal University (Humanities and Management Sciences*, *8*(2), 237-267.

Allen, D., & Katamaya, A. (2016). Relative second language proficiency and the giving and receiving of written peer feedback. *System*, *56*, 96–106. doi:10.1016/j.system.2015.12.002

Allen, M., Mabry, E., Mattrey, M., Bourhis, J., Titsworth, S., & Burrell, N. (2004). Evaluating the effectiveness of distance learning: A comparison using meta-analysis. *Journal of Communication*, *54*(3), 402–420. doi:10.1111/j.1460-2466.2004.tb02636.x

Alshuraidah, A., & Storch, N. (2019). Investigating a collaborative approach to peer feedback. *ELT Journal*, *73*(2), 166–174. doi:10.1093/elt/ccy057

Altenberg, E. P. (2005). The perception of word boundaries in a second language. *Second Language Research*, *21*(4), 325–358. doi:10.1191/0267658305sr250oa

Altstaedter, L. L., & Doolittle, P. (2014). Students' perceptions of peer feedback. *Argentinian Journal of Applied Linguistics*, *2*(2), 60–76.

Alvarez-Marinelli, H., Blanco, M., Lara-Alecio, R., Irby, B. J., Tong, F., Stanley, K., & Fan, Y. (2016). Computer assisted English language learning in Costa Rican elementary schools: An experimental study. *Computer Assisted Language Learning, 29*(1), 103–126. doi:10.1080/09588221.2014.903977

American Council on Teaching of Foreign Languages. (2011). *21st Century Skills map*. Retrieved from https://www.actfl.org/sites/default/files/pdfs/21stCenturySkillsMap/p21_worldlanguagesmap.pdf

Amores, M. J. (1997). A new perspective on peer-editing. *Foreign Language Annals, 30*(4), 513–522. doi:10.1111/j.1944-9720.1997.tb00858.x

Aquil, R. (2012). Listening English connected speech: A problem and solutions. *Arab World English Journal, 3*(2), 329-364.

Aquil, R. (2012). Listening English connected speech: A problem and solutions. *Arab World English Journal, 3*(2), 329–364.

Arbaugh, J. (2000b). Virtual classroom versus physical classroom: An exploratory study of class discussions patterns and student learning in an asynchronous Internet-based MBA course. *Journal of Management Education, 24*(2), 207–227. doi:10.1177/105256290002400206

Arbaugh, J., & Berbunan-Fich, R. (2006). An investigation of epistemological and social dimensions of teaching in online learning environments. *Academy of Management Learning & Education, 5*(4), 435–447. doi:10.5465/amle.2006.23473204

Ashcroft, R. J., Garner, J., & Hadingham, O. (2018). Incidental vocabulary learning through watching movies. *Australian Journal of Applied Linguistics, 1*(3), 135–147. doi:10.29140/ajal.v1n3.89

Avery, P., & Ehrlich, S. (1992). *Teaching American English Pronunciation*. Oxford University Press.

Ayan, E., & Demirel, E. (2017). Analyzing Research Tendencies of ELT Researchers and Trajectory of English Language Teaching and Learning in the last Five Years. *International Journal of Curriculum and Instruction, 9*(2), 89–106.

Aydın, S. (2016). A Qualitative Research on Foreign Language Teaching Anxiety. *The Qualitative Report, 21*(4), 629-642. https://nsuworks.nova.edu/tqr/vol21/iss4/2

Bachmann, L. F. (1990). *Fundamental considerations in language testing*. Oxford University Press.

Bahari, A. (2019). FonF practice model from theory to practice: CALL via focus on form approach and non-linear dynamic motivation to develop listening and speaking proficiency. *Computers & Education, 130*, 40–58. doi:10.1016/j.compedu.2018.11.009

Baker, P. (2006). *Using Corpora in Discourse Analysis*. Continuum.

Balas, J. L. (2005). Blogging is so last year: Now podcasting is hot. *Computers in Libraries, 10*, 29–32.

Bani-Salameh, Z. A., Kabilan, M. K., & Bani-Salalmeh, L. (2011). Utilising multiple media ESP programme in enhancing flight attendants' safety knowledge and problem solving skills. *British Journal of Educational Technology, 42*(6), 1003–1015. doi:10.1111/j.1467-8535.2010.01124.x

Barcelos, A. (2003). Researching beliefs about SLA: A critical review. In P. Kalaja & A. M. F. Barcelos (Eds.), *Beliefs about SLA: New research approaches*. Kluwer Academic Publishers. doi:10.1007/978-1-4020-4751-0_1

Barnes, D. (1976). *From communication to curriculum*. Harmondsworth, England: Penguin.

Barrera-Pardo, D. (2008). The reality of stress timing. *ELT Journal, 62*(1), 11–17. doi:10.1093/elt/ccm073

Barrot, J. S. (2016). Examining the teaching beliefs and practices of experienced ESL teachers: A sociocognitive-transformative perspective. *The Southeast Asian Journal of English Language Studies, 22*(1), 153–163. doi:10.17576/3L-2016-2201-12

Başal, A., Gülözer, K., & Demir, İ. (2015). Use of video and audio in EFL listening test. *Journal of Education and Training Studies, 3*(6), 83-89. . doi:10.11114/jets.vi6.1001

Başal, A., Gülizer, K., & Demir, I. (2015). Use of Video and Audio Texts in EFL Listening Test. *Journal of Education and Training Studies*, *3*(6), 83–89. doi:10.11114/jets.v3i6.1001

Basal, A., Yilmaz, S., Tanrıverdi, A., & Sari, L. (2016). Effectiveness of mobile applications in vocabulary teaching. *Contemporary Educational Teaching*, *7*(1), 47–59. doi:10.30935/cedtech/6162

Basaran, S., & Dilber, E. (2013). Effects of captioned TV shows on language learners motivation and perception. *Respectus Philologicus*, *23*(28), 83–96. doi:10.15388/RESPECTUS.2013.23.28.7

Başer, Z. (October, 2011). *A critical analysis of student to student interaction tests in assessment of speaking*. Paper presented at the 1st International Congress on Curriculum and Instruction, Eskişehir, Turkey.

Batty, A. O. (2015). A comparison of video- and audio-mediated listening tests with many-facet Rasch modeling and differential distractor functioning. *Language Testing*, *32*(1), 3–20. doi:10.1177/0265532214531254

Baxter, P., & Jack, S. (2008). Qualitative case study methodology: Study design and implementation for novice researchers. *Qualitative Report*, *13*(4), 544–559.

Bayyurt, Y. (2013). Current Perspectives on Sociolinguistics and English Language Education. *The Journal of Language Teaching and Learning*, *3*(1), 69–78.

Bayyurt, Y., & Karataş, N. B. (2011). Needs Analysis: The first step in content development for mobile learning. *Boğaziçi Journal of Education*, *28*(1), 13–22.

Becker, S. R., & Sturm, J. L. (2017). Effects of audiovisual media on L2 listening comprehension: A preliminary study in French. *CALICO Journal*, *34*, 147–177.

Beggren, J. (2015). Learning from giving feedback: A study of secondary level students. *ELT Journal*, *69*(1), 58–70. doi:10.1093/elt/ccu036

Bernard, R. M., Borokhovski, E., Schmid, R. F., & Tamim, R. M. (2018). Gauging the effectiveness of educational technology integration in education: what the best-quality meta-analyses tell us. *Learning, Design, and Technology: An International Compendium of Theory, Research, Practice, and Policy*, 1-25.

Best, K., Jones-Katz, L., Smolarek, B., Stolzenburg, M., & Williamson, D. (2014). Listening to our students: An exploratory practice study of ESL writing students' views of feedback. *TESOL Journal*, *6*(2), 332–357. doi:10.1002/tesj.152

Bird, S. A., & Williams, J. N. (2002). The effect of bimodal input on implicit and explicit memory: An investigation into the benefits of within-language subtitling. *Applied Psycholinguistics*, *23*(4), 509–533. doi:10.1017/S0142716402004022

Bitchener, J., & Ferris, D. R. (2012). *Written corrective feedback in second language acquisition and writing*. New York: Routledge. doi:10.4324/9780203832400

Blau, E. K. (1990). The effect of syntax, speed, and pauses on listening comprehension. *TESOL Quarterly*, *24*(4), 746–753. doi:10.2307/3587129

Bonk, W. J. (2000). Second language lexical knowledge and listening comprehension. *International Journal of Listening*, *14*(1), 14–31. doi:10.1080/10904018.2000.10499033

Borenstein, M., Hedges, L. V., Higgins, J. P., & Rothstein, H. R. (2011). Publication bias. In *Introduction to meta-analysis*. John Wiley & Sons.

Borg, S. (2003). Teacher cognition in language teaching: A review of research on what language teachers think, know, believe, and do. *Language Teaching*, *36*(2), 81–109. doi:10.1017/S0261444803001903

Borg, S. (2006). *Teacher cognition and language education*. Continuum.

Borg, S. (2007). Research engagement in English Language Teaching. *Teaching and Teacher Education*, *23*(5), 731–747. doi:10.1016/j.tate.2006.03.012

Borg, S. (2019). *Language teacher cognition: Perspectives and debates*. Springer.

Borg, S., & Burns, A. (2008). Integrating grammar in adult TESOL classrooms. *Applied Linguistics*, *29*(3), 456–482. doi:10.1093/applin/amn020

Bosker, H. R., Quené, H., Sanders, T., & de Jong, N. H. (2014). The Perception of Fluency in Native and Nonnative Speech. *Language Learning*, *64*(3), 579–614. doi:10.1111/lang.12067

Botero, G. G., Restrepo, M. A. B., Zhu, C., & Questier, F. (2019). Complementing in-class language learning with voluntary out-of-class MALL. Does training in self-regulation and scaffolding make a difference? *Computer Assisted Language Learning*, 1–27. doi:10.1080/09588221.2019.1650780

Boulton, A., & Cobb, T. (2017). Corpus use in language learning: A meta-analysis. *Language Learning*, *67*(2), 348–393. doi:10.1111/lang.12224

Bower, J. V., & Rutson-Griffiths, A. (2016). The relationship between the use of spaced repetition software with a TOEIC word list and TOEIC score gains. *Computer Assisted Language Learning*, *29*(7), 1238–1248. doi:10.1080/09588221.2016.1222444

Bradley, L. (2014). Peer-reviewing in an intercultural wiki environment: Student interaction and reflections. *Computers and Composition*, *34*, 80–95. doi:10.1016/j.compcom.2014.09.008

Branch, R. M., & Kopcha, T. J. (2013). Instructional Design Models. In Handbook of Research on Educational Communications and Technology (pp. 77-87). Doi:10.1007/978-1-4614-3185-5_7

Brett, P. (1997). A comparative study of the effects of the use of multiple media on listening comprehension. *System*, *25*(1), 39–53. doi:10.1016/S0346-251X(96)00059-0

Brewster, J., Ellis, G., & Girard, D. (1992). *The Primary English teacher"s guide*. Penguin.

Bromley, D. B. (1990). Academic contributions to psychological counselling: A philosophy of science for the study of individual cases. *Counselling Psychology Quarterly*, *3*(3), 299–307. doi:10.1080/09515079008254261

Brooks, L. (2009). Interacting in pairs in a test of oral proficiency: Co-constructing a better performance. *Language Testing*, *26*(3), 341–366. doi:10.1177/0265532209104666

Brouwer, S. (2010). *Processing strongly reduced forms in casual speech* (PhD Dissertation). Radbaud University. Retrieved from https://repository.ubn.ru.nl/handle/2066/90675

Brown & Kondo-Brown. (Eds.). (2006). Perspectives on teaching connected speech to second language speakers. University of Hawaii Press.

Brown, E. (Ed.). (2001). Mobile learning explorations at the Stanford Learning Lab. *Speaking of Computers*, *55*. Retrieved September 24, 2014, from http://cgi.stanford.edu/~dept-ctl/tomprof/posting.php?ID=289

Brown, J. D. (2006). Authentic communication: Why is it important' ta teach reduced forms? *Authentic Communication: Proceedings of the 5th Annual JALT Pan-SIG Conference*, 13-24. Retrieved June 6, 2007 from http://jalt.org/pansig/2006/HTML/Brown.htm

Brown, J. D. (2006). Authentic communication: Whyzit importan' ta teach reduced forms? *Authentic Communication: Proceedings of the 5th Annual JALT Pan-SIG Conference*, 13-24. Retrieved June 6, 2007 from http://jalt.org/pansig/2006/HTML/Brown.htm

Brown, H. D. (1987). *Principles of language learning and teaching* (2nd ed.). Prentice-Hall.

Brown, H. D., & Abeywickrama, P. (2019). *Language assessment: Principles and classroom practices*. Pearson.

Brown, J. D. (2005). *Testing in Language Programs: A comprehensive guide to English language assessment*. McGraw-Hill.

Brown, J. D., & Hilferty, A. (1986a). Listening for reduced forms. *TESOL Quarterly*, *20*(4), 759–763. doi:10.2307/3586525

Brown, J. D., & Hilferty, A. (1986b). The effectiveness of teaching reduced forms for listening comprehension. *RELC Journal*, *17*(2), 59–70. doi:10.1177/003368828601700204

Brown, J. D., & Hilferty, A. (2006). The effectiveness of teaching reduced forms for listening comprehension. In J. D. Brown & K. Kondo-Brown (Eds.), *Perspectives on Teaching Connected Speech to Second Language Speakers* (pp. 51–58). University of Hawai'i, National Foreign Language Resource Center.

Brusilovsky, P. (1996). Methods and techniques of adaptive hypermedia. *User Modeling and User-Adapted Interaction*, *6*(2-3), 87–129. doi:10.1007/ BF00143964

Brusilovsky, P. (2007). Adaptive Navigation Support. In P. Brusilovski, A. Kobsa, & W. Nejdl (Eds.), *The Adaptive Web: Methods and Strategies of Web Personalization* (Vol. 4321, pp. 263–290). Springer-Verlag.

Brusilovsky, P. (2012). Adaptive Hypermedia for Education and Training. In P. J. Durlach & A. M. Lesgold (Eds.), *Adaptive Technologies for Training and Education* (pp. 46–68). CUP.

Brusilovsky, P., Eklund, J., & Schwarz, E. (1998). Web-based education for all: A tool for development adaptive courseware. *Computer Networks and ISDN Systems*, *30*(1), 291–300.

Brusilovsky, P., & Millán, E. (2007). User models for adaptive hypermedia and adaptive educational systems. In P. Brusilovski, A. Kobsa, & W. Nejdl (Eds.), *The Adaptive Web: Methods and Strategies of Web Personalization* (Vol. 4321, pp. 3–53). Springer-Verlag.

Buck, G. (2001). *Assessing Listening*. Cambridge University Press. doi:10.1017/CBO9780511732959

Buck, G., & Tatsuoka, K. (1998). Application of the rule-space procedure to language testing: Examining attributes of a free response listening test. *Language Testing*, *15*(2), 119–157. doi:10.1177/026553229801500201

Burley-Allen, M. (1995). *Listening: The forgotten skill*. John Wiley & Sons, Inc.

Burston, J. (2014). Twenty years of MALL project implementation: A meta-analysis of learning outcomes. *ReCALL*, 1–17.

Çağıltay, K., & Göktaş, Y. (Eds.). (2016). Öğretim Teknolojilerinin Temelleri: Teoriler, Araştırmalar, Eğilimler. Pegem Akademi.

Cahill, R. (2006). Teaching reduced interrogative forms to low-level EFL students in Japan. In J.D. Brown & K. Brown, (Eds.), Perspectives on teaching connected speech to second language speakers (pp. 99-125). Honolulu, HI: University of Hawai'i, National Foreign Language Resource Center.

Caimi, A., & Mariotti, C. (2015). Beyond the book: The use of subtitled audiovisual material to promote content and language integrated learning in higher education. In J. Díaz-Cintas & J. Neves (Eds.), *Audiovisual translation: Taking stock* (pp. 230–243). Cambridge Scholars Publishing.

Çakmak, F., & Erçetin, G. (2018). Effects of gloss type on text recall and incidental vocabulary learning in mobile-assisted L2 listening. *ReCALL, 30*(1), 24–47. doi:10.1017/S0958344017000155

Canagarajah, S. (2014). In search of a new paradigm for teaching as an international language. *TESOL Quarterly, 4*(4), 767–785. doi:10.1002/tesj.166

Cao, X., Yamashita, N., & Ishida, T. (2016). How non-native speakers perceive listening comprehension problems: Implications for adaptive support technologies. In T. Yoshino, G. Zurita, T. Yuizono, N. Baloian, G-D. Chen, & T. Inoue (Eds.), *Collaboration Technologies and Social Computing - 8th International Conference, CollabTech 2016, Proceedings* (pp. 89-104). Springer Verlag. 10.1007/978-981-10-2618-8_8

Card, D., Polin, L., Parra, J., Rhoads, J. B., & Sartori, T. (2006). Can you hear me now: The return of voice to distance learning. Presented at the *Fifth International Conference on Web-based education of the International Association of Science and Technology for Development*, Puerto Vallarta, Mexico.

Cardenas-Claros, M., & Gruba, P. (2013). Decoding the "CoDe": A framework for conceptualizing and designing help options in computer-based second language listening. *ReCALL, 25*(2), 250–271. doi:10.1017/S0958344013000049

Carson, J. G., & Nelson, G. (1994). Writing groups: Cross-cultural issues. *Journal of Second Language Writing, 3*(1), 17–30. doi:10.1016/1060-3743(94)90003-5

Carson, J. G., & Nelson, G. L. (1996). Chinese students' perceptions of ESL peer feedback group interaction. *Journal of Second Language Writing, 5*(1), 1–19. doi:10.1016/S1060-3743(96)90012-0

Cartney, P. (2010). Exploring the use of peer assessment as a vehicle for closing the gap between feedback given and feedback used. *Assessment & Evaluation in Higher Education, 35*(5), 551–564. doi:10.1080/02602931003632381

Cauldwell, R. T. (1996). Direct encounters with fast speech on CD-Audio to teach listening. *System, 24*(4), 521–528. doi:10.1016/S0346-251X(96)00046-2

Cavus, N., & Ibrahim, D. (2017). Learning English using children's stories in mobile devices. *British Journal of Educational Technology, 48*(2), 625–641. doi:10.1111/bjet.12427

Cazden, C. (1988). *Classroom discourse: The language of teaching and learning.* Porltsmouth, NH: Heinemann.

Celce-Murcia, M. (1995). Discourse analysis and the teaching of listening. In G. Cook & B. Seidlhofer (Eds.), *Principle and practice in applied linguistics* (pp. 363–377). Oxford University Press.

Celce-Murcia, M., Brinton, D. M., & Goodwin, J. M. (2011). *Teaching pronunciation: A reference guide* (2nd ed.). Cambridge University Press.

Chang, C. C., Lei, H., & Tseng, J. S. (2011). Media presentation mode, English listening comprehension and cognitive load in ubiquitous learning environments: Modality effect or redundancy effect? *Australasian Journal of Educational Technology, 27*(4), 633–654. doi:10.14742/ajet.942

Chang, C. C., Warden, C. A., Liang, C., & Chou, P. N. (2018). Performance, cognitive load, and behaviour of technology-assisted English listening learning: From CALL to MALL. *Journal of Computer Assisted Learning, 34*(2), 105–114. doi:10.1111/jcal.12218

Chang, C. F. (2012). Peer review via three modes in an EFL writing course. *Computers and Composition, 29*(1), 63–78. doi:10.1016/j.compcom.2012.01.001

Chang, C. K., Chen, G. D., & Hsu, C. K. (2011). Providing adequate interactions in online discussion forums using few teaching assistants. *The Turkish Online Journal of Educational Technology, 10*(3), 193–202.

Chang, C.-S. (2009). Gains to L2 listeners from reading while listening versus listening only in comprehending short stories. *System, 37*(4), 652–663. doi:10.1016/j.system.2009.09.009

Chang, M. M., & Lin, M. C. (2013). Strategy-oriented web-based English instruction—A meta-analysis. *Australasian Journal of Educational Technology, 29*(2), 203–216. doi:10.14742/ajet.67

Charles, T. J. (2017). *The role of captioned video in developing speech segmentation for learners of English as a second language* (PhD thesis). University of York.

Chaudron, C. (1984). The effects of feedback on students' composition revisions. *RELC Journal, 15*(2), 1–14. doi:10.1177/003368828401500201

Cha, Y. J., & Park, L. E. (2010). An analysis of synchronous interaction and its influence on EFL writers' revisions. *Multimedia-Assisted Language Learning, 13*(2), 9–36. doi:10.15702/mall.2010.13.2.9

Chen, A. (2015). The impacts of listening strategy instruction on strategy use and listening performance of EFL learners. *International Journal on Studies in English Language and Literature, 3*(5), 75–87.

Chen, C.-H., Hung, H.-T., & Yeh, H.-C. (2021). Virtual reality in problem-based learning contexts: Effects on the problem-solving performance, vocabulary acquisition and motivation of English language learners. *Journal of Computer Assisted Learning, 37*(3), 851–860. doi:10.1111/jcal.12528

Chen, H. (2011). Developing and evaluating SyntoLearn, a fully automatic video and transcript synchronization tool for EFL learners. *Computer Assisted Language Learning, 24*(2), 117–130. doi:10.1080/09588221.2010.526947

Chen, I., & Chang, C. (2011). Content presentation modes in mobile language listening tasks: English proficiency as a moderator. *Computer Assisted Language Learning, 24*(5), 451–470. doi:10.1080/09588221.2011.577749

Chen, J., Belkada, S., & Okamoto, T. (2004). How a web-based course facilitates acquisition of English for academic purposes. *Language Learning & Technology, 8*, 33–49.

Chen, T. (2016). Technology-supported peer feedback in ESL/EFL writing classes: A research synthesis. *Computer Assisted Language Learning, 29*(2), 365–397. doi:10.1080/09588221.2014.960942

Chen, Z., Chen, W., Jia, J., & An, H. (2020). The effects of using mobile devices on language learning: A meta-analysis. *Educational Technology Research and Development, 68*(4), 1769–1789. doi:10.100711423-020-09801-5

Cheon, J., & Grant, M. (2009). Web-based assessment tool for communication and active listening skill development. *TechTrends, 53*(6), 24–34. doi:10.100711528-009-0340-4

Chinnery, G. M. (2006). Emerging technologies going to the MALL: Mobile assisted language learning. *Language Learning & Technology, 10*, 9–16.

Chiu, Y. H. (2013). Computer-assisted second language vocabulary instruction: A meta-analysis. *British Journal of Educational Technology, 44*(2), E52–E56. doi:10.1111/j.1467-8535.2012.01342.x

Chiu, Y., Kao, C., & Reynolds, B. (2012). The relative effectiveness of digital game-based learning types in English as a foreign language setting: A meta-analysis. *British Journal of Educational Technology, 43*(4), E104–E107. doi:10.1111/j.1467-8535.2012.01295.x

Cho, K., & Schunn, C. D. (2007). Scaffolded writing and rewriting the discipline: A web-based reciprocal peer review system. *Computers & Education, 48*(3), 409–426. doi:10.1016/j.compedu.2005.02.004

Christensen, L., & Johnson, B. (2004). *Educational research: quantitative, qualitative and mixed approaches* (2nd ed.). Pearson Education.

Clark, H. H., & Fox Tree, J. E. (2002). Using Uh and Um in spontaneous speaking. *Cognition, 84*(1), 73–111. doi:10.1016/S0010-0277(02)00017-3 PMID:12062148

Clifford, R. (1998). Mirror, mirror, on the wall: Reflections on computer assisted language learning. *CALICO Journal, 16*(1), 1–11.

Cobb, T., & Boulton, A. (2015). Classroom applications of corpus analysis. In D. Biber & R. Reppen (Eds.), *The Cambridge handbook of corpus linguistics* (pp. 478–497). Cambridge University Press. doi:10.1017/CBO9781139764377.027

Cohen, J. (1988). *Statistical power analysis for the behavioral sciences* (2nd ed.). Erlbaum.

Coniam, D. (2001). The use of audio or video comprehension as an assessment instrument in the certification of English language teachers: A case study. *System, 29*(1), 1–14. doi:10.1016/S0346-251X(00)00057-9

Connor, U., & Asenavage, K. (1994). Peer response groups in ESL writing classes: How much impact on revision? *Journal of Second Language Writing, 3*(3), 257–276. doi:10.1016/1060-3743(94)90019-1

Conrad, L. (1985). Semantic versus syntactic cues in listening comprehension. *Studies in Second Language Acquisition, 7*(1), 59–69. doi:10.1017/S0272263100005155

Cook, G., & Seidlhofer, B. (1995). An applied linguist in principle and practice. In G. Cook & B. Seidlhofer (Eds.), *Principle and practice in applied linguistics* (pp. 1–26). Oxford University Press.

Cook, G., & Seidlhofer, B. (Eds.), *Principle and practice in applied linguistics* (pp. 363–377). Oxford University Press.

Cooper, S., Dale, C., & Spencer, S. (2009). A tutor in your back pocket: Reflections on the use of iPods and podcasting in an undergraduate popular music programme. *British Journal of Music Education, 26*(1), 85–97. doi:10.1017/S0265051708008280

Copley, J. (2007). Audio and video podcasts for lectures for campus-based students production and evaluation of student use. *Innovations in Education and Teaching International, 44*(4), 387–399. doi:10.1080/14703290701602805

Corgan, R., Hammer, V., Margolies, M., & Crossley, C. (2004). Making your online course successful. *Business Education Forum, 58*(3), 51–53.

Corrales, K., & Maloof, C. (2011). Student perceptions on how content based instruction supports learner development in a foreign language context. *Zona Próxima, 15*, 40–53.

Creswell, J. W. (2012). *Educational Research: Planning, conducting and evaluating quantitative and qualitative research.* Pearson Education, Inc.

Creswell, J. W. (2013). *Qualitative inquiry and research design: Choosing among five approaches.* Thousand Oaks, California: Sage Publications, Inc.

Crinon, J. (2012). The dynamics of writing and peer review at primary school. *Journal of Writing Research, 4*(2), 121–154. doi:10.17239/jowr-2012.04.02.2

Cross, J. (2011). Comprehending news videotexts: The influence of the visual content. *Language Learning & Technology, 15*, 44–68.

Cross, J. (2013). Promoting autonomous listening to podcasts: A case study. *Language Teaching Research*, *18*(1), 8–32. doi:10.1177/1362168813505394

Danan, M. (1992). Reversed subtitling and dual coding theory: New directions for foreign language instruction. *Language Learning*, *42*(4), 497–527. doi:10.1111/j.1467-1770.1992.tb01042.x

Danan, M. (2004). Captioning and subtitling: Undervalued language learning strategies. *Meta*, *49*(1), 67–77. doi:10.7202/009021ar

Davis, K. (1995). Qualitative theory and methods in applied linguistics research. *TESOL Quarterly*, *29*(3), 427–453. doi:10.2307/3588070

Davis, L. (2009). The influence of interlocutor proficiency in a paired oral assessment. *Language Testing*, *26*(3), 367–396. doi:10.1177/0265532209104667

De Guerrero, M., & Villamil, O. (1994). Social-cognitive dimensions of interaction in L2 peer revision. *Modern Language Journal*, *78*(4), 484–496. doi:10.1111/j.1540-4781.1994.tb02065.x

De la Fuente, M. J. (2014). Learners' attention to input during focus on form listening tasks: The role of mobile technology in the second language classroom. *Computer Assisted Language Learning*, *27*(3), 261–276. doi:10.1080/09588221.2012.733710

Dehghani, A. P., & Jowkar, M. (2005). The impact of computer-video projector on EFL learners' listening comprehension. *Academic Research International*, *3*, 106–112.

Diab, N. M. (2010). Effects of peer- versus self-editing on students' revision of language errors in revised drafts. *System*, *38*(1), 85–95. doi:10.1016/j.system.2009.12.008

DiGiovanni, E., & Nagaswani, G. (2001). Online peer review: An alternative to face-to-face? *ELT Journal*, *55*(3), 263–272. doi:10.1093/elt/55.3.263

Donato, R. (2004). Aspects of collaboration in pedagogical discourse. *Annual Review of Applied Linguistics*, *24*, 284–302. doi:10.1017/S026719050400011X

Doolittle, P. E., & Hicks, D. (2003). Constructivism as a theoretical foundation for the use of technology in social studies. *Theory and Research in Social Education*, *31*(1), 72–104. doi:10.1080/00933104.2003.10473216

Driscoll, M. P. (2000). *Psychology of learning for instruction* (2nd ed.). Needham Heights, MA: Allyn & Bacon.

Ducasse, A. M., & Brown, A. (2009). Assessing paired orals: Raters' orientation to interaction. *Language Testing*, *26*(3), 423–443. doi:10.1177/0265532209104669

Ducate, L., & Lomicka, L. (2009). Podcasting: An effective tool for honing language students' pronunciation? *Language Learning & Technology*, *13*, 66–86.

Duman, G., Orhon, G., & Gedik, N. (2014). Research trends in mobile assisted language learning from 2000 to 2012. *ReCALL, FirstView*, 1-20.

Dunlap, J. C. (2005). Workload reduction in online courses: Getting some shuteye. *Performance Improvement*, *44*(5), 18–25. doi:10.1002/pfi.4140440507

Dunlop, J. C., & Grabinger, R. S. (1996). Rich environment for the active learning in the higher education. In B. G. Wilson (Ed.), *Constructing learning environments: Case studies in instructional design* (pp. 65–82). Englewood Cliiffs, NJ: Educational Technology Publications.

Dyson, A. H., & Freedman, S. W. (1991). Writing. In *Handbook of research on teaching the English language arts*. Macmillan.

East, M., & King, C. (2012). L2 learners' engagement with high stakes listening tests: Does technology have a beneficial role to play? *CALICO Journal*, *29*(2), 208–223. doi:10.11139/cj.29.2.208-223

Eddington, D., & Elzinga, D. (2008). The phonetic, Context of American English Flapping: Quantitative Evidence. *Language and Speech*, *52*(3), 245–266. doi:10.1177/0023830908098542 PMID:19626926

Eggen, P. D., & Kauchak, D. P. (2016). *Educational psychology: Windows on Classrooms*. Pearson Education.

Ekşi, G. Y. (2012). Peer review versus teacher feedback in process writing: How effective? *IJAES*, *13*(1), 33–48.

Ellis, R. (2008). *The study of Second Language Acquisition*. Oxford University Press.

Elwood, J. A., & Bode, J. (2014). Student preferences vis-à-vis teacher feedback in university EFL writing classes in Japan. *System*, *42*, 333–343. doi:10.1016/j.system.2013.12.023

Emerick, M. R. (2019). Explicit teaching and authenticity in L2 listening instruction: University language teachers' beliefs. *System*, *80*, 107–119. doi:10.1016/j.system.2018.11.004

Eraut, M. (1988). What has happened to learning design. In H. Mathias, N. Rushby, & R. Budgett (Eds.), *Aspects of Educational Technology XXI: Designing New Systems and Technologies for Learning*. Kogan Page.

Erbanova, P. (2014). *Aspects of connected speech in English: Assessing students' progress after pronunciation training* (Master's Thesis). Masaryk University. Retrieved from https://is.muni.cz/th/c4mfj/Master_s_thesis.pdf

Erçetin, G. (2011). Pedagogical issues in developing mobile assisted language learning materials. *Boğaziçi Journal of Education*, *28*(1), 22–31.

Erdölen, Y. E. (2020, November 8). *ABD Seçim ve Kampanya Değerlendirmesi|Gülfem Saydan Sanver & Yunus Emre Erdölen | Üsküdar Motoru #3* [daktilo1984 Youtube Channel]. Retrieved from https://www.youtube.com/watch?v=2LMbpvI1Xfg&feature=youtu.be

Ertmer, P. A., Richardson, J. C., Belland, B., Camin, D., Connolly, P., Coulthard, G., Lei, K., & Mong, C. (2007). Using peer feedback to enhance the quality of student online postings: An exploratory study. *Journal of Computer-Mediated Communication*, *12*(2), 412–433. doi:10.1111/j.1083-6101.2007.00331.x

Etemadi, A. (2012). Effects of bimodal subtitling of English movies on content comprehension and vocabulary recognition. *International Journal of English Linguistics*, *2*(1), 239–248. doi:10.5539/ijel.v2n1p239

Ezza, E. S. (2010). Arab EFL learners' writing dilemma at tertiary level. *English Language Teaching*, *3*(4), 33–39. doi:10.5539/elt.v3n4p33

Fallahkhair, S., Pemberton, L., & Griffiths, R. (2007). Development of a cross-platform ubiquitous language learning service via mobile phone and interactive television. *Journal of Computer Assisted Learning*, *23*(4), 312–325. doi:10.1111/j.1365-2729.2007.00236.x

Ferris, D. (2002). *Treatment of error in second language student writing*. The University of Michigan Press.

Ferris, D. (2006). Does error feedback help student writers? New evidence on the short- and long- term effect of written error correction. In K. Hyland & F. Hyland (Eds.), *Peer feedback in second language writing: Context and issues* (pp. 81–104). Cambridge University Press. doi:10.1017/CBO9781139524742.007

Ferris, D. R. (2003). *Response to student writing: Implications for second language students*. Lawrence Erlbaum Associates. doi:10.4324/9781410607201

Ferris, D. R., Liu, H., Sinha, A., & Senna, M. (2013). Written corrective feedback for individual L2 writers. *Journal of Second Language Writing*, *22*(3), 307–329. doi:10.1016/j.jslw.2012.09.009

Feyten, C. M. (1991). The power of listening ability: An overlooked dimension in language acquisition. *Modern Language Journal*, *75*(2), 173–180. doi:10.1111/j.1540-4781.1991.tb05348.x

Field, J. (2008). *Listening in the language classroom*. Cambridge University Press.

Field, J. (2008). Revising segmentation hypotheses in first and second language listening. *System*, *36*(1), 35–51. doi:10.1016/j.system.2007.10.003

Filatova, O. (2016). More than a word cloud. *TESOL Journal*, *7*(2), 438–448. doi:10.1002/tesj.251

Fitze, M. (2006). Discourse and participation in ESL face-to-face and written electronic conferences. *Language Learning & Technology*, *10*(1), 67–86.

Flowerdew, J., & Miller, L. (2005). *Developing listening skills through technology. In Second Language Listening: Theory and Practice*. Cambridge University Press.

Flower, L., & Hayes, J. R. (1980). The cognition of discovery: Defining a rhetorical problem. *College Composition and Communication*, *31*(1), 21–32. doi:10.2307/356630

Foot, M. C. (1999). Relaxing in pairs. *ELT Journal*, *53*(1), 36–41. doi:10.1093/elt/53.1.36

Freedle, R., & Kostin, I. (1999). Does the text matter in a multiple-choice test of comprehension? The case for the construct validity of TOEFL's minitalks. *Language Testing*, *16*(1), 2–32.

Freedman, S. W. (1992). Outside-in and inside-out: Peer response groups intwo ninth-grade classes. *Research in the Teaching of English*, *26*(1), 71–107.

Fulford, A., & Zhang, S. (1998). Assessing distance learners' satisfaction with instruction: A quantitative and qualitative measure. *American Journal of Distance Education*, *12*(3), 4–27. doi:10.1080/08923649809527002

Galaczi, E. D. (2008). Peer-peer interaction in a speaking test: The case of the First Certificate in English Examination. *Language Assessment Quarterly*, *5*(2), 89–119. doi:10.1080/15434300801934702

Gan, Z., Davison, C., & Hamp-Lyons, L. (2009). Topic negotiation in peer group oral assessment situations: A conversation analytic approach. *Applied Linguistics*, *30*(3), 315–334. doi:10.1093/applin/amn035

Gao, Q., & Liu, L. (2013). Investigating college English teachers' beliefs in listening teaching and their relationships with teaching practices. *Foreign language word*. Retrieved 1 May, 2021 from https://en.cnki.com.cn/Article_en/CJFDTotal-WYJY201302004.htm

Gao, X., & Ma, Q. (2011). Vocabulary learning and teaching beliefs of pre-service and in-service teachers in Hong Kong and mainland China. *Language Awareness*, *20*(4), 327–342. doi:10.1080/09658416.2011.579977

Gaskell, M. G., & Snoeren, N. D. (2008). The impact of strong assimilation on the perception of connected speech. *Journal of Experimental Psychology. Human Perception and Performance*, *34*(6), 1–29. doi:10.1037/a0011977 PMID:19045997

Gere, A. R., & Abbott, R. D. (1985). Talking about writing: The language of writing groups. *Research in the Teaching of English*, *10*(4), 362–381.

Gilles, R., & Snook, L. (Eds.). (1990). *A student guide to freshman composition* (11th ed.). University of Arizona Bellwether Press.

Gjevori, E. (2020, November 3). *The View from Turkey regarding US elections.* Retrieved from https://www.trtworld.com/magazine/the-view-from-turkey-regarding-us-elections-41143

Goctu, R. (2017). Using movies in EFL classrooms. *European Journal of Language and Literature Studies, 3*(2), 121–124. doi:10.26417/ejls.v8i1.p121-124

Goh, C. (2013). Second language listening comprehension: Process and pedagogy. In M. Celce-Murica (Ed.), *Teaching English as a second or foreign language* (pp. 72–89). Heinle & Heinle.

Goh, C. C. (2000). A cognitive perspective on language learners' listening comprehension prob lems. *System, 28*(1), 55–75. doi:10.1016/S0346-251X(99)00060-3

Goldberg, A., Russell, M., & Cook, A. (2003). The effect of computers on student writing: A meta-analysis of studies from 1992 to 2002. *The Journal of Technology, Learning, and Assessment, 2*, 1–51.

Gow, D. W. Jr, & Im, A. M. (2004). A cross-linguistic examination of assimilation context effects. *Journal of Memory and Language, 51*(2), 279–296. doi:10.1016/j.jml.2004.05.004

Graham, S., Santos, D., & Francis-Brophy, E. (2014). Teacher beliefs about listening in a foreign language. *Teaching and Teacher Education, 40*, 44–60. doi:10.1016/j.tate.2014.01.007

Graham, S., Santos, D., & Vanderplank, R. (2010). Strategy clusters and sources of knowledge in French L2 listening comprehension. *International Journal of Innovation in Language Learning and Teaching, 4*(1), 1–20. doi:10.1080/17501220802385866

Grgurović, M., Chapelle, C. A., & Shelley, M. C. (2013). A meta-analysis of effectiveness studies on computer technology-supported language learning. *ReCALL, 25*(2), 165–198. doi:10.1017/S0958344013000013

Grgurović, M., & Hegelheimer, V. (2007). Help options and multiple media listening: Students' use of subtitles and the transcript. *Language Learning & Technology, 11*, 45–66.

Griffiths, R. (1990). Facilitating listening comprehension through rate-control. *RELC Journal, 21*(1), 55–65. doi:10.1177/003368829002100104

Griffiths, R. (1992). Speech rate and listening comprehension: Further evidence of the relationship. *TESOL Quarterly, 26*(2), 385–390. doi:10.2307/3587015

Gruba, P. (1993). A comparison study of audio and video in language testing. *JALT Journal, 15*, 85–88.

Gruba, P. (2006). Playing the videotext: A media literacy perspective on video-mediated L2 listening. *Language Learning & Technology, 10*, 77–92.

Guardado, M., & Shi, L. (2007). ESL students' experiences of online peer feedback. *Computers and Composition, 24*(4), 443–461. doi:10.1016/j.compcom.2007.03.002

Gupta, K. C.-L. (2020). Researcher-teacher collaboration in adopting critical Content and Language Integrated Learning (CLIL): Processes, challenges and outcomes. *Trabalhos em Lingüística Aplicada, 59*(1), 42–77. doi:10.1590/010318136014125912020

Habertürk. (2020, November 4). *Türkiye'nin Nabzı, ABD Seçim Özel.* Habertürk TV Youtube Channel.

Habertürk. (2020, November 5). *HT 360, Donald Trump görevi Joe Biden'a devretmeyi reddedebilir mi?* [Habertürk TV Youtube Channel]. Retrieved from https://www.youtube.com/watch?v=3DbOt5f79Jk

Habertürk. (2020, November 6). *HT 360, Kim bu mahcup Trumpçılar?* [Habertürk TV Youtube Channel]. Retrieved from https://www.youtube.com/watch?v=1Pgy12jjyrg

Hall, G., & Cook, G. (2012). Own-language use in language teaching and learning. *Language Teaching, 45*(3), 271–308. doi:10.1017/S0261444812000067

Hall, G., & Cook, G. (2013). *Own-language use in ELT: Exploring global practices and attitudes.* British Council.

Hanjani, A. M., & Li, L. (2014). Exploring L2 writers' collaborative revision interactions and their writing performance. *System, 44*, 101–114. doi:10.1016/j.system.2014.03.004

Hansen, J. G., & Liu, J. (2005). Guiding principles for effective peer response. *ELT Journal, 59*(1), 31–38. doi:10.1093/elt/cci004

Hasan, A. S. (2000). Learners' perceptions of listening comprehension problems. *Culture and Curriculum, 13*(2), 137-153. . doi:10.1080/07908310008666595

Hattie, J., & Timperley, H. (2007). The power of feedback. *Review of Educational Research, 77*(1), 81–112. doi:10.3102/003465430298487

Hayati, A., & Mohmedi, F. (2011). The effect of films with and without subtitles on listening comprehension of EFL learners. *British Journal of Educational Technology, 42*(1), 181–192. doi:10.1111/j.1467-8535.2009.01004.x

Hayes, J. R. (2012). Modeling and remodeling writing. *Written Communication, 29*(3), 369–388. doi:10.1177/0741088312451260

Hedgecock, J., & Lefkowitz, N. (1992). Collaborative oral/aural revision in foreign language writing instruction. *Journal of Second Language Writing, 1*(2), 255–276. doi:10.1016/1060-3743(92)90006-B

Henrichen, L. E. (1984). Sandhi-variation: A filter of input for learners of ESL. *Language Learning, 34*(3), 103–126. doi:10.1111/j.1467-1770.1984.tb00343.x

Herrington, A., & Cadman, D. (1991). Peer review and revising in an anthropology course: Lessons for learning. *College Composition and Communication, 42*(2), 184–199. doi:10.2307/358198

Herrington, J., & Kervin, L. (2007). Authentic learning supported by technology: Ten suggestions and cases of integration in classroom. *Educational Media International, 44*(3), 219–236. doi:10.1080/09523980701491666

Herron, C., York, H., Corrie, C., & Cole, S. P. (2006). A comparison study of the effects of a story-based video instructional package versus a text-based instructional package in the intermediate-level foreign language classroom. *CALICO Journal, 23*(2), 281–307. doi:10.1558/cj.v23i2.281-307

Higgins, R., Hartley, P., & Skelton, A. (2002). The conscientious consumer: Reconsidering the role of assessment feedback in student learning. *Studies in Higher Education, 27*(1), 53–64. doi:10.1080/03075070120099368

Hochart, J. (1998). Improving listening and speaking skills in English through the use of authoring systems. *ReCALL, 10*(2), 18–24. doi:10.1017/S0958344000003700

Hoey, E. M., & Kendrick, K. H. (2018). Conversation analysis. In A. M. B. de Groot & P. Hagoort (Eds.), *Research methods in psycholinguistics and the neurobiology of language: A practical guide* (pp. 151–173). Wiley Blackwell.

Hojeij, Z., & Baroudi, S. (2018). Student perceptions on peer feedback training using a blended method: A UAE case. *Issues in Educational Research*, *28*(3), 655–678. http://www.iier.org.au/iier28/hojeij.pdf

Ho, M. C. (2015). The effects of face-to-face and computer-mediated peer review on EFL writers' comments and revisions. *Australasian Journal of Educational Technology*, *31*(1), 1–15. doi:10.14742/ajet.495

Ho, M., & Savignon, S. (2007). Face-to-face and computer-mediated peer review in EFL writing. *CALICO Journal*, *24*(2), 269–290. doi:10.1558/cj.v24i2.269-290

Hoogendyk, F. W. C., & Arifin, Z. & Novita, D. (2014). Increasing students' vocabulary by using bimodal subtitling in short movie. *Jurnal Pendidikan dan Pembelajaran (Universitas Tanjungpura)*, *3*(11), 1-14.

Hsieh, Y. (2019). Effects of video captioning on EFL vocabulary learning and listening comprehension. *Computer Assisted Language Learning*, 1–23.

Hsu, C. K., Hwang, G. J., & Chan, C. K. (2014). An automatic caption filtering and partial hiding approach to improving the English listening comprehension of EFL students. *Journal of Educational Technology & Society*, *17*, 270–283.

Hsu, C. K., Hwang, G. J., Chang, Y. T., & Chang, C. K. (2013). Effects of Video Caption Modes on English Listening Comprehension and Vocabulary Acquisition Using Handheld Devices. *Journal of Educational Technology & Society*, *16*, 403–414.

Hsu, C.-K., Hwank, G.-J., Chang, Y.-T., & Chang, C.-K. (2013). Effects of video caption modes on English listening comprehension and vocabulary acquisition using handheld devices. *Journal of Educational Technology & Society*, *16*(1), 403–414.

Hubbard, P., & Levy, M. (2006). The scope of CALL education. In P. Hubbard & M. Levy (Eds.), *Teacher Education in CALL* (pp. 3–20). John Benjamins., doi:10.1075/lllt.14.04hub

Hu, G. W. E., & Lam, S. T. E. (2010). Issues of cultural appropriateness and pedagogical efficacy: Exploring peer review in a second language writing class. *Instructional Science*, *38*(4), 371–384. doi:10.100711251-008-9086-1

Hui, W. W., Hu, P. H., Clark, T. K., Tam, K. Y., & Milton, J. J. (2008). Technology-assisted learning: A longitudinal field study of knowledge category, learning effectiveness and satisfaction in language learning. *Journal of Computer Assisted Learning*, *24*(3), 245–259. doi:10.1111/j.1365-2729.2007.00257.x

Hwang, G. J., Hsu, T. C., Lai, C. L., & Hsueh, C. J. (2017). Interaction of problem-based gaming and learning anxiety in language students' English listening performance and progressive behavioral patterns. *Computers & Education*, *106*, 26–42. doi:10.1016/j.compedu.2016.11.010

Hwang, G. J., & Tsai, C. C. (2011). Research trends in mobile and ubiquitous learning: A review of publications in selected journals from 2001 to 2010. *British Journal of Educational Technology*, *42*(4), E65–E70. doi:10.1111/j.1467-8535.2011.01183.x

Hwang, W. Y., Huang, Y. M., Shadiev, R., Wu, S. Y., & Chen, S. L. (2014). Effects of using mobile devices on English listening diversity and speaking for EFL elementary students. *Australasian Journal of Educational Technology*, *30*(5), 503–516. doi:10.14742/ajet.237

Hwang, W. Y., Shih, T. K., Ma, Z. H., Shadiev, R., & Chen, S. Y. (2016). Evaluating listening and speaking skills in a mobile game-based learning environment with situational contexts. *Computer Assisted Language Learning*, *29*(4), 639–657. doi:10.1080/09588221.2015.1016438

Hwu, F. (2004). On the applicability of the input-enhancement hypothesis and input processing theory in multiple media CALL: The case of Spanish preterite and imperfect instruction in an input application. *CALICO Journal*, *21*(2), 317–338. doi:10.1558/cj.v21i2.317-338

Hyland, F. (2000). ESL Writers and Feedback: Giving more autonomy to students. *Language Teaching Research*, *4*(1), 33–54. doi:10.1177/136216880000400103

Hyland, K., & Hyland, F. (2006). Feedback on second language students' writing. *Language Teaching*, *39*(2), 83–101. doi:10.1017/S0261444806003399

Idsardi, W. J., & Sung, E.-K. (2003). Flaps in American English and Korean: An acoustic and perceptual study. Generative Grammar in a Broader Perspective. *Proceedings of the 4th GLOW in Asia*, 187-208.

Indahsari, D. (2020). Using podcast for EFL students in language learning. *JEES*, *5*(2), 103–108. doi:10.21070/jees.v5i2.767

Information about Teleparty (n.d.). http://www.netflixpart.com

Irgin, P. (2021). Testing and assessment of listening, listening test task types, and sample test items. In S. Çelik, H. Çelik, & C. Coombe (Eds.), *Language assessment and test preparation in English as a foreign language (EFL) education* (pp. 91–115). Vizetek.

Irgin, P., & Erten, I. H. (2020). Exploring the role of strategy instruction: Young learners' listening performance and strategy use. *Eurasian Journal of Applied Linguistics*, *6*(3), 415–441.

Ito, Y. (2001). Effect of reduced forms on ESL learners' input-intake process. *Second Language Studies 20*(1), 99-124. Retrieved June 6, 2007 from https://www.hawaii.edu/sls/uhwpesl/20(1)/Ito.pdf

Ito, Y. (2006). The comprehension of English reduced forms by second language learners and its effect on input-intake process. In J. D. Brown & K. Kondo-Brown (Eds.), *Perspectives on teaching connected speech to second language speakers* (pp. 67–81). University of Hawai'i National Foreign Language Resource Center.

Jaber, R. A., & Dollar, Y. K. (2019). A meta-analysis of the effect of bimodal subtitling on vocabulary learning among adult EFL learners. *Focus on ELT Journal*, *1*(1), 43–52. doi:10.14744/felt.2019.00001

Jacobs, A., Fricke, M., & Kroll, J. F. (2016). Cross-Language Activation Begins During Speech Planning and Extends Into Second Language Speech. *Language Learning*, *66*(2), 324–353. doi:10.1111/lang.12148 PMID:27773945

Jacobs, G. M., Curtis, A., Braine, G., & Juan, S. Y. (1998). Feedback on student writing: Taking the middle path. *Journal of Second Language Writing*, *7*(3), 307–317. doi:10.1016/S1060-3743(98)90019-4

Jahin, J. H. (2012). The effect of peer reviewing on writing apprehension and essay writing ability of prospective EFL teachers. *The Australian Journal of Teacher Education*, *37*(1), 60–84. doi:10.14221/ajte.2012v37n11.3

Jakobsdóttir, S., & Hooper, S. (1995). Computer-assisted foreign language learning: Effects of text, context, and gender on listening comprehension and motivation. *Educational Technology Research and Development*, *43*(4), 43–59. doi:10.1007/BF02300490

Jehn, K. A., & Shah, P. P. (1997). Interpersonal relationships and task performance: An examination of mediation processes in friendship and acquaintance groups. *Journal of Personality and Social Psychology*, *72*(4), 775–790. doi:10.1037/0022-3514.72.4.775

Jensen, E. D., & Vinther, T. (2003). Exact repetition as input enhancement in second language acquisition. *Language Learning*, *53*(3), 373–428. doi:10.1111/1467-9922.00230

Jia, C., & Hew, K. F. T. (2019). Supporting lower-level processes in EFL listening: The effect on learners' listening proficiency of a dictation program supported by a mobile instant messaging app. *Computer Assisted Language Learning*, 1–28. doi:10.1080/09588221.2019.1671462

Kanellopoulou, C. (2019). Film subtitles as a successful vocabulary learning tool. *Open Journal of Modern Linguistics*, *9*(02), 145–152. doi:10.4236/ojml.2019.92014

Kang, T. (2015). The effectiveness of multiple media tools in L2 listening: A meta-analysis. In V. Turel (Ed.), *Intelligent design of interactive multimedia listening software* (pp. 255–284). IGI Global. doi:10.4018/978-1-4666-8499-7.ch010

Kao, C. W. (2014). The effects of digital game-based learning task in English as a foreign language contexts: A meta-analysis. *Education Journal*, *42*, 113–141.

Karimi, M. N., & Nazari, M. (2017). The congruity/incongruity of EFL teachers' beliefs about listening instruction and their listening instructional practices. *The Australian Journal of Teacher Education*, *42*(2), 62–80. doi:10.14221/ajte.2017v42n2.5

Kaufman, J. H., & Schunn, C. D. (2011). Students' perceptions about peer assessment for writing: Their origin and impact on revision work. *Instructional Science*, *39*(3), 387–406. doi:10.100711251-010-9133-6

Kayi-Aydar, H. (2013). Scaffolding language learning in an academic ESL classroom. *ELT Journal*, *67*(3), 324–335. doi:10.1093/elt/cct016

Keh, C. L. (1990). Feedback in the writing process: A model and methods for implementation. *ELT Journal*, *44*(4), 294–394. doi:10.1093/elt/44.4.294

Kennedy, D. M., & McNaught, C. (1997). Design elements for interactive multimedia. *Australian Journal of Educational Technology*, *13*(1), 1–22.

Kessler, G., & Bukowski, D. (2010). Developing collaborative autonomous learning abilities in computer mediated language learning: Attention to meaning among students in wiki space. *Computer Assisted Language Learning*, *23*(1), 41–58. doi:10.1080/09588220903467335

Khalil, E. (2018). The efficacy of peer feedback in Turkish EFL students' writing performance. *Journal of Literature and Art Studies*, *8*(6), 920–931.

Kim, H. S. (2015). Using Authentic Videos to Improve EFL Students' Listening Comprehension. *International Journal of Contents*, *11*(4), 15–24. doi:10.5392/IJoC.2015.11.4.015

Kırkgöz, Y. (2009). Globalization and English Language Policy in Turkey. *Educational Policy*, *23*(5), 663–684. doi:10.1177/0895904808316319

Kissling, E. M. (2012). *The effect of phonetic instruction on adult learners' perception and production* (Doctoral dissertation). Georgetown University. Retrieved from https://repository.library.georgetown.edu/handle/10822/557711

Köksal, D. (2005). *Çeviri eğitimi: Kuram ve uygulama*. Nobel.

Köksal, D. (2005). *Çeviri Eğitimi: Kuram ve Uygulama*. Nobel.

Kondo, M., Ishikawa, Y., Smith, C., Skamoto, K., Shimomura, H., & Wada, N. (2012). Mobile assisted language learning in university EFL courses in Japan: Developing attitudes and skills for self-regulated learning. *ReCALL*, *24*(2), 169–187. doi:10.1017/S0958344012000055

Ko, S., & Rossen, S. (2001). *Teaching online: A practical guide*. Houghton Mifflin.

Koschmann, T. (1996). Paradigm shifts and instructional technology: An introduction. In T. Koschmann (Ed.), *CSCL: Theory and practice of an emerging paradigm* (pp. 1–23). Lawrence Erlbaum Associates.

Kremmel, B., Brunfaut, T., & Alderson, J. C. (2015). Exploring the role of phraseological knowledge in foreign language reading. *Applied Linguistics*, *1*(24), amv070. Advance online publication. doi:10.1093/applin/amv070

Krych-Appelbaum, M., & Musial, J. (2007). Students' perception of value of interactive oral communication as part of writing course papers. *Journal of Instructional Psychology*, *34*(3), 131–136.

Kuhl, P. K., Conboy, B. T., Coffey-Corina, S., Padden, D., Rivera-Gaxiola, M., & Nelson, T. (2008). Phonetic Learning as a pathway to language: New data and native language magnet theory expanded (NLM-e). *Philosophical Transactions of the Royal Society of London. Series B, Biological Sciences*, *363*(1493), 979–1000. doi:10.1098/rstb.2007.2154 PMID:17846016

Kukulska-Hulme, A., & Shield, L. (2008). An overview of mobile assisted language learning: From content delivery to supported collaboration and interaction. *ReCALL*, *20*(3), 271–289. doi:10.1017/S0958344008000335

Kumaravadivelu, B. (2012). *Language Teacher Education for a Global Society: A Modular Model for Knowing, Analyzing, Recognizing, Doing and Seeing*. Routledge. doi:10.4324/9780203832530

Kuyyogsuy, S. (2019). Students' attitudes toward peer feedback: Paving a way for students' English writing improvement. *English Language Teaching*, *12*(7), 107–119. doi:10.5539/elt.v12n7p107

Lam, R. (2010). A peer review training workshop: Coaching students to give and evaluate peer feedback. *TESL Canada Journal*, *27*(2), 114–127. doi:10.18806/tesl.v27i2.1052

Lantolf, J. (Ed.). (2000). *Sociocultural theory and second language learning*. Oxford University Press.

Lantolf, J., & Thorne, S. (2006). *Sociocultural theory and the genesis of second language development*. Oxford University Press.

Larson-Hall, J. (2009). *A guide to doing statistics in second language research using SPSS*. Routledge. doi:10.4324/9780203875964

Le Mon, R. (1988). Cultural differences affecting the international transfer of computerised language learning. *System*, *16*(1), 37–40. doi:10.1016/0346- 251X(88)90008-5

Lee, I. (2017). *Classroom writing assessment and feedback in L2 school contexts*. Springer. doi:10.1007/978-981-10-3924-9

Lee, M. (2015). Peer feedback in second language writing: Investigating junior secondary students' perspectives on inter-feedback and intra-feedback. *System*, *55*, 1–10. doi:10.1016/j.system.2015.08.003

Lei, Z. (2017). Salience of student written feedback by peer revision in EFL writing class. *English Language Teaching*, *10*(12), 151–157. doi:10.5539/elt.v10n12p151

Leki, I. (1990). Potential problems with peer responding in ESL writing classes. *The CATESOL Journal*, *3*, 5–19.

Levak, N., & Son, J. B. (2017). Facilitating second language learners' listening comprehension with Second Life and Skype. *ReCALL*, *29*(2), 200–218. doi:10.1017/S0958344016000215

Leveridge, A., & Yang, J. (2013). Testing learner reliance on caption supports in second language listening comprehension multiple media environments. *ReCALL*, *25*(2), 199–214. doi:10.1017/S0958344013000074

Levis, G. M., & Levis, J. (2010). Authentic speech and teaching sentence focus In J. Levis, & K. LeVelle (Eds.), *Proceedings of 1st Pronunciation in Second Language Learning and Teaching Conference* (pp. 135-144). Ames, IA: Iowa State University.

Liang, M. (2010). Using synchronous online peer response groups in EFL writing: Revision-related discourse. *Language Learning & Technology*, *14*(1), 45–64.

Liao, C. Y., & Yeldham, M. (2015). Taiwanese high school EFL teachers' perceptions of listening instruction. *The Asian Journal of Applied Linguistics*, *2*(2), 92–101.

Li, L. (2020). *Language teacher cognition a sociocultural perspective*. Palgrave Macmillan. doi:10.1057/978-1-137-51134-8

Lin, H. (2015). A meta-synthesis of empirical research on the effectiveness of computer-mediated communication (CMC) in SLA. *Language Learning & Technology*, *19*, 85–117.

Lin, S., Liu, E., & Yuan, S. (2001). Web-based peer assessment: Feedback for students with various thinking styles. *Journal of Computer Assisted Learning*, *17*(4), 420–432. doi:10.1046/j.0266-4909.2001.00198.x

Lin, S., Winaitham, W., & Saitakham, K. (2008). *The use of websites for practicing listening skills of undergraduate students. A case study at Suranaree University of Technology*. Online Submission.

Lin, W. C., Huang, H.-T., & Liou, H.-C. (2013). The effects of text-based SCMC on SLA: A metaanalysis. *Language Learning & Technology*, *17*, 123–142.

Liou, H. C., & Peng, C. Y. (2009). Training effects on computer-mediated peer review. *System*, *37*(3), 514–525. doi:10.1016/j.system.2009.01.005

Lipsey, M. W., & Wilson, D. B. (2001). *Practical meta-analysis*. Sage.

Little, C. W., Clark, J. C., Tani, N. E., & Connor, C. M. (2018). Improving writing skills through technology-based instruction: A meta-analysis. *Review of Education*, *6*(2), 183–201. doi:10.1002/rev3.3114

Liu, G. Z., Chen, J. Y., & Hwang, G. J. (2018). Mobile-based collaborative learning in the fitness center: A case study on the development of English listening comprehension with a context-aware application. *British Journal of Educational Technology*, *49*(2), 305–320. doi:10.1111/bjet.12581

Liu, J., & Hansen, J. (2002). *Peer response in second language writing classrooms*. University of Michigan Press. doi:10.3998/mpub.8952

Liu, J., & Sadler, R. W. (2003). The effect and affect of peer review in electronic versus traditional modes on L2 writing. *Journal of English for Academic Purposes*, *2*(3), 193–227. doi:10.1016/S1475-1585(03)00025-0

Liu, N., & Carless, D. (2006). Peer feedback: The learning element of peer assessment. *Teaching in Higher Education*, *11*(3), 279–290. doi:10.1080/13562510600680582

Liu, T. Y. (2009). A context-aware ubiquitous learning environment for language listening and speaking. *Journal of Computer Assisted Learning*, *25*(6), 515–527. doi:10.1111/j.1365-2729.2009.00329.x

Li, W., & Renandya, W. A. (2012). The effective approaches to teaching listening: Chinese EFL teachers' perspectives. *The Journal of Asia TEFL*, *9*(4), 79–111.

Lockhart, C., & Ng, P. (1995). Analyzing talk in ESL peer response groups: Stances, functions, and content. *Language Learning*, *45*(4), 605–655. doi:10.1111/j.1467-1770.1995.tb00456.x

Lodge, K. (2009). *A Critical Introduction to Phonetics*. Continuum.

Long, D. R. (1990). What you don't know can't help you. *Studies in Second Language Acquisition, 12*(1), 65–80. doi:10.1017/S0272263100008743

Long, M., & Porter, P. (1985). Group work, interlanguage talk and second language *acquisition. TESOL Quarterly, 19*(2), 305–325. doi:10.2307/3586827

Lord, G. (2008). Podcasting communities and second language pronunciation. *Foreign Language Annals, 41*(2), 364–379. doi:10.1111/j.1944-9720.2008.tb03297.x

Lou, Y., Abrami, P. C., & d'Apollonia, S. (2001). Small group and individual learning with technology: A meta-analysis. *Review of Educational Research, 71*(3), 449–521. doi:10.3102/00346543071003449

Lundstrom, K., & Baker, W. (2009). To give is better than to receive: The benefits of peer review to the reviewer's own writing. *Journal of Second Language Writing, 18*(1), 30–43. doi:10.1016/j.jslw.2008.06.002

Lu, R., & Bol, L. (2007). A comparison of anonymous versus identifiable e-Peer review on college student writing performance and the extent of critical feedback. *Journal of Interactive Online Learning, 6*(2), 100–115.

Lynch, M. M. (2002). *The online educator: A guide to creating the virtual classroom.* Routledge Falmer. doi:10.4324/9780203458556

Macaro, E., Graham, S., & Woore, R. (2015). *Improving foreign language teaching: Towards a research-based curriculum and pedagogy.* Routledge. doi:10.4324/9781315771229

Macaro, E., Vanderplank, R., & Graham, S. (2005). A systematic review of the role of prior knowledge in unidirectional listening comprehension. In *Research Evidence in Education Library.* EPPI-Centre, Social Science Research Unit, Institute of Education, University of London.

MacGregor, R. C. (1993). Are we recognising the organisational impact on educational soft- ware design? *Australian Journal of Educational Technology, 9*(1), 59–68.

Mackey, A., & Gass, S. (2005). *Second language research: Methodology and design.* Lawrence Erlbaum.

Manchón, R. M. (2012). Multiple perspectives on L2 writing development. In R. M. Manchón (Ed.), *L2 writing development: Multiple perspectives* (pp. 17–46). Walter de Gruyter. doi:10.1515/9781934078303.1

Mandasari, B. (2016). An analysis of teachers' beliefs toward authentic materials in teaching listening. *Teknosastik, 14*(1), 19–25. doi:10.33365/ts.v14i1.82

Mangelsdorf, K. (1992). Peer reviews in the ESL composition classroom: What do the students think? *ELT Journal, 46*(3), 274–285. doi:10.1093/elt/46.3.274

Mangelsdorf, K., & Schlumberger, A. (1992). ESL student response stances in a peer-review task. *Journal of Second Language Writing, 1*(3), 235–254. doi:10.1016/1060-3743(92)90005-A

Mardani, M., & Najmabadi, A. (2016). The Effect of movies with different types of subtitles on incidental English vocabulary learning of Iranian High School EFL learners. *Iranian Journal of Research in English Language Teaching (= Research in English Language Pedagogy), 4*(1), 52-63.

Markham, P. L., Peter, L. A., & McCarthy, T. J. (2001). The effects of native language vs. target language captions on foreign language students' DVD video comprehension. *Foreign Language Annals, 34*(5), 439–445. doi:10.1111/j.1944-9720.2001.tb02083.x

Marks, J. (1999). Is stress timing real? *ELT Journal, 52*(3), 191–199. doi:10.1093/elt/53.3.191

Matsuzawa, T. (2006). Comprehension of English reduced forms by Japanese business people and the effectiveness of instruction. In J. D. Brown & K. Kondo Brown (Eds.), *Perspectives on teaching connected speech to second language speakers* (pp. 59–66). University of Hawaii Press. National Foreign Language Resource Center.

Matthew, G. (2020). The effect of adding same-language subtitles to recorded lectures for non-native, English speakers in e-learning environments. *Research in Learning Technology*, *28*(0), 1–16. doi:10.25304/rlt.v28.2340

Matthew, R. (2008). Adventures in podcasting. *PS, Political Science & Politics*, *41*(03), 585–593. doi:10.1017/S1049096508080797

Matthews, J., Cheng, J., & O'Toole, J. M. (2015). Computer-mediated input, output and feedback in the development of L2 word recognition from speech. *ReCALL*, *27*(3), 321–339. doi:10.1017/S0958344014000421

Mavridi, S., & Xerri, D. (Eds.). (2020). *English for 21st century skills*. Express Publishing. https://www.expresspublishing.co.uk/en/content/english-21st-century-skills-0

Mayer, R. E. (2009). Multimedia learning (2nd ed.). Cambridge University Press. doi:10.1017/CBO9780511811678

Mayer, R. E. (2001). *Multimedia Learning*. Cambridge University Press. doi:10.1017/CBO9781139164603

Mayer, R. E. (2005). Cognitive theory of multimedia learning. In R. E. Mayer (Ed.), *The Cambridge Handbook of Multimedia Learning*. Cambridge University Press. doi:10.1017/CBO9780511816819.004

Mayer, R. E., Heiser, J., & Lonn, S. (2001). Cognitive constraints on multimedia learning: When presenting more material results in less understanding. *Journal of Educational Psychology*, *93*(1), 187–198. doi:10.1037/0022-0663.93.1.187

Mayer, R. E., & Moreno, R. (2003). Nine ways to reduce cognitive load in multimedia learning. *Educational Psychologist*, *38*(1), 43–52. doi:10.1207/S15326985EP3801_6

Maykut, P., & Morehouse, R. (1994). *Beginning qualitative research: A philosophic and practical guide*. The Falmer Press.

May, L. (2009). Co-constructed interaction in a paired speaking test: The rater's perspective. *Language Testing*, *26*(3), 397–421. doi:10.1177/0265532209104668

McBride, K. (2011). The effect of rate of speech and distributed practice on the development of listening comprehension. *Computer Assisted Language Learning*, *24*(2), 131–154. doi:10.1080/09588221.2010.528777

McCarthy, M. (2010). Spoken fluency revisited. *English Profile Journal*, *1*(1), 1–15. doi:10.1017/S2041536210000012

McConnell, D. (2002). The experience of collaborative assessment in e-learning. *Studies in Continuing Education*, *24*(1), 73–92. doi:10.1080/01580370220130459

McGroarty, M. E., & Zhu, W. (1997). Triangulation in classroom research: A study of peer revision. *Language Learning*, *47*(1), 1–43. doi:10.1111/0023-8333.11997001

McMinn, S. W. J. (2008). *Podcasting possibilities: Increasing time and motivation in the language learning classroom*. Retrieved November 24, 2012, from http://www.eife-l.org/publications/proceedings/ilf08/contributions/improving-quality-of-learning-with-technologies/McMinn.pdf/view

Mecartty, F. H. (2000). Lexical and grammatical knowledge in reading and listening comprehension by foreign language learners of Spanish. *Applied Language Learning*, *11*, 323–348.

Meinardi, M. (2009). Speed bumps for authentic listening material. *ReCALL*, *21*(3), 302–318. doi:10.1017/S0958344009990048

Melenca, M. (2001). *Teaching connected speech rules to Japanese speakers of English so as to avoid a staccato speech rhythm* (Master's thesis). Concordia University. Retrieved from https://spectrum.library.concordia.ca/1389/

Mendonca, C. O., Johnson, K. E., & Mendonca, C. O. (1994). Peer review negotiations: Revision activities in ESL writing instruction. *TESOL Quarterly, 28*(4), 745–769. doi:10.2307/3587558

Meskill, C. (1996). Listening skill development through multimedia. *Journal of Educational Multimedia and Hypermedia, 5*(2), 179–201.

Michael, P. H., & Webb, S. (2017). The effects of captions on EFL learners' comprehension of English-language television programs. *CALICO Journal, 34*, 20–38.

Miles, M. B., & Huberman, A. M. (1994). *Qualitative data analysis: An expanded sourcebook*. Sage.

Miller, D. B. (2006). Podcasting at the University of Connecticut: Enhancing the educational experience. *Campus Technology*. Retrieved November 24, 2012, from https://campustechnology.com/

Min, H. T. (2005). Training students to become successful peer reviewers. *System, 33*(2), 293–308. doi:10.1016/j.system.2004.11.003

Min, H. T. (2006). The effects of trained peer review on EFL students' revision types and writing quality. *Journal of Second Language Writing, 15*(2), 118–141. doi:10.1016/j.jslw.2006.01.003

Min, H. T. (2016). Effect of teacher modeling and feedback on EFL students' peer review skills in peer review training. *Journal of Second Language Writing, 31*, 43–57. doi:10.1016/j.jslw.2016.01.004

Mirzaei, M. S., Meshgi, K., Akita, Y., & Kawahara, T. (2017). Partial and synchronized captioning: A new tool to assist learners in developing second language listening skill. *ReCALL, 29*(2), 178–199. doi:10.1017/S0958344017000039

Mitterer, H., Kim, S., & Cho, T. (2013). Compensation for complete assimilation in speech perception: The case of Korean labial- to – velar assimilation. *Journal of Memory and Language, 69*(1), 59–83. doi:10.1016/j.jml.2013.02.001

Mitterer, H., & McQueen, J. M. (2009). Foreign subtitles help but native-language subtitles harm foreign speech perception. *PLoS One, 4*(11), e7785. Advance online publication. doi:10.1371/journal.pone.0007785 PMID:19918371

Mizumoto, A., & Chujo, K. (2015). A meta-analysis of data-driven learning approach in the Japanese EFL classroom. *English Corpus Studies, 22*, 1–18.

Monteiro, K., & Kim, Y. (2020). The effect of input characteristics and individual differences on L2 comprehension of authentic and modified listening tasks. *System, 94*, 102–336. doi:10.1016/j.system.2020.102336

Morley, J. (2001). Aural comprehension instruction: Principles and practice. In M. Celce-Murcia (Ed.), *Teaching English as a second or foreign language* (pp. 69–86). Heinle & Heinle.

Mory, E. H. (2004). Feedback research review. In D. H. Jonassen (Ed.), *Handbook of research on educational communications and technology* (pp. 745–783). Lawrence Erlbaum.

Nah, K. C. (2011). Optimising the use of wireless application protocol (WAP) sites for listening activities in a Korean English as a foreign language (EFL) context. *Computer Assisted Language Learning, 24*(2), 103–116. doi:10.1080/09588221.2010.526946

Nah, K. C., White, P., & Sussex, R. (2008). The potential of using a mobile phone to access the Internet for learning EFL listening skills within a Korean context. *ReCALL, 20*(3), 331–347. doi:10.1017/S0958344008000633

Narciss, S. (2013). Designing and evaluating tutoring feedback strategies for digital learning environments on the basis of the interactive tutoring Feedback Model. *Digital Education Review, 23*, 7-26. Retrieved from http://greav.ub.edu/der

Narciss, S. (2008). Feedback strategies for interactive learning tasks. In J. M. Spector, M. D. Merrill, J. G. van Merrienboer, & M. P. Driscoll (Eds.), *Handbook of research on educational communications and technology* (3rd ed., pp. 125–144). Lawrence Erlbaum Associates.

Narciss, S. (2012). Feedback in instructional contexts. In N. M. Seel (Ed.), *Encyclopedia of the sciences of learning* (pp. 1285–1289). Springer. doi:10.1007/978-1-4419-1428-6_282

Nation, I. S. P., & Newton, J. (2009). *Teaching ESL/EFL listening and speaking*. Routledge.

Nelson, G. I., & Carson, J. G. (2006). Cultural issues in peer response: 'revisiting culture. In K. Hyland & F. Hyland (Eds.), *Feedback in second language writing: Context and issues* (pp. 42–59). Cambridge University Press. doi:10.1017/CBO9781139524742.005

Nelson, G. L., & Carson, J. G. (1998). ESL students' perceptions of effectiveness in peer feedback groups. *Journal of Second Language Writing, 7*(2), 113–131. doi:10.1016/S1060-3743(98)90010-8

Nelson, G. L., & Murphy, J. M. (1992). An L2 writing group: Task and social dimensions. *Journal of Second Language Writing, 1*(3), 171–193. doi:10.1016/1060-3743(92)90002-7

Nicholson, A. Y. W., & Ngai, J. Y. K. (1996). Managing the development and production of interactive multimedia courseware in education. *Australian Journal of Educational Technology, 12*(1), 35–45.

Nicol, D. J., & Macfarlane-Dick, D. (2006). Formative assessment and self-regulated learning: A model and seven principles of good feedback practice. *Studies in Higher Education, 31*(2), 199–218. doi:10.1080/03075070600572090

Nikolou, S., & Darra, M. (2018). The use and development of podcasting as a technological tool in secondary education in Greece: A case study. *International Education Studies, 11*(11), 109–109. doi:10.5539/ies.v11n11p109

Nikolova, O. R. (2002). Effects of students' participation in authoring of multimedia materials on student acquisition of vocabulary. *Language Learning & Technology, 6*(1), 100–122.

Norris, J., & Ortega, L. (2006). The value and practice of research synthesis for language learning. In J. Norris & L. Ortega (Eds.), *Synthesizing research on language learning and teaching* (pp. 3–52). Benjamins. doi:10.1075/lllt.13.04nor

Norusis, M. J. (2008). *SPSS 16.0 guide to data analysis*. Prentice Hall.

Notar, C. E., Wilson, J. D., & Ross, K. G. (2002). Distant learning for the development of higher-level cognitive skills. *Education, 122*, 642–650.

O'Bryan, A., & Hegelheimer, V. (2007). Integrating CALL into the classroom: The role of podcasting in an ESL listening strategies course. *ReCALL, 19*(2), 162–180. doi:10.1017/S0958344007000523

O'Bryan, A., & Hegelheimer, V. (2009). Using a mixed methods approach to explore strategies, metacognitive awareness and the effects of task design on listening development. *Canadian Journal of Applied Linguistics, 12*(1), 9–37.

Ockey, G. J. (2007). Construct implications of including still image or video in computer-based listening tests. *Language Testing, 24*(4), 517–537. doi:10.1177/0265532207080771

Otake, T., Hatano, G., Cutler, A., & Mehler, J. (1993). Mora or Syllable? Speech segmentation in Japanese. *Journal of Memory and Language, 32*(2), 258–278. doi:10.1006/jmla.1993.1014

Oxford, R. L. (1993). Progress in Tertiary Content-Based ESL Instruction. *TESL Canada Journal, 11*(1), 75–97. doi:10.18806/tesl.v11i1.626

Palloff, R. M., & Pratt, K. (2013). *Lessons from the virtual classroom: The realities of online teaching* (2nd ed.). Jossey-Bass.

Pan, Y.-X., Jiang, D.-N., Yao, L., Picheny, M., & Qin, Y. (2010). *Effects of automated transcription quality on non-native speakers' comprehension in real-time computer-mediated communication.* Paper presented at CHI 2010: Sound and Speech, Atlanta, GA.

Pardo-Ballester, C. (2012). CALL evaluation: Students' perception and use of LoMasTv. *CALICO Journal, 29*(3), 532–547. doi:10.11139/cj.29.3.532-547

Pardo-Ballester, C. (2016). Using video in web-based listening tests. *New Approaches in Educational Research, 5*(2), 91–98. doi:10.7821/naer.2016.7.170

Pardo, N. A., & Téllez, M. F. (2009). ELT materials: The key to fostering effective teaching and learning settings. *PROFILE Issues in Teachers Professional Deveopment, 11*(2), 171–186.

Parry, T. S., & Meredith, R. A. (1984). Videotape vs. Audiotape for Listening Comprehension Tests: An Experiment. *OMLTA Journal*, 47-53.

Paulus, T. M. (1999). The effect of peer and student feedback on student writing. *Journal of Second Language Writing, 8*(3), 265–289. doi:10.1016/S1060-3743(99)80117-9

Pearson Student Mobile Device Survey 2013. (2013). Retrieved September 19, 2014, from https://www.pearsoned.com/wp-content/uploads/Pearson-Student-Mobile-Device-Survey-2013-National-Report-on-College-Students-public-release.pdf

Peterson, M. (2010). Computerized games and simulations in computer-assisted language learning: A meta-analysis of research. *Simulation & Gaming, 41*(1), 72–93. doi:10.1177/1046878109355684

Peterson, P. W. (2001). Skills and strategies for proficient listening. In M. Celce-Murica (Ed.), *Teaching English as a second or foreign language* (pp. 87–100). Heinle & Heinle.

Pham, T. H. T., & Gillies, R. (2010). Designing a culturally appropriate format of formative peer assessment for Asian students: The case of Vietnamese students. *International Journal of Educational Reform, 19*(2), 72–85. doi:10.1177/105678791001900201

Phipps, S., & Borg, B. (2009). Exploring tensions between teachers' grammar teaching beliefs and practices. *System, 37*(3), 380–390. doi:10.1016/j.system.2009.03.002

Plonsky, L., & Oswald, F. L. (2012). How to do a meta-analysis. In A. Mackey & S. M. Gass (Eds.), *Research methods in second language acquisition: A practical guide* (pp. 275–295). Basil Blackwell. doi:10.1002/9781444347340.ch14

Polat, M. (2020). A Rasch Analysis of Rater Behaviour in Speaking Assessment. *International Online Journal of Education & Teaching, 7*(3), 1126–1141.

Polishook, M. (2005). Handheld composing: Reconceptualizing artistic practice with PDAs. In A. Kukulska-Hulme & J. Traxler (Eds.), *Mobile learning: A handbook for educators and trainers* (pp. 133–138). Routledge.

Rahimi, M. (2013). Is training students' reviewers worth its while? A study of how training influences the quality of students' feedback and writing. *Language Teaching Research, 17*(1), 67–89. doi:10.1177/1362168812459151

Razak, N. A., & Saeed, M. A. (2014). Collaborative writing revision process among learners of English as a foreign language (EFL) in an online community of practice (CoP). *Australasian Journal of Educational Technology, 30*(5), 580–599. doi:10.14742/ajet.786

Razı, S. (2014). *Plagiarism detectors in undergraduate academic writing assignments*. Paper presented at the 14th International Languge, Literature and Stylistics Symposium, İzmir, Turkey.

Razı, S. (2016). Open and anonymous peer review in a digital online environment compared in academic writing context. In C. Goria, O. Speicher, & S. Stollhans (Eds.), Innovative language teaching and learning at university: enhancing participation and collaboration (pp. 49-56). Dublin, Ireland: Research-publishing.net. doi:10.14705/rpnet.2016.000404

Razı, S. (2015). Cross-checked problems in undergraduate academic writing. In K. Dikilitaş, R. Smith, & W. Trotman (Eds.), *Teacher-researchers in action* (pp. 147–161). IATEFL.

Regan, M., Nash, J., & Licata, D. (2000). *Mobile learning*. Retrieved August 13, 2011from http://sll.stanford.edu/projects/mobilelearning

Renandya, W. A., & Farrell, T. S. C. (2011). Teacher, the tape is too fast: Extensive listening in ELT. *ELT Journal*, *65*(1), 52–59. doi:10.1093/elt/ccq015

Richards, J. (2016). *Teaching Speaking #3 – Talk in Interaction* [Blog post]. Retrieved from https://www.cambridge.org/elt/blog/2016/02/03/teaching-speaking-2-talk-interaction-ii/

Richardson, J., & Swan, K. (2003). Examining social presence in online courses in relation to students' perceived learning and satisfaction. *Journal of Asynchronous Learning Networks*, *7*, 68–88.

Ridgway, T. (2000). Listening strategies—I beg your pardon'. *ELT Journal*, *54*(2), 179–185. doi:10.1093/elt/54.2.179

Riehl, A. K. (2003). American English Flapping: Perceptual and acoustic evidence against paradigm uniformity with phonetic features. *Working Papers of the Cornell Phonetics Laboratory*, *15*, 271-337.

Rinehart, D., & Chen, S. J. (2012). *The benefits of a cycle of corrective feedback on L2 writing: improving Taiwanese technological university students' English writing proficiency*. Lambert Academic Publishing.

Roach, P. (2009). English Phonetics and Phonology (4th ed.). Academic Press.

Roberts, J. A., Yaya, L. H. P., & Manolis, C. (2014). The invisible addiction: Cell-phone activities and addiction among male and female college students. *Journal of Behavioral Addictions*, *3*(4), 1–12. doi:10.1556/JBA.3.2014.015 PMID:25595966

Robinson, J. M. (2002). In search of fairness: An application of multi-reviewer anonymous peer review in a large class. *Journal of Further and Higher Education*, *26*(2), 183–192. doi:10.1080/03098770220129451

Rogerson, M. (2006). Don'cha Know? A survey of ESL teachers' perspectives on reduced form instruction.Second Language Studies. In J. D. Brown & K. Kondo Brown, (Eds.), Perspectives on teaching connected speech to second language speakers (pp. 85-97). Honolulu, HI: University of Hawaii Press. National Foreign Language Resource Center.

Rollinson, P. (1998). *Peer response and revision in an ESL writing group: a case study* (Unpublished PhD thesis). Universidad Autónoma de Madrid.

Rollinson, P. (2005). Using peer feedback in ESL writing class. *ELT Journal*, *59*(1), 23–30. doi:10.1093/elt/cci003

Rosa, M. (2002). Don'cha know? A survey of ESL teachers' perspectives on reduced forms instruction. *Second Language Studies*, *21*(1), 49–78.

Rosell-Aguilar, F. (2007). Top of the Pods: In search of podcasting 'pedagogy' for language learning. *Computer Assisted Language Learning*, *20*(5), 471–492. doi:10.1080/09588220701746047

Rost, M. (2011). *Teaching and researching Listening*. Longman.

Roussel, S. (2011). A computer assisted method to track listening strategies in second language learning. *ReCALL*, *23*(2), 98–116. doi:10.1017/S0958344011000036

Rubin, J. (1995). An overview to a guide for the teaching of second language listening. In D. J. Mendelsohn & J. Rubin (Eds.), *A Guide for the Teaching of Second Language Listening* (pp. 7–11). Dominie Press, Inc.

Ruecker, T. (2011). The potential of dual-language cross-cultural peer review. *ELT Journal*, *65*(4), 398–407. doi:10.1093/elt/ccq067

Ruegg, R. (2015). The relative effects of peer and teacher feedback on improvement in EFL students' writing ability. *Linguistics and Education*, *29*, 73–82. doi:10.1016/j.linged.2014.12.001

Saeed, M. A., Ghazali, K., & Aljaberi, M. A. (2018). A review of previous studies on ESL/EFL learners' interactional feedback exchanges in face-to face and computer-assisted peer review of writing. *International Journal of Educational Technology in Higher Education*, *15*(6), 1–25. doi:10.118641239-017-0084-8

Sah, F. M., & Shah, P. M. (2020). Teachers' beliefs and practices in teaching listening. *Creative Education*, *11*(02), 182–195. doi:10.4236/ce.2020.112013

Saito, H., & Fujita, T. (2004). Characteristics and user acceptance of peer rating in EFL writing classrooms. *Language Teaching Research*, *8*(1), 31–54. doi:10.1191/1362168804lr133oa

Saka, N. E., & Zengin, B. (2020). Use of Video/Listening Content Transcripts in English for Coronavirus Awareness and Information, *Virology & Immunological Journal*, *4*(1), 1–9. doi:10.23880/vij-16000230

Sato, M. (2013). Beliefs about peer interaction and peer corrective feedback: Efficacy of classroom intervention. *Modern Language Journal*, *97*(3), 611–630. doi:10.1111/j.1540-4781.2013.12035.x

Savignon, S. J., & Roithmeier, W. (2004). Computer-mediated communication: Texts and strategies. *CALICO Journal*, *21*(2), 265–290. doi:10.1558/cj.v21i2.265-290

Saville, N., & Hargreaves, P. (1999). Assessing speaking in the revised FCE. *ELT Journal*, *53*(1), 42–51. doi:10.1093/elt/53.1.42

Saygin, A. P., & Wilson, S. M. (2002). *Paradigm reanalysis and the representation of morphologically complex words in Turkish.* Paper presented at the 38th Annual Meeting of the Chicago Linguistic.

Sayın, Z., & Seferoğlu, S. S. (2015). Çevrim-içi öğrenmeye yeni bir bakış açısı: Kitlesel açık çevrimiçi dersler. In B. Akkoyunlu, A. İşman, & H. F. Odabaşı (Eds.), *Eğitim teknolojileri okumaları* (pp. 525–538). TOJET-The Turkish Online Journal of Educational Technology.

Schegloff, E. A. (2007). *Sequence Organization in Interaction: A Primer in Conversation Analysis I.* Cambridge University Press. doi:10.1017/CBO9780511791208

Schultz, K., Jones-Walker, C. E., & Chikkatur, A. P. (2015). Listening to students, negotiating beliefs: Preparing teachers for urban classrooms. *Curriculum Inquiry*, *38*(2), 155–187. doi:10.1111/j.1467-873X.2007.00404.x

Seferoğlu, G., & Uzakgören, S. (2004). Equipping learners with listening strategies in English language classes. *Hacettepe Üniversitesi Eğitim Fakültesi Dergisi*, *27*, 223–231.

Şendağ, S., Gedik, N., & Toker, S. (2018). Impact of repetitive listening, listening-aid and podcast length on EFL podcast listening. *Computers & Education*, *125*, 273–283. doi:10.1016/j.compedu.2018.06.019

Sengupta, S. (1998). From text revision to text improvement: A story of secondary school composition. *RELC Journal*, *29*(1), 110–137. doi:10.1177/003368829802900107

Settler, J., & Jenkins, J. (2005). Pronunciation. State-of-the-Art Review Article. *Language Teaching, 38*(1), 1–17. doi:10.1017/S026144480500251X

Şevik, M. (2017). Does bimodal subtitling effect listening comprehension in EFL contexts. In D. Köksal (Ed.), *Researching ELT: Classroom methodology and beyond* (pp. 55–64). Peter Lang International Academic Publisher.

Sha, G. (2010). Using TTS voices to develop audio materials for listening comprehension: A digital approach. *British Journal of Educational Technology, 41*(4), 632–641. doi:10.1111/j.1467-8535.2009.01025.x

Sharifi, M., Rostami AbuSaeedi, A., Jafarigohar, M., & Zandi, B. (2018). Retrospect and prospect of computer assisted English language learning: A meta-analysis of the empirical literature. *Computer Assisted Language Learning, 31*(4), 413–436. doi:10.1080/09588221.2017.1412325

Sharples, M. (2000). The design of personal mobile technologies for lifelong learning. *Computers & Education, 34*(3-4), 177–193. doi:10.1016/S0360-1315(99)00044-5

Shepherd, M. A., Simonian, J. R., & Trussler, L. C. (2017). The Subtitle effect: How subtitle-facilitated perceptual learning can make subtitles (seem) less necessary. *Communication Research Reports, 34*(4), 350–358. doi:10.1080/08824096.2017.1356712

Shin, M.-H., & Sung, J. (2019). The Effect of Teaching Suprasegmental Features on Korean University Students' Listening Comprehension. *International Journal of Advanced Science and Technology, 28*(3), 1 - 7. Retrieved from http://sersc.org/journals/index.php/IJAST/article/view/256

Shin, D. (1998). Using video-taped lectures for testing academic language. *International Journal of Listening, 12*, 56–79. doi:10.1080/10904018.1998.10499019

Shrum, J. L., & Glisan, E. (2005). *Teacher's handbook: Contextualized language instruction* (3rd ed.). Heinle.

Shute, V. J. (2008). Focus on formative feedback. *Review of Educational Research, 78*(1), 153–189. doi:10.3102/0034654307313795

Shute, V. J., & Zapata-Rivera, D. (2012). Adaptive educational systems. In J. D. Paula & M. L. Alan (Eds.), *Adaptive technologies for training and education* (pp. 7–27). Cambridge University Press., doi:10.1017/ CBO9781139049580.004

Şimşek, A. (2016). Öğretim tasarımı ve modelleri. In K. Çağıltay & Y. Göktaş (Eds.), Öğretim Teknolojilerinin Temelleri: Teoriler, Araştırmalar, Eğilimler (pp. 105-122). Pegem Akademi.

Skehan, P. (2001). Tasks and language performance. In M. Bygate, P. Skehan, & M. Swain (Eds.), *Research pedagogic tasks: second language learning, teaching, and testing*. Longman.

Smits, M. S., Boon, J., Sluijsmans, D. M., & Van Gogh, T. (2006). Content and timing of feedback in a web-based learning environment: Effects on learning as a function of prior knowledge. *Interactive Learning Environments, 16*(2), 183–193. doi:10.1080/10494820701365952

Sommers, E., & Lawrence, S. (1992). Women's ways of talking in teacher-directed and student-directed peer response groups. *Linguistics and Education, 4*(1), 1–36. doi:10.1016/0898-5898(92)90018-R

Sorden, S. D. (2005). A cognitive approach to instructional design for multimedia learning. *Information Science Journal, 8*, 263–279. doi:10.28945/498

Speck, B. W. (2000). Grading students' classroom writing: Issues and strategies (ASHE-ERIC Higher Education Report Vol. 27, No. 3). The George Washington University, Graduate School of Education and Human Development.

Stanley, J. (1992). Coaching student writers to be effective peer evaluators. *Journal of Second Language Writing, 1*(3), 217–233. doi:10.1016/1060-3743(92)90004-9

Stephenson, M., & Hall, G. (2021). Organizing talk in group speaking tests: Learning from high-scoring students. *ELT Journal, 75*(1), 42–54. doi:10.1093/elt/ccaa052

Stipek, D. J., Givvin, K. B., Salmon, J. M., & MacGyvers, V. L. (2001). Teachers' beliefs and practices related to mathematics instruction. *Teaching and Teacher Education, 17*(2), 213–226. doi:10.1016/S0742-051X(00)00052-4

Stoller, F. L. (2004). Content-based instructions: Perspectives on curriculum planning. *Annual Review of Applied Linguistics, 24*(1), 261–283. doi:10.1017/S0267190504000108

Sueyoshi, A., & Hardison, D. M. (2005). The role of gestures and facial cues in second language listening comprehension. *Language Learning, 55*(4), 661–699. doi:10.1111/j.0023-8333.2005.00320.x

Suh, S. S., Kim, S. W., & Kim, N. J. (2010). Effectiveness of MMORPG-based instruction in elementary English education in Korea. *Journal of Computer Assisted Learning, 26*(5), 370–378. doi:10.1111/j.1365-2729.2010.00353.x

Sung, E.-K. (2003). Acoustics and perception of flaps in American English and Korean. *Language Research, 39*(4), 933–958.

Sung, Y. T., Chang, K. E., & Yang, J. M. (2015). How effective are mobile devices for language learning? A meta-analysis. *Educational Research Review, 16*, 68–84. doi:10.1016/j.edurev.2015.09.001

Suvorov, R. (2011). The effects of context visuals on L2 listening comprehension. *University of Cambridge ESOL Examinations Research Notes, 45*, 2–8.

Sze, P. M. (2007). Developing students' listening and speaking skills through ELT podcasts. *Education Journal, 34*, 115–134.

Tai, Y. (2012). Contextualizing a MALL: Practice design and evaluation. *Journal of Educational Technology & Society, 12*, 220–230.

Tarantino, J. (2016). Examining the effects of independent MALL on vocabulary recall and listening comprehension: An exploratory case study of preschool children. *CALICO Journal, 33*, 260–277.

Taylor, A. M. (2009). CALL-based versus paper-based glosses: Is there a difference in reading comprehension? *CALICO Journal, 27*(1), 147–160. doi:10.11139/cj.27.1.147-160

Taylor, G. (2005). Perceived strategies of students watching captioned video processing. *Foreign Language Annals, 38*(3), 422–427. doi:10.1111/j.1944-9720.2005.tb02228.x

The Cambridge Framework for Life Competences, Introductory Guide for Teachers and Educational Managers. (n.d.). Cambridge University Press. Retrieved from: https://www.cambridge.org/tr/files/6615/9741/7863/CambridgeLifeCompetencies_Introduction_Booklet_digitalPDF_082020.pdf

The Turkish Coalition of America. (2007, June 8). *Patara: Its role in US history.* Retrieved from https://www.tc-america.org/files/Patara-Brochure.pdf

The Turkish coalition of America. (2007, June 8). *The Lycian confederation and the American constitution* TCA Issues & Information – TCA Issue Paper-9. Retrieved from https://www.tc-america.org/issues-information/us-turkey-relations/the-lycian-confederation-and-the-american-constitution-153.htm

Thompson, G., & McKinley, J. (2018). Integration of content and language learning. The TESOL Encyclopedia of English language, 1-13. doi:10.1002/9781118784235.eelt0634

Toda, T. (2006). Focus on form in teaching connected speech. In J. D. Brown & K. Kondo-Brown (Eds.), *Perspectives of Teaching Connected Speech to Second Language Speakers* (pp. 189–2015). University of Hawaii, National Foreign Language Resource Center.

Topping, K. (1998). Peer assessment between students in colleges and universities. *Review of Educational Research, 68*(3), 249–276. doi:10.3102/00346543068003249

Toro, J. M., Sebastian-Galles, N., & Mattys, S. L. (2009). The rule of perceptual salience during the segmentation of connected speech. *The European Journal of Cognitive Psychology*, *21*(5), 786–800. doi:10.1080/09541440802405584

Tottie, G. (2011). Uh and Um as sociolinguistic markers in British English. *International Journal of Corpus Linguistics*, *16*(2), 173–197. doi:10.1075/ijcl.16.2.02tot

Trappes-Lomax, H. (2004). Discourse analysis. In A. Davies & C. Elder (Eds.), *The Handbook of Applied Linguistics* (pp. 133–164). Blackwell Publishing. doi:10.1002/9780470757000.ch5

Trofimovich, P., & Isaacs, T. (2012). Disentangling accent from comprehensibility. *Bilingualism: Language and Cognition*, *15*(4), 905–916. doi:10.1017/S1366728912000168

Tsai, S. C. (2015). Implementing courseware as the primary mode of task-based ESP instruction: A case study of EFL students. *Computer Assisted Language Learning*, *28*(2), 171–186. doi:10.1080/09588221.2013.818554

Tsui, A. B. M., & Ng, M. (2000). Do secondary L2 writers benefit from peer comments. *Journal of Second Language Writing*, *9*(2), 147–170. doi:10.1016/S1060-3743(00)00022-9

Tucker, B. V. (2007). *Spoken word recognition of reduced American flap* (Doctoral Dissertation). The University of Arizona. Retrieved from https://sites.ualberta.ca/~bvtucker/Papers/Tucker_2007_Dissertation.pdf

Tunison, S., & Noonan, B. (2001). On-line learning: Secondary students' first experience. *Canadian Journal of Education*, *26*(4), 495–514. doi:10.2307/1602179

Türel, V. (2012). Design of feedback in interactive multimedia language learning environments. *Linguistik Online, 54*(4), 35-49. Retrieved June 1, 2021, http://www.linguistik-online.de/54_12/tuerel.pdf

Turel, V. (2018). Adaptive Hypermedia in Education. In M. Khosrow-Pour (Ed.), Encyclopaedia of information science and technology (4th ed.). (Vols. 1–10; pp. 2357–2368). IGI-Global. doi:10.4018/978-1-5225-2255-3

Turel, V. (2014). The use and design of supplementary visuals for the enhancement of listening skills in hypermedia. In C. M. Akrivopoulou & N. Garipidis (Eds.), *Human rights and the impact of ICT in the public sphere: Participation, democracy, and political autonomy* (pp. 268–291). IGI-Global. doi:10.4018/978-1-4666-6248-3.ch016

Turel, V. (2014b). Factors affecting listening. In C. M. Akrivopoulou & N. Garipidis (Eds.), *Human rights and the impact of ICT in the public sphere: Participation, democracy, and political autonomy* (pp. 310–331). IGI-Global. doi:10.4018/978-1-4666-6248-3.ch018

Turel, V. (2014c). Design of cultural differences in hypermedia environments. *International Journal of Human Rights and Constitutional Studies*, *2*(2), 150–170. doi:10.1504/IJHRCS.2014.062766

Turel, V. (2014d). Priority of listening materials for autonomous intermediate language learners. In C. M. Akrivopoulou & N. Garipidis (Eds.), *Human rights and the impact of ICT in the public sphere: Participation, democracy, and political autonomy* (pp. 292–309). IGI-Global. doi:10.4018/978-1-4666-6248-3.ch017

Turel, V. (2015a). Hypermedia and its' role in learning. In M. Khosrow-Pour (Ed.), *Encyclopaedia of information science and technology* (3rd ed., pp. 2495–2505). IGI-Global. doi:10.4018/978-1-4666-5888-2.ch243

Turel, V. (2015b). Introduction: Listening, multimedia and optimum design. In V. Turel (Ed.), *Intelligent Design of Interactive Multimedia Listening Software* (pp. xx–xxxiiv). IGI-Global.

Turel, V., & Kılıç, E. (2014). The inclusion and design of cultural differences in Interactive multimedia environments. In C. M. Akrivopoulou & N. Garipidis (Eds.), *Human rights and the impact of ICT in the public sphere: Participation, democracy, and political autonomy* (pp. 245–267). IGI-Global. doi:10.4018/978-1-4666-6248-3.ch015

Turel, V., & McKenna, P. (2013). Design of language learning software. In *Computer-assisted foreign language teaching and learning: Technological advances* (pp. 188–209). IGI Global. doi:10.4018/978-1-4666-2821-2

Turel, V., & McKenna, P. (2015a). Design of multimedia listening software: Unfamiliar items, glossary, aural texts and speech rate. In V. Turel (Ed.), *Intelligent design of interactive multimedia listening software* (pp. 170–197). IGI-Global. doi:10.4018/978-1-4666-8499-7.ch007

Turel, V., & McKenna, P. (2015b). Design of multimedia listening software: Instructions, tasks, texts and self-assessment tests. In V. Turel (Ed.), *Intelligent design of interactive multimedia listening software* (pp. 142–169). IGI-Global. doi:10.4018/978-1-4666-8499-7.ch006

Underhill, N. (1987). *Testing spoken language: A handbook for oral testing techniques*. Cambridge University Press.

Underwood, P., & Wallace, M. (2012). *The effects of instruction of English speech on low proficiency EFL students*. Retrieved from https://www.academia.edu/854671/The_Effects_of_Instruction_in_Reduced_Forms_of_English_Speech_on_Low-proficiency_EFL_University_Students

Underwood, P. R., Bridges, S., & Hattori, T. (2010). *Speak Easy: Spoken English for Japanese Learners*. Suita Mediaisland.

Ur, P. (1984). *Teaching listening comprehension*. Cambridge University.

Utami, D. N. (2016). The EFL teachers' beliefs and their teaching practices. *Journal of Languages and Literature*, *10*(2), 135–144. doi:10.19105/ojbs.v10i2.974

Vahdany, F., & Azaraki, A. M. (2015). The effect of watching captioned movies on Iranian learners' vocabulary learning at pre-intermediate level. *Educational Technology*, *4*, 30619–30633.

Van den Berg, B. A. M., Admiraal, W. F., & Pilot, A. (2006). Designing student peer assessment in higher education: Analysis of written and oral peer feedback. *Teaching in Higher Education*, *11*(2), 135–147. doi:10.1080/13562510500527685

Van der Pol, J., Van den Berg, B. A. M., Admiral, W. F., & Simons, P. R. J. (2008). The nature, reception, and use of online peer feedback in higher education. *Computers & Education*, *51*(4), 1804–1817. doi:10.1016/j.compedu.2008.06.001

Van Popta, E., Kral, M., Camp, G., Martens, R. L., & Simons, P. R. J. (2017). Exploring the value of peer feedback in online learning for the provider. *Educational Research Review*, *20*, 24–34. doi:10.1016/j.edurev.2016.10.003

Vandergrift, L. (2004). Listening to learn or learning to listen? *Annual Review of Applied Linguistics*, *24*, 3–25. doi:10.1017/S0267190504000017

Vandergrift, L. (2007). Recent developments in second and foreign language listening comprehension research. *Language Teaching*, *40*(3), 191–210. doi:10.1017/S0261444807004338

Vandergrift, L., & Goh, C. C. M. (2012). Listening in multiple media environments. In *Teaching and Learning Second Language Listening* (pp. 217–238). Routledge, Taylor & Francis Group.

Vandergrift, L., & Goh, C. C. M. (2012). *Teaching and learning second language listening: Metacognition in Action*. Taylor & Francis. doi:10.4324/9780203843376

Vandergrift, L., & Tafaghodtari, M. H. (2010). Teaching L2 learners how to listen does make a difference: An empirical study. *Language Learning, 60*(2), 470–497. Advance online publication. Retrieved May 5, 2021, from. doi:10.1111/j.1467-9922.2009.00559.x

Vanderplank, R. (2016). *Captioned media in foreign language learning and teaching: Subtitles for the deaf and hard-of-hearing as tools for language learning*. Palgrave MacMillan.

Verdugo, D. R., & Belmonte, I. A. (2007). Using digital stories to improve listening comprehension with Spanish young learners of English. *Language Learning & Technology, 11*, 87–101.

Villamil, O. S., & De Guerrero, M. C. M. (1996). Peer revision in the L2 classroom: Social-cognitive activities, mediating strategies, and aspects of social behaviour. *Journal of Second Language Writing, 5*(1), 51–75. doi:10.1016/S1060-3743(96)90015-6

Villamil, O. S., & De Guerrero, M. C. M. (2006). Sociocultural theory: A framework for understanding of socio-cognitive aspects of social behavior. In K. Hyland & F. Hyland (Eds.), *Feedback in second language writing: Context and issues* (pp. 23–41). Cambridge University Press. doi:10.1017/CBO9781139524742.004

Vygotsky, L. S. (1978). *Mind in society: The development of higher psychological processes*. Harvard University Press.

Wagner, E. (2007). Are they watching? Test-taker viewing behavior during an L2 video listening test. *Language Learning & Technology, 11*, 67–86.

Wagner, E. (2010). The effect of the use of video texts on ESL listening test-taker performance. *Language Testing, 27*(4), 493–513. doi:10.1177/0265532209355668

Wagner, E. (2013). An investigation of how the channel of input and access to test questions affect l2 listening test performance. *Language Assessment Quarterly, 10*(2), 178–195. doi:10.1080/15434303.2013.769552

Wang, W. Q. (2014). Students' perceptions of rubric-referenced feedback on EFL writing: A longitudinal inquiry. *Assessing Writing, 19*, 80–96. doi:10.1016/j.asw.2013.11.008

Warshauer, M., & Ware, P. (2006). Automated writing evaluation: Defining the classroom research agenda. *Language Teaching Research, 10*(2), 157–180. doi:10.1191/1362168806lr190oa

Weinberg, A., Knoerr, H., & Vandergrift, L. (2011). Creating podcasts for academic listening in French: Student perceptions of enjoyment and usefulness. *CALICO Journal, 28*(3), 588–605. doi:10.11139/cj.28.3.588-605

Weinstein, N. (2001). *Whaddaya Say?* (2nd ed.). Pearson Education.

Weir, C. J. (2005). *Language testing and validation*. Palgrave Macmillan. doi:10.1057/9780230514577

Weissberg, R. (2006). Scaffolded feedback: theoretical conversations with advanced L2 writers. In K. Hyland & F. Hyland (Eds.), *Feedback in second language writing: Contexts and issues* (pp. 246–265). Cambridge University Press. doi:10.1017/CBO9781139524742.015

Wette, R. (2015). Teacher-led collaborative modelling in academic L2 writing courses. *ELT Journal, 69*(1), 71–80. doi:10.1093/elt/ccu043

Wilberschied, L., & Berman, P. M. (2004). Effect of using photos from authentic video as advance organizers on listening comprehension in an FLES Chinese class. *Foreign Language Annals, 37*(4), 534–543. doi:10.1111/j.1944-9720.2004.tb02420.x

Wilson, J. J. (2010). *How to teach listening*. Pearson Longman.

Winke, P., Gass, S., & Sydorenko, T. (2010). The effects of captioning videos used for foreign language listening activities. *Language Learning & Technology*, *14*, 65–86.

Wolvin, A. D., & Coakley, C. G. (1985). *Listening*. Willian C. Brown Company.

Wong, S. W. L., Mok, P. P. K., Chung, K. K.-H., Leung, V. W. H., Bishop, D. V. M., & Chow, B. W.-Y. (2015). Perception of Native English Reduced Forms in Chinese Learners: Its Role in Listening Comprehension and Its Phonological Correlates. *TESOL Quarterly*. . doi:10.1002/tesq.273

Woottipong, K. (2014). Effect of using video materials in the teaching of listening skills for university students. *International Journal of Linguistics*, *6*(4), 200–212. doi:10.5296/ijl.v6i4.5870

Wu, W. C. V., Petit, E., & Chen, C. H. (2015). EFL writing revision with blind expert and peer review using a CMC open forum. *Computer Assisted Language Learning*, *28*(1), 58–80. doi:10.1080/09588221.2014.937442

Wu, W. H., Wu, Y. C., Chen, C. Y., Kao, H. Y., Lin, C. H., & Huang, S. H. (2012). Review of trends from mobile learning studies: A meta-analysis. *Computers & Education*, *59*(2), 817–827. doi:10.1016/j.compedu.2012.03.016

Yang, J.-C., Lin, Y.-L., & Chung, C.-I. (2009). Developing a film-based learning system with English verbal reduced forms for supporting English listening comprehension. In M. Chang, R. Kuo, Kinshuk, G.-D. Chen, & M. Hirose (Eds.), Learning by playing: Gamebased education system design and development (Vol. 5670, pp. 74–84). Berlin: Springer Berlin Heidelberg. doi:10.1007/978-3-642-03364-3_9

Yang, H. Y. (2014a). Does multimedia support individual differences?–EFL learners' listening comprehension and cognitive load. *Australasian Journal of Educational Technology*, *30*(6), 699–713. doi:10.14742/ajet.639

Yang, H. Y. (2014b). The effects of advance organizers and subtitles on EFL learners' listening comprehension skills. *CALICO Journal*, *31*(3), 345–373. doi:10.11139/cj.31.3.345-373

Yang, J., & Chang, P. (2014). Captions and reduced forms instruction: The impact on EFL students' listening comprehension. *ReCALL*, *26*(1), 44–61. doi:10.1017/S0958344013000219

Yang, M., Badger, R., & Yu, Z. (2006). A comparative and teacher feedback in a Chinese EFL writing class. *Journal of Second Language Writing*, *15*(3), 179–200. doi:10.1016/j.jslw.2006.09.004

Yasin, B., Mustafa, F., & Permatasari, R. (2017). How Much Videos Win over Audios in Listening Instruction for EFL Learners. *Turkish Online Journal of Educational Technology-TOJET*, *17*, 92–100.

Yeh, C.-C. (2013). An investigation of a podcast learning project for extensive listening. *Language Education in Asia*, *4*(2), 135–149. doi:10.5746/LEiA/13/V4/I2/A04/Yeh

Yeldham, M. (2018). Viewing L2 captioned videos: What's in it for the listener? *Computer Assisted Language Learning*, *31*(4), 367–389.

Yeldham, M., & Gruba, P. (2014). Toward an instructional approach to developing interactive second language listening. *Language Teaching Research*, *18*(1), 33–53. doi:10.1177/1362168813505395

Yin, R. K. (2003). *Case study research: Design and methods* (3rd ed.). Sage.

Yun, J. (2011). The effects of hypertext glosses on L2 vocabulary acquisition: A meta-analysis. *Computer Assisted Language Learning*, *24*(1), 39–58. doi:10.1080/09588221.2010.523285

Yu, S., & Lee, I. (2015). Understanding EFL students' participation in group peer feedback of L2 writing: A case study from an activity theory perspective. *Language Teaching Research*, *19*(5), 572–593. doi:10.1177/1362168814541714

Zacharias, N. T. (2005). Teachers' beliefs about internationally-published materials: A survey of tertiary English teachers in Indonesia. *RELC Journal*, *36*(1), 23–37. doi:10.1177/0033688205053480

Zamel, V. (1995). College composition and communication. *National Council of Teachers of English, 46*(4), 506-521.

Zengin, B. (2019). Use of film scripts and their translations in teaching English as a foreign language. *International Online Journal of Education & Teaching*, *6*(4), 944–958. https://iojet.org/index.php/IOJET/article/view/718

Zengin, B., Başal, A., & Yükselir, C. (2019). Investigation into the perceptions of English teachers and instructors on the use of literature in English language teaching. *The Reading Matrix: An International Online Journal*, *19*(1), 155–166.

Zengin, B., & Çubukçu, F. (2013). English language learners' attitudes to backseat TV systems and subtitled movies. *Turkish Studies – International Periodical for the Languages, Literature and History of Turkish or Turkic*, *8*(8), 1561–1573.

Zengin, B., & Cubukcu, F. (2013). English language learners' attitudes to backseat TV systems and subtitled movies. *Turkish Studies – International Periodical for the Languages. Literature and History of Turkish or Turkic*, *8*(8), 1561–1573. doi:10.7827/TurkishStudies.5353

Zengin, B., Doğan, D., & Çubukçu, F. (2015). English major students' attitudes towards movies and series as language resources. In V. Türel (Ed.), *Intelligent design of interactive multimedia listening software* (pp. 276–295).

Zengin, B., Doğan, D., & Çubukçu, F. (2015). English major students' attitudes towards movies and series as language resources. In V. Turel (Ed.), *Intelligent Design of Interactive Multimedia Listening Software* (pp. 276–295). IGI Global. doi:10.4018/978-1-4666-8499-7.ch011

Zengin, B., & Yetkin Aker, D. (2016). Education Policy and the Effect of Subtitle Options in Learning Process: A Case of Turkish Students as English Language Learners. *Humanitas*, *4*(7), 485–498.

Zengin, B., & Yetkin Aker, D. (2016). Education policy and the effect of subtitle options in learning process: A case of Turkish students as English language learners. *Humanitas. A Journal of Social Sciences*, *4*(7), 485–498.

Zhang, S. (1995). Reexamining the affective advantage of peer feedback in the ESL writing class. *Journal of Second Language Writing*, *4*(3), 209–222. doi:10.1016/1060-3743(95)90010-1

Zhan, L. (2016). Written teacher feedback: Student perceptions, teacher perceptions, and actual teacher performance. *English Language Teaching*, *9*(8), 73–84. doi:10.5539/elt.v9n8p73

Zhao, H. (2010). Investigating use and understanding of peer and teacher feedback on writing: A comparative study in a Chinese English writing classroom. *Assessing Writing*, *15*(1), 3–17. doi:10.1016/j.asw.2010.01.002

Zhao, H. (2014). Investigating teacher-supported peer assessment for EFL writing. *ELT Journal*, *68*(2), 155–168. doi:10.1093/elt/cct068

Zhao, Y. (1997). The effects of listeners' control of speech rate on second language comprehension. *Applied Linguistics*, *18*(1), 49–68. doi:10.1093/applin/18.1.49

Zhao, Y. (2003). Recent developments in technology and language learning: A literature review and meta-analysis. *CALICO Journal*, *21*(1), 7–27. doi:10.1558/cj.v21i1.7-27

Zhu, W. (2001). Interaction and feedback in mixed peer response groups. *Journal of Second Language Writing*, *10*(4), 251–276. doi:10.1016/S1060-3743(01)00043-1

Zhu, W., & Mitchell, D. (2012). Participation in peer response as activity: An examination of peer response stances from an activity theory perspective. *TESOL Quarterly*, *46*(2), 362–386. doi:10.1002/tesq.22

Zielinski, B. W. (2007). The listener: No longer the silent partner in reduced intelligibility. *System*, *36*(1), 69–84. doi:10.1016/j.system.2007.11.004

About the Contributors

Vehbi Turel has a First Class Degree in Teaching of English as a Foreign Language, M.Ed. in Educational Technology and TESOL, and a Ph.D. on Design of Interactive Multimedia Language (Listening) Software. He is a professional software developer and is mainly interested in teaching, research and creating CALL materials in English, Kurdish and Turkish. He has thirty-four years teaching experience and has been working as a co-editor, editor-in-chief, editorial board member, author and reviewer for several refereed international journals and book publishers. He teaches and publishes internationally.

* * *

Zeynep Başer is an Assistant Professor at the English Translation and Interpretation Department at Kırıkkale University in Turkey. She graduated from Middle East Technical University Department of English Language Teaching in 2010. She completed her Masters at Middle East Technical University, Department of English Language Teaching in 2012 with her Master's thesis on practices and professional development of a newly appointed English teacher and the factors that influence them. She was part of the psychology and linguistics group. IN 2018 she completed her PhD in Middle East Technical University, Cognitive Sciences Department with her dissertation study on the effect of sentence processing and structural preparation in the mother tongue and second language. Between 2010 and 2018, she worked at Kırıkkale University, School of Foreign Languages. Her research interests are mainly psycholinguistics, experimental linguistics, cognitive sciences, and foreign language education.

Kaine Gulozer is an assistant professor at the Department of Foreign Language Education, Education Faculty of Bartın University. She has extensive experience in language teaching as well as organizing ERASMUS programs as a part of partnership of Erasmus projects. She relies on excellent networks to research, teacher training and practice. Dr. Gulozer teaches a range of undergraduate level of courses such as "Teacher Training", "Material Development in Education", "Instructional Technology," "Project Development and Management". Additionally, she delivers postgraduate level of courses such as "Academic writing for publication purposes" and "Entrepreneurship" in Lifelong program of the faculty.

Pelin Irgin is an assistant professor of English Language Education at TED University in Turkey. She received her PhD degree in English Language Teaching at Hacettepe University, Turkey and completed her doctorate thesis research at University of Reading, UK. Her research interests are language learning and cognition, EFL listening comprehension and listening strategies, and L2 writing. She is currently serving on the editorial boards of international publications. She is also the community manager of TESOL_SPLIS Steering Committee of 2020-2022.

Işıl Kaçar is a teacher educator and a faculty member at the Department of Foreign Language Education at Middle East Technical University in Ankara, Turkey. Her research interests are pre-service teacher education, (e-)mentoring, academic writing, technology integration into English Language Teaching. and virtual exchange /telecollaboration.

Tingting Kang serves as the Multilanguage Learner Support Specialist at Lafayette College. She received her Ph.D. degree in Applied Linguistics from Northern Arizona University in 2016 and her M.A. degree in Teaching English to Speakers of Other Languages from Murray State University in 2011. At Lafayette College, she is in charge of the English for Academic Purpose (EAP) center. The purpose of the EAP center is to equip ESL learners with the reading, writing, and oral communication skills and the knowledge of US culture necessary to succeed in an English-speaking college setting. Her research interests include computer-assisted language learning, second language listening and speaking skills, corpus linguistics, and language assessment.

Zeynep Kocoglu is a full-time faculty member in the department of English Language Education, Faculty of Education of Yeditepe University. Her current teaching areas include computer-assisted language learning, materials development, use of literature in Language teaching and testing.

Elif Tokdemir Demirel is an Associate Professor and the Chair of the English Translation and Interpretation Department at Kırıkkale University in Turkey. She completed her MA in Bilkent University, MA TEFL Program and her PhD at Middle East Technical University, English Language Teaching Department. She previously worked at Karadeniz Technical University, Department of English Language and Literature. She is a member of the Digital Humanities Turkey Research Group and a Board Member of the European Writing Centers Association. Her research interests are second language writing, Corpus linguistics and particularly learner corpora studies, digital humanities, mobile learning and teaching technologies in ELT and classroom interaction.

Buğra Zengin has worked in the field of English language teaching for over twenty years. He holds his BA in English language and literature, his M.A. degree in English literature and his PhD degree in English language teaching. His research interests include applied linguistics with ELT focus in general, use of movie scripts, use of transcripts, use of Google search skills, augmented reality and memory strategies in foreign language education, and recently political literacy in particular. He has a book of mnemonics for vocabulary learning of Turkish learners of English and a model for the English tense. He also generated mnemonics for Japanese.

Index

T

U

V

W

www.ingramcontent.com/pod-product-compliance
Lightning Source LLC
LaVergne TN
LVHW062144300926
849161LV00010B/2616